NH

Energy Crisis
Policy Response

edited by
Peter N. Nemetz

The Institute for Research on Public Policy
L'Institut de recherches politiques

Montreal

ISBN 0 920380 49 2

Legal Deposit Third Quarter
Bibliothèque nationale du Québec

This issue is published in collaboration with
The Journal of Business Administration
University of British Columbia

The Institute for Research on Public Policy/L'Institut de recherches politiques
2149 Mackay Street
Montreal, Quebec
H3G 2J2

Founded in 1972, THE INSTITUTE FOR RESEARCH ON PUBLIC POLICY is a national organization whose independence and autonomy are ensured by the revenues of an endowment fund, which is supported by the federal and provincial governments and by the private sector. In addition, the Institute receives grants and contracts from governments, corporations, and foundations to carry out specific research projects.

The *raison d'être* of the Institute is threefold:

— To act as a catalyst within the national community by helping to facilitate informed public debate on issues of major public interest

— To stimulate participation by all segments of the national community in the process that leads to public policy making

— To find practical solutions to important public policy problems, thus aiding in the development of sound public policies

The Institute is governed by a Board of Directors, which is the decision-making body, and a Council of Trustees, which advises the board on matters related to the research direction of the Institute. Day-to-day administration of the Institute's policies, programmes, and staff is the responsibility of the president.

The Institute operates in a decentralized way, employing researchers located across Canada. This ensures that research undertaken will include contributions from all regions of the country.

Wherever possible, the Institute will try to promote public understanding of, and discussion on, issues of national importance, whether they be controversial or not. It will publish its research findings with clarity and impartiality. Conclusions or recommendations in the Institute's publications are solely those of the author, and should not be attributed to the Board of Directors, Council of Trustees, or contributors to the Institute.

The president bears final responsibility for the decision to publish a manuscript under the Institute's imprint. In reaching this decision, he is advised on the accuracy and objectivity of a manuscript by both Institute staff and outside reviewers. Publication of a manuscript signifies that it is deemed to be a competent treatment of a subject worthy of public consideration.

Publications of the Institute are published in the language of the author, along with an executive summary in both of Canada's official languages.

iv

Doris Anderson, O.C.
Former President, Advisory Council on the
Status of Women, Toronto
Dr. Roger Blais, P.Eng.
Managing Director, Centre québécois
d'innovation industrielle, Montreal
Robert W. Bonner, Q.C.
Chairman, British Columbia Hydro &
Power Authority, Vancouver
Professor John L. Brown
Faculty of Business Administration &
Commerce, University of Alberta,
Edmonton
George Cooper,
Halifax
James S. Cowan,
Stewart, MacKeen & Covert, Halifax
Dr. Mark Eliesen
Director of Research, New Democratic
Party, Ottawa
W.A. Friley
President, Skyland Oil, Calgary
Dr. Donald Glendenning,
President, Holland College, Charlottetown
Judge Nathan Green
The Law Courts, Halifax
Dr. Leon Katz, O.C.
Department of Physics, University of
Saskatchewan, Saskatoon
Tom Kierans
Chairman, Ontario Economic Council,
Toronto
Dr. Leo Kristjanson
President, University of Saskatchewan,
Saskatoon
Allen T. Lambert, O.C.
Vice-President, Toronto-Dominion Bank,
Toronto
Terry Mactaggart
President, Niagara Institute,
Niagara-on-the-Lake
Dr. John McCallum,
Faculty of Administrative Studies,
University of Manitoba, Winnipeg
Professor William A.W. Neilson
Faculty of Law, University of Victoria
R.D. Nolan, P. Eng.,
General Manager, Neill & Gunter Ltd.,
Fredericton
Robert Olivero,
President, Management House Ltd.,
St. John's
Marilyn L. Pilkington
Osgoode Hall Law School, Toronto
Eldon D. Thompson,
President, Telesat Canada, Vanier

Dr. Israel Unger,
Department of Chemistry, University of
New Brunswick, Fredericton
Philip Vineberg, O.C., Q.C.
Phillips, Vineberg, and Associates,
Montreal
Dr. Norman Wagner
President, University of Calgary
Ida Wasacase
Director, Saskatchewan Indian Federated
College, University of Regina
Professor Paul Weiler
Mackenzie King Professor,
Harvard University, Cambridge
Dr. John Tuzo Wilson, C.C., O.B.E.
Director General, Ontario Science Centre,
Toronto
Rev. Lois Wilson
Moderator, United Church of Canada,
Kingston
Ray Wolfe, C.M.
Chairman and President, The Oshawa
Group, Toronto

Ex Officio Members
Dr. Robert E. Bell, C.C.
President, Royal Society of Canada
Dr. Pierre Bois
President, Medical Research Council
W. Bruce Brittain
President, Institute of Public
Administration of Canada
Dr. Owen Carrigan
Representing the Canada Council
Dr. Alan Earp
President, Association of Universities &
Colleges of Canada
Dr. Claude Fortier, O.C.
Chairman, Science Council of Canada
Dr. Larkin Kerwin, C.C.
President, National Research Council
Dr. David Slater
Chairman, Economic Council of Canada
Professor André Vachet
Representing the Social Science Federation
of Canada

Institute Management

Gordon Robertson	President
Louis Vagianos	Executive Director
David MacDonald, P.C.	Fellow in Residence
Raymond Breton	Director, Ethnic and Cultural Diversity Program
John M. Curtis	Director, International Economics Program
Rowland J. Harrison	Director, Natural Resources Program
Ian McAllister	Director, Regional Employment Opportunities Program
William T. Stanbury	Director, Regulation and Government Intervention Program
Zavis P. Zeman	Director, Technology and Society Program
Donald Wilson	Director, Conference and Seminars Program
Dana Phillip Doiron	Director, Communications Services
Ann C. McCoomb	Associate Director, Communications Services
Tom Kent	Editor, *Policy Options Politiques*

Foreword

The oil crisis of 1973—74 has been described as a turning point in post-war history. The OPEC oil embargo delivered a powerful economic and political jolt to the entire world. It led to a drastic shift in wealth among nations. It slowed the high level of post-war economic growth. It undermined the hopes and plans of the poor developing countries. In Canada and other industrialized countries, increases in the price of oil, coupled with constrained production, have generated a sense of urgency about the need to adapt to the reality of depleting non-renewable resources.

Prospective slow-downs in the flow of crude oil have cast shadows in the shape of recurring energy crises. Price, policy, and innovation have combined to bring about economies in consumption, with great variation among countries in the degree to which these have been achieved. There is general recognition of the risk to the world's economies unless new energy strategies are developed. In most modern states, such strategies attempt to generate a shift away from a heavy reliance on imported oil towards a more balanced energy system. This calls for the application of resource management techniques to assess the risks, benefits, priorities, and potentials of the different energy options.

Contributors to this book attempt to use various managerial approaches in the formulation of energy policies. Among the remedies put forth, there is little agreement as to which policies will best achieve a balanced energy system. While some experts argue that Canadian energy policy should emphasize intensive development of coal, others claim that it ought to strive for greater reliance on electricity, and still others contend that the transition to "soft energy paths" is a preferable policy approach. This collection of essays offers discussion on a broad range of policy responses on these and other energy questions, examining not only technical and economic possibilities, but political and institutional alternatives as well.

Gordon Robertson
President
May 1981

Avant-propos

On a qualifié la crise du pétrole de 1973 – 1974 d'un des points tournants de la période d'après-guerre. L'embargo pétrolier exercé par l'OPEP a asséné un rude coup économique et politique au monde entier. Il a provoqué un ralentissement de la croissance économique d'après-guerre, tout en ébranlant les espoirs et les projets des pays en voie de développement. Au Canada et dans d'autres pays industrialisés, les augmentations du prix du pétrole alliées à la limitation de la production ont fait prendre conscience de l'urgence de s'adapter à l'épuisement des ressources non renouvelables.

Les éventuels ralentissements de l'approvisionnement en brut font planer l'ombre de crises de l'énergie périodiques. Les prix, les politiques et l'innovation ont entraîné des économies de consommation, leur importance variant énormément d'un pays à l'autre. On admet généralement que si l'on n'élabore pas de nouvelles stratégies énergétiques, les économies de la planète courent de graves dangers. Dans la plupart des États modernes, de telles stratégies visent à provoquer un passage d'une forte dépendance du pétrole importé à un régime énergétique mieux équilibré. Cela nécessite un recours aux techniques de gestion des ressources en vue d'évaluer les risques, les avantages, les priorités et les possibilités des différentes options.

Les auteurs qui ont collaboré à ce volume font appel à diverses conceptions de la gestion dans le but de formuler des approches du problème de l'énergie. Parmi les remèdes proposés, on n'arrive pas à s'entendre sur les mesures les plus aptes à procurer un système énergétique équilibré. Tandis que certains experts soutiennent que la politique énergétique canadienne devrait être axée sur la mise en valeur intensive du charbon, d'autres prétendent qu'elle devrait faire une plus grande place à l'électricité et d'autres encore affirment que l'on devrait privilégier une transition aux « énergies douces ». Ce recueil présente un examen d'une vaste gamme d'orientations possibles face à ces questions énergétiques (et à d'autres), et traite non seulement des possibilités techniques et économiques mais aussi des choix politiques et institutionnels.

Gordon Robertson
Président
Mai 1981

Table of Contents

ix

Notes on the Authors

Scott E. ATKINSON
Scott E. Atkinson is a senior research economist with the American Petroleum Institute in Washington, D.C. He received a B.A. from Williams College, an M.A. in economics from the University of Northern Colorado, and a Ph.D. in economics from the University of Colorado. His principal areas of research are energy and environmental economics.

Leo Denis BARRY
Currently Minister of Mines and Energy for Newfoundland and Labrador, Leo Barry has a B.Sc. and B.A. from Memorial University, an LL.B. from Dalhousie Law School, and an LL.M. from Yale. A member of the Newfoundland Bar, Mr. Barry practised law in St. John's and lectured in political science at Memorial University from 1968 to 1972. In March 1972, he was elected a Member of the Newfoundland House of Assembly and became Deputy Speaker at its next sitting. Mr. Barry was Minister of Mines and Energy for Newfoundland from December 1972 until September 1975. From 1975 until his recent return to the Newfoundland Cabinet, Mr. Barry held several positions including chairman of the Newfoundland Labour Relations Board and associate professor at Dalhousie Law School.

Norbert BERKOWITZ
Dr. Norbert Berkowitz graduated with a B.Sc. (Hons.) and Ph.D. in chemistry from the University of London, and came to Canada in 1952 to join the staff of the Alberta Research Council. He currently divides his time between the University of Alberta, Edmonton, where he is professor of fuel science in the Department of Mineral Engineering, and Alberta's Energy Resources Conservation Board, Calgary, on which he serves as a vice-chairman. Dr. Berkowitz has lectured extensively in the Americas, Europe, and Asia, and has authored over one hundred and thirty papers on fuel science topics as well as a textbook (*An Introduction to Coal Technology*), which was recently published by Academic Press, New York.

Ernst R. BERNDT
Ernst R. Berndt received his Ph.D. in economics from the University of Wisconsin in 1972. Currently he is a professor of applied economics in the Sloan School of Management at the Massachusetts Institute of Technology.

Peter BLAIR
Dr. Blair is an assistant professor of regional science and public policy at the University of Pennsylvania. His primary research interests are in modelling

and simulation of energy systems and energy-environmental policy analysis. Dr. Blair received a B.S. in engineering from Swarthmore College and an M.S., M.Sc., and Ph.D. from the University of Pennsylvania.

Thomas CASSEL
Dr. Cassel is a principal of Technecon Analytic Research, Inc. in Philadelphia. He is a consultant to public sector agencies, corporations, and private investors in fields of investment decision analysis and project feasibility—primarily dealing with energy technologies. Dr. Cassel received his B.Sc., M.Sc., and Ph.D. from the University of Pennsylvania.

Donald N. DEWEES
Donald N. Dewees is a professor of economics in the Department of Political Economy and a professor at the Faculty of Law at the University of Toronto. He is also an associate at the Institute for Policy Analysis. He received an LL.B. degree from the Harvard Law School, and a Ph.D. in economics, also from Harvard. His research interests include urban transportation, environmental economics, and topics in the area of law and economics. Recent studies have included analysis of energy conservation opportunities in consumer goods, and the evaluation of alternative policies for pollution control. Professor Dewees is currently serving as the director of research for the Royal Commission on Matters of Health and Safety arising from the use of asbestos in Ontario.

Robert EDELSTEIN
Dr. Edelstein, a professor of finance at the Wharton School, University of Pennsylvania, is active in public and private sector research and consulting in real estate finance, urban economics, public finance, and environmental and energy economics. Dr. Edelstein received his A.B., A.M., and Ph.D. in economics from Harvard University.

G. Michael FOLIE
G. Michael Folie is associate professor of economics at the University of New South Wales, Sydney, Australia. After graduating from the University of Melbourne, he received his Ph.D. in engineering from the University of Southampton and the M.Sc. in economics from the London School of Economics. Dr. Folie has acted as consultant to numerous private and public organizations. His research interests are now in the area of energy policy and resource economics in which he has published numerous papers and monographs. Recent publications include papers in *Energy Systems and Policy, Applied Mathematical Modelling*, and *Australian Quarterly*.

Robert HALVORSEN

Robert Halvorsen is an associate professor of economics at the University of Washington in Seattle, Washington. He received a B.B.A. from the University of Michigan and a M.B.A., M.P.A., and Ph.D. in economics from Harvard University. His principal areas of research are natural resources and environmental economics.

Nancy HASSIG

Nancy Hassig serves as a project manager for the Electric Utility Rate Design Study at the Electric Power Research Institute (EPRI) in Palo Alto, California. Dr. Hassig came to EPRI in 1978 as a consultant while completing her dissertation at Stanford University in the area of electricity pricing and revenue reconciliation. Previous work experience includes financial accounting manager at Hewlett-Packard Corporation and operations research analyst at Memorex Corporation. Dr. Hassig has taught computer science at San Jose State University and West Valley College. Her educational credits include a Ph.D. in industrial engineering from Stanford University, and a M.S. in business administration from the University of Santa Clara, and a M.S. in statistics from Stanford University.

Jonathan R. KESSELMAN

Jonathan R. Kesselman received his Ph.D. degree in economics in 1972 from the Massachusetts Institute of Technology. Currently he is professor of economics at the University of British Columbia.

Fred H. KNELMAN

Fred Knelman is professor of science and human affairs at Concordia University in Montreal. He received a Ph.D. in chemical engineering from the University of London and a D.I.C. from the Imperial College of Science and Technology in 1953. Dr. Knelman has held faculty appointments at McGill and York Universities and has been visiting professor at the University of California (Santa Barbara) and the California Institute of the Arts. His publications include *Energy Conservation* (Background Study for the Science Council of Canada, 1975), *Nuclear Energy: The Unforgiving Technology* (1976), and *Anti-Nation: Transition to Sustainability* (1979).

Amory B. LOVINS

Amory B. Lovins is an American consultant physicist working as a team with his wife and colleague Hunter in fifteen countries. Educated at Harvard and Oxford, he resigned a Junior Research Fellowship of Merton College, Oxford in 1971 to serve as British representative of Friends of the Earth, Inc. Twice appointed Regents' Lecturer in the University of California (Berkeley, energy policy, 1978, and Riverside, economics, 1980), he was 1979 Grauer Lecturer at the University of British Columbia, and has advised a wide range of governments and international institutions. The most recent of his eight books, co-authored by Hunter Lovins, is *Energy/War: Breaking the Nuclear Link* (Friends of the Earth (San Francisco), 1980, and Harper & Row (New York), 1981).

J. Robert MALKO

J. Robert Malko has resumed his position as chief economist at the Wisconsin Public Service Commission after serving as programme manager of the Electric Utility Rate Design Study at the Electric Power Research Institute since 1978. From 1975 to 1977 he served as chief economist at the Wisconsin Public Service Commission and chairman of the Staff Subcommittee on Economics of the National Association of Regulatory Utility Commissioners. Dr. Malko has taught public utility economics at the University of Wisconsin at Madison and Stanford University. He has carried out consulting assignments for state and local governments; he has appeared as an expert witness on electricity pricing issues before various regulatory commissions; and he has written several articles on electricity pricing. Dr. Malko received the B.S. degree in mathematics and economics from Loyola College (Baltimore, Maryland) and the M.S. and Ph.D. degree in economics from Purdue University.

John MELVIN

John Melvin graduated in mechanical engineering from McGill University in 1950. After gaining experience as a diesel-electric locomotive service engineer and an oil refinery and chemical plant maintenance engineer, he joined the Chalk River Nuclear Laboratories of Atomic Energy of Canada Ltd. in 1954. Since then he has participated in the design and development of experimental facilities, nuclear fuel and process equipment, reliability engineering, and R&D planning and budgeting. He is currently engaged in a study of the scope for electricity/oil substitution. He is a member of the Canadian Society for Mechanical Engineering and the Association of Professional Engineers of Ontario.

Peter N. NEMETZ

Peter Nemetz received a B.A. in economics and political science from the University of British Columbia and an A.M. and Ph.D. in economics from Harvard University. He is currently an assistant professor in the Policy Analysis Division of the Faculty of Commerce and Business Administration at the University of British Columbia. Recent publications include: co-author, "The Role of Effluent Monitoring in Environmental Control," *Water, Air and Soil Pollution* 10, No. 4 (1979); editor, *Energy Policy: The Global Challenge* (The Institute for Research on Public Policy, 1979); *Economic Incentives for Energy Conservation at the Consumer Level* (Consumer and Corporate Affairs Canada, 1979 and 1980); "System

Solutions to Urban Wastewater Control,'' *Journal of Environmental Economics and Management* 7, No. 2 (1980); editor, *Resource Policy: International Perspectives* (The Institute for Research on Public Policy, 1980); co-author, ''The Use of Biological Criteria in Environmental Policy,'' *Water Resources Bulletin* 16, No. 6 (1980); and co-author, ''The Biology-Policy Interface: Theories of Pathogenesis, Benefit Valuation and Public Policy Formation,'' *Policy Sciences* 13 (1981):125–38.

Dennis RAY
Dennis Ray is currently pursuing a Ph.D. in transportation and public utilities from the School of Business, University of Wisconsin. After earning a B.Sc. degree from the University of New Mexico and gaining four years of engineering experience, he began graduate study while working on electricity pricing reform issues at the Wisconsin Public Service Commission. His recent work includes a study of the viability of lifeline pricing, an analysis of the industrial response to time-of-day rates, and an assessment of the economic and institutional barriers to the diffusion of waste heat recovery systems.

Ernest SIDDALL
Mr. Siddall is a special assistant to the Vice-President, Design & Development, AECL Engineering Company. He received the degree of B.Sc. (Eng) with first class honours from London University in 1939. After service in the Royal Corps of Signals in the British Army, he had experience in the telephone industry and in the instrumentation aspects of high explosive research before coming to Canada in 1951. He was a member of the design team of Canada's first aircraft flight simulator. With Atomic Energy of Canada Limited from 1954 onwards, he contributed to important advances in the safety systems of reactors. He took part in the design of the Douglas Point, RAPP (India), Pickering A, and the early stages of the Bruce A reactors and supervised the evolution of control by digital computer in these projects.

Alistair M. ULPH
Alistair M. Ulph is currently lecturer in the Department of Economics, University of Southampton, England. He graduated in economics from the University of Glasgow and received the B.Phil. in management studies from Oxford University. Mr. Ulph's research interests are in environmental and resource economics, with particular emphasis on energy economics. He has published widely in these fields, and recent papers have appeared in *Energy Economics, The Economic Record*, and the *Scottish Journal of Political Economy*.

G.C. WATKINS
G.C. Watkins was born and educated in England and is a graduate of the University of Leeds. Currently he is president of the economic consulting firm of DataMetrics Limited and holds an appointment as adjunct professor of economics at the University of Calgary.

Leonard WAVERMAN
Leonard Waverman is a professor of economics in the Department of Political Economy and Faculty of Law at the University of Toronto; a director of the Policy and Economic Analysis Program of the Institute for Policy Analysis; a past, part-time member of the Ontario Energy Board; and an executive council member of the International Association of Energy Economists. Professor Waverman is the author of articles and books in natural resource economics and public utility economics. His present research interests include issues of incremental costing in regulated industries and economic analysis of administrative law.

D.G.T. WILLIAMS
D.G.T. Williams is reader in public law and president of Wolfson College in the University of Cambridge. He has been a member of the Royal Commission on Environmental Pollution since 1976, of the Commission on Energy and the Environment, and of the Council on Tribunals since 1972. He was a member of the Clean Air Council, 1971–79.

Introduction and Executive Summary

N^A

Energy Crisis—Policy Response

by
Peter N. Nemetz

INTRODUCTION

Almost a decade has passed since the global apperception of the crisis in energy supply. Considering the nature of the threat posed to the survival of modern industrialized society, the formulation and implementation of appropriate policy responses have been characterized by an alarming amount of inertia and indecision. This inability to respond effectively to such a serious challenge can be explained by a number of factors, foremost among which are the complex interdependence of energy issues with political, economic, environmental, and social systems, and the existence of three distinct energy problems distinguished by their own characteristic time frames, potential range of solutions, and incumbent risks.

The first of these problems, the expected dislocation of petroleum supply in the next two decades, is merely a prologue to the second and more intractable problem of identifying appropriate transition fuels in the post-petroleum era, and the third problem of developing truly long-term energy technologies that can effectively remove the constraint of energy supply from man's endeavours (Nemetz, 1979).

The evolution of Canadian energy policy since 1973 and the radical transformation in the assessment of Canada's energy capabilities and constraints have been delineated by a succession of government policy documents. From confident assurances of plentiful supplies of petroleum and natural gas (Canada, 1973), the government has progressed through uneasy hopes for self-reliance (Canada, 1976), to renewed optimism for the early achievement of total energy self-sufficiency (Canada, 1980).

The most recent of these major policy statements entitled *The National Energy Program* was issued by the Canadian federal government in October 1980. Three principal approaches have been adopted in an attempt to free Canada from reliance on foreign petroleum supplies by 1990: (1) a reduction in demand for oil by continuing price increases, (2) the encouragement of increased development of domestic conventional and non-conventional

petroleum supplies through a package of selective price increases, tax
incentives, and grants, and (3) the encouragement of interfuel substitution
away from petroleum to natural gas, electricity, coal, and renewable energy
sources.

One of the distinguishing characteristics of the Canadian and indeed
international debate over current issues of energy policy is the juxtaposition
of two disparate developments—the emerging consensus on the nature of the
energy problem confronting the world and a wide and potentially irreconcilable difference of views on appropriate remedies. The papers in this volume
mirror this diversity.

A similarity of energy problems among nations is insufficient for the
development of a consensus on desirable policy responses. Canada and
Australia face comparable problems of energy supply. Both countries are
well endowed with a wide range of energy sources, yet both face increasing
dependence on petroleum imports in the absence of remedial government
activity. What is the nature of an appropriate policy response? While Ottawa
is solidly committed to rapidly releasing Canada from dependence on
external petroleum supplies, *Alistair Ulph* and *Michael Folie,* in their
analysis of the Australian situation, conclude that

> ... to concentrate attention on self-sufficiency is to misdirect the debate on energy
> policy. The specific problems that arise in connection with self-sufficiency—
> balance-of-payments difficulties and security of supply—can be handled in a fairly
> straightforward manner, though not by trying to maintain high levels of self-
> sufficiency. This can lead into issues concerning the equitable sharing of welfare
> losses, and the ability of markets to provide adequate signals for risky activities with
> long lead times. However, these latter issues would be important even if there were
> no international trade, and it is on the latter set of questions that policy makers and
> advisers should concentrate their efforts.

In "Missing Elements in Canadian Energy Policy," *Norbert Berkowitz,*
one of Canada's leading experts on coal development, provides a sober
technological assessment of the prospects for rapid, national petroleum
self-sufficiency. The author presents a strong case for broadening the scope
of energy policy to include more intensive development of Canadian coal. As
Berkowitz warns,

> ... grave uncertainty over how much oil sands and heavy oil development
> programmes can in fact be speeded up makes it in my view imperative to base a
> significant part of future syncrude production on coal.

In a critique of current Canadian plans for expanded development of the
tar sands and frontier petroleum, *Fred Knelman* states that

> [c]onventional wisdom, unfortunately, is tied to conventional sources. Bad energy,
> like bad money, drives out the good. Moreover, bad energy is both costly and
> uncertain. The billions of dollars required for tar sands, pipelines, and frontier and
> offshore sources will surely cut into the money we desperately require for necessary

social services, let alone for conservation. Thus bad energy drives out both good energy and good money.

Interfuel substitution is essential to the development of any viable national energy policy, yet the concept is exceptionally broad, encompassing fuels and technologies from both the ''hard'' and ''soft'' end of the energy spectrum. Agreement on the appropriate manifestation of this policy remedy is not always forthcoming. One of the most controversial elements of interfuel substitution concerns the role of electricity in meeting future energy needs. Complicating the dispassionate assessment of this option is its invariable association with the issue of nuclear power.

John Melvin argues that ''both economics and technology point to a major role for electricity, at least in the Canadian context. . . . The essential first step is to complete the substitution of electricity for imported oil during the next ten to fifteen years. This is shown to be technologically feasible and economically beneficial.''

In a paper that addresses the safety issues associated with the use of uranium and coal as energy sources, *Ernest Siddall* also lends his support to the increased use of electricity and nuclear power. In an approach similar to Wildavsky's ''Richer Is Safer'' (1980), the author describes the ''benefits in respect to safety [that] may result indirectly from the development of low-cost energy.''

In contrast to the position of Melvin and Siddall, *Amory Lovins* articulates the case against greater reliance on energy from electricity. In describing the current financial state of the United States utility industry, Lovins argues that the ''economic options for profitably selling energy, especially electrical energy, are constrained by the thermodynamic laws governing its conversion and distribution.'' In asking whether further central electrification in the United States makes economic sense, the author concludes in the negative, '' . . . if we do what is economically rational, that is, cost minimizing, risk minimizing, and profit maximizing.'' Lovins foresees, however, a ''strong potential role of the utilities as bankers for the energy transition [to softer energy paths] that is already irresistibly underway.''

The debate on the future role of electrical energy will not be resolved easily, as it is based not only on divergent national experiences, but also on broader philosophical differences concerning the interrelation between energy and the structure and functioning of society.

Despite the existence of potentially irreconcilable differences of opinion on some energy-related issues, it is reassuring to observe the gradual evolution of agreement on other vital questions of energy policy. Foremost among these is the increasing support for market mechanisms, especially price, as one of the indispensible instruments of government energy policy. *Donald Dewees* concludes that

...governments may play an important role in energy conservation. That role, however, is to improve the operation of markets, not to replace them. We cannot hold the price of energy below social cost and achieve the same conservation results as if those prices reflected social cost, no matter what policies are adopted. A serious energy conservation programme requires energy prices at social costs, information programmes for a few major energy-consuming products, perhaps an energy excise tax for some products and, in isolated cases, co-operative minimum efficiency standards that are not technology forcing. This is not massive regulation but mini-regulation.

One particularly important aspect of pricing policy is the reform of rate structures for such fuels as natural gas and electricity. Historical pricing patterns have promoted the excessive use of energy and indirectly led to the creation of an industrial base that is especially vulnerable to increasing energy prices. The design of electricity rates has become an important issue for regulatory commissions and utilities in North America

[b]ecause of the energy dilemma, general inflation, increases in system peak demand, capital shortages, and growing concerns for environmental and consumer interests (Malko, Ray, and Hassig).

In their paper on innovative electricity pricing in the U.S. Midwest, *Robert Malko, Dennis Ray,* and *Nancy Hassig* attempt

... to provide regulatory commissions, electric utilities and other interested groups with information concerning recent legislative developments and implementation activities relating to time-of-day pricing of electricity. Instead of "reinventing the wheel," electric utilities and regulatory commissions should seriously examine and benefit from existing time-of-day pricing programmes that have made progress in formulating, applying, and analysing these innovative and potentially useful rate structures.

The central question of economic efficiency that emerges from the study of general energy price levels and specific rate structures has equal relevance to the design of taxing mechanisms and the control of regulated industry. *Ernst Berndt, Jonathan Kesselman,* and *Campbell Watkins* address "the question of whether regulated utilities should be required to 'flow through' the tax benefits of accelerated depreciation to their customers." The authors examine this controversial issue using the criterion of economic efficiency. They focus particularly on the alternative of "'normalization', by which the taxes charged for rate-setting purposes are those payable if straight-line rather that accelerated depreciation were required." In assessing the concept of normalization, the authors conclude that

... tax normalization corresponds more closely to the competitive standard than does tax flow-through. Moreover, normalization will tend to avoid inducing inefficient behaviour elsewhere in the economy.

Clearly, government must address a broad range of social, economic, and political issues. To protect diverse and divergent interests, a package of evaluative criteria must be utilized in resolving major issues of public policy. While economic efficiency *per se* represents only one of several guides to governmental action, its importance in the design and implementation of energy policy is paramount in view of the central role of energy provision in our economic system and the massive capital requirements for energy projects in the coming decades. For this reason, it is particularly important that government avoid the inadvertent introduction of distortions into the pricing of energy products.

An early and perceptive assessment of the effect of regulation on the efficiency of regulated industries was provided by Averch and Johnson in their pathbreaking article of 1962. In a contribution to this volume, *Scott Atkinson* and *Robert Halvorsen* adopt a similar theoretical approach in weighing the effects of fuel adjustment clauses on input choice in U.S. electric utilities. The authors conclude that

> ... a fuel adjustment clause can be expected to result in the use of more than the cost-minimizing amount of fuel relative to capital and labour in periods of increasing fuel prices.

Atkinson and Halvorsen have identified a potentially important distortion induced by governmental regulatory responses to recent significant increases in the general price level and the cost of fuel in particular.

It is imperative that policies developed to address the momentous challenges of energy production and utilization be well designed, well studied, and free from lacunae and contradictions that can lead to failure or counter-productive results. It is here, in particular, that the academic community has a vital role to play, assisting government in the formulation and implementation of rational, effective, and efficient energy policies. It is especially important when theoretical contributions have immediate relevance to the policy-making process. An example of this nexus between academe and governmental policy making is provided by *Peter Blair, Thomas Cassel,* and *Robert Edelstein* who examine optimal, multi-period investment decision making for electric generation activities.

> The theoretical thrust of the presentation is directed toward incorporating various assumptions about the world, including uncertainty, into the analytic structure of the problem-solving procedure.

Of importance is the fact that the concepts developed by the authors "are being applied and operationalized in research being conducted under the auspices of the United States Department of Energy."

While there are frequently marked similarities in the nature of the energy problems faced by many countries, there are other energy-related issues associated with specific differences in the constitutional structure of governments. Two papers in this volume address an issue of increasing importance to Canada as an archetypal federal state—the constitutional aspects of the production of energy resources and the distribution of related revenues. *Leo Barry* provides a legal study of

> ... the ways in which Canada's constitutional structure may affect schemes to give Canada improved electrical planning, generation, and transmission on a regional and national basis. . . . the main question . . . , as in so many other areas of Canadian constitutional law, concerns the division of jurisdiction between the federal and provincial governments.

Conclusions are drawn by the author that

> ... indicate a role . . . for both federal and provincial governments but identify areas of constitutional uncertainty and possible drastic shifts in the federal-provincial balance of power if co-operative arrangements are not developed between the two levels of government.

While Barry has identified an area of potential future conflict within the Canadian confederation, *Leonard Waverman* examines an issue that has already caused considerable political friction within Canada—the distribution of resource rents generated by the petroleum and natural gas industries. Waverman's conclusion is forceful and unambiguous:

> ... there is sound justification for a significant federal share of resource rents. Part of the sector's well-being is due to past federal subsidies. Moreover, in a federation concerned with equity of income distribution, it is unacceptable that resource-rich areas attempt to tax their way out of Confederation.

CONCLUSION

The diverse offerings of this volume present a convincing demonstration that the process of energy policy formation and implementation is exceptionally complex. To borrow from *David Williams'* description of environmental control and energy policy in the United Kingdom,

> ... the issues are scientific and technical, political and administrative, social and economic, and legal both at the national and international levels.

What distinguishes energy from other major and interdependent issues of public policy is its unique role as the *sine qua non* of modern, industrialized civilization. The pervasive role of energy in our social, economic, and political order mandates that the study of energy policy alternatives receive a singular and unprecedented urgency and pre-eminence in the deliberations of modern government and society.

SELECTED REFERENCES

Averch, H. and Johnson, L.L. (1962) "Behavior of the Firm Under Regulatory Constraint," *American Economic Review* 52 (December): 1052-69.

Canada, Department of Energy, Mines and Resources. (1973) *An Energy Policy for Canada: Phase 1*. Ottawa: Information Canada.

Canada, Department of Energy, Mines and Resources. (1976) *An Energy Strategy for Canada: Policies for Self-Reliance*. Ottawa: Minister of Supply and Services Canada.

Canada, Department of Energy, Mines and Resources. (1980) *The National Energy Program*. Ottawa: Minister of Supply and Services Canada.

Nemetz, Peter N., ed. (1979) *Energy Policy: The Global Challenge*. Montreal: The Institute for Research on Public Policy.

Wildavsky, Aaron. (1980) "Richer is Safer." *The Public Interest* 60 (Summer): 23-39.

Introduction et abrégé

La crise de l'énergie et la politique gouvernementale

par
Peter N. Nemetz

INTRODUCTION

Près d'une décennie s'est écoulée depuis que l'on a pris conscience, à l'échelle mondiale, d'une crise des approvisionnements en énergie. Compte tenu de la menace que pose cette crise à la survie même de la société industrielle moderne, l'élaboration et la mise en oeuvre de lignes de conduite officielles ont témoigné d'une inertie et d'une indécision alarmantes. De nombreux facteurs expliquent cette incapacité de réagir efficacement à un tel défi. Les plus importants sont l'interdépendance des questions énergétiques et des systèmes politique, économique, écologique et social, ainsi que l'existence de trois différents problèmes énergétiques ayant chacun son propre échéancier, son éventail de solutions possibles et ses risques inhérents.

Le premier de ces problèmes, soit le bouleversement des approvisionnements en pétrole prévu au cours des vingt prochaines années, n'est que le prologue du deuxième et plus difficile problème que constitue l'identification de combustibles de transition en vue de l'ère postpétrolière, et du troisième problème que représente la mise au point de réelles techniques énergétiques à longue échéance qui puissent libérer les entreprises humaines des contraintes imposées par l'approvisionnement en énergie (Nemetz, 1979).

Plusieurs documents d'orientation du gouvernement ont tracé l'évolution de la politique énergétique du Canada depuis 1973, et la transformation radicale de l'évaluation des possibilités et des limites du Canada en matière d'énergie. Partant d'assurances confiantes d'approvisionnements abondants en pétrole et en gaz naturel (Canada, 1973), en passant par des espoirs incertains d'autonomie (Canada, 1976), le gouvernement manifeste aujourd'hui un optimisme renouvelé face à une autosuffisance énergétique prochaine et complète (Canada, 1980).

C'est en octobre 1980 que le gouvernement fédéral canadien publiait le dernier de ces importants énoncés de principe intitulé *Le programme énergétique national*. On a opté pour trois grandes approches dans le but de libérer le Canada, dès 1990, de sa dépendance des approvisionnements étrangers : 1) une réduction de la demande de pétrole par une augmentation constante des prix; 2) l'incitation à l'exploitation accrue des sources de pétrole, classiques ou non, par un ensemble d'augmentations sélectives des prix, de stimulants fiscaux et de subventions; et 3) l'encouragement au remplacement du pétrole par le gaz naturel, l'électricité, le charbon et les énergies renouvelables.

Le débat canadien et, à vrai dire, international sur les questions énergétiques est marqué par la juxtaposition de deux tendances disparates : la naissance d'un consensus sur la nature du problème énergétique mondial, et une différence d'opinions profonde et possiblement irrémédiable sur les remèdes. Les textes de ce volume témoignent de cette diversité.

La ressemblance des problèmes énergétiques d'un pays à l'autre ne suffit pas à assurer le consensus sur les politiques à adopter pour y remédier. Le Canada et l'Australie font face à des problèmes comparables d'approvisionnement en énergie. Les deux pays disposent d'une vaste gamme de sources d'énergie, mais ils sont tous deux aux prises avec une dépendance croissante des importations de pétrole par manque de mesures correctives des gouvernements. Quelle serait la nature d'une réaction appropriée? Tandis qu'Ottawa est fermement résolu à rendre le Canada indépendant des approvisionnements en pétrole étrangers, *Alistair Ulph* et *Michael Folie* concluent, dans leur analyse de la situation australienne, que :

> (…) un débat sur la politique énergétique qui insiste sur l'autosuffisance fait fausse route. Les problèmes que soulève la question de l'autosuffisance—la balance des paiements et la certitude des approvisionnements—peuvent se régler sans trop de détours mais sans toutefois tenter d'atteindre une pleine autosuffisance. Cela peut occasionner des problèmes concernant le partage des pertes de bien-être et la capacité des marchés d'émettre les avertissements requis dans le cas d'activités de risque à long délai d'exécution. Toutefois, ces dernières questions seraient importantes même s'il n'y avait pas de commerce international, et c'est là que les décisionnaires et les conseillers devraient concentrer leurs efforts.

Dans « Missing Elements in Canadian Energy Policy », *Norbert Berkowitz*, un des plus grands experts canadiens en matière de mise en valeur du charbon, présente une évaluation technique réfléchie des chances d'atteindre rapidement l'autosuffisance en pétrole. L'auteur développe un solide argument en faveur de l'élargissement de la portée de la politique énergétique en vue d'une mise en valeur plus poussée du charbon canadien. Comme nous en prévient M. Berkowitz,

> (…) l'incertitude prononcée face à la possibilité d'accélérer la mise en oeuvre des programmes de mise en valeur des sables pétrolifères et de l'huile lourde nous oblige,

à mon avis, à fonder une bonne part de la production future de brut synthétique sur le charbon.

Dans une critique des plans canadiens visant à l'exploitation accrue des sables pétrolifères et du pétrole des régions vierges, *Fred Knelman* affirme que :

> (...) malheureusement, l'opinion courante est fondée sur les sources classiques. En énergie comme en matière d'argent, le mauvais chasse le bon. En outre, la mauvaise énergie est à la fois chère et peu fiable. Les milliards qui seront engloutis dans les sables bitumineux, les pipelines et le pétrole des régions vierges ainsi que celui au large des côtes le seront certainement au détriment des services sociaux essentiels et encore davantage des économies d'énergie. Ainsi, la mauvaise énergie chasse à la fois la bonne énergie et le bon argent.

Le remplacement des énergies est essentiel à l'élaboration de toute politique énergétique viable; ce concept est cependant très vaste et comprend à la fois des énergies « dures » et des énergies « douces ». On n'arrive pas toujours à s'entendre sur l'expression concrète d'un tel remède. Une des questions les plus controversées en matière de remplacement des énergies est celle du rôle futur de l'électricité. Son inévitable relation avec le nucléaire rend difficile une évaluation impartiale de cette option.

John Melvin soutient que l'« économie et la technologie permettent toutes deux de prévoir un rôle important pour l'électricité, du moins dans le contexte canadien (...) Il s'agit en premier lieu de remplacer, d'ici dix ou quinze ans, le pétrole importé par l'électricité. Cela s'avère techniquement possible et économiquement avantageux. »

Dans un texte qui porte sur la sûreté du recours au charbon et à l'uranium comme sources d'énergie, *Ernest Siddall* accorde lui aussi son appui à l'utilisation accrue de l'électricité et de l'énergie nucléaire. À l'aide d'une approche analogue à celle de M. Wildavsky dans « Richer Is Safer » (1980), l'auteur décrit les « avantages indirects, en matière de sûreté, que pourrait procurer la mise en valeur des énergies bon marché ».

À l'opposé de MM. Melvin et Siddall, *Amory Lovins* expose les arguments contre un recours plus poussé à l'électricité. Dans une description de l'état financier actuel des entreprises de service public américaines, M. Lovins soutient que « les lois thermodynamiques qui régissent la conversion et la distribution de l'énergie, et en particulier de l'énergie électrique, limitent les occasions économiques de la vendre à profit ». L'auteur conclut que « (...) si nous faisons ce qui est économiquement rationnel, soit la minimisation des coûts et des risques, et l'optimisation des profits », l'électrification centralisée n'a aucun sens économique aux États-Unis. Selon M. Lovins, cependant, les entreprises de service public « pourraient éventuellement jouer un rôle important en tant que banquiers de la transition énergétique (vers des énergies plus douces) qui est déjà irrémédiablement en cours ».

Il ne sera pas facile de conclure le débat sur le rôle futur de l'énergie électrique, car il s'alimente non seulement des expériences nationales divergentes mais aussi de plus profondes différences philosophiques concernant les rapports entre l'énergie, d'une part, et la structure et le fonctionnement de la société, d'autre part.

Malgré l'existence de certaines différences d'opinion possiblement irréconciliables sur certaines questions relatives à l'énergie, il est rassurant de constater l'apparition graduelle d'un accord sur les autres questions essentielles de politique énergétique. On remarque avant tout l'accroissement de l'appui aux mécanismes commerciaux, et en particulier les prix, comme instruments indispensables de toute politique énergétique. *Donald Dewees* conclut que :

> (...) les gouvernements peuvent jouer un rôle important dans les économies d'énergie. Ce rôle, cependant, est d'améliorer le fonctionnement des marchés et non pas de les remplacer. Nous ne pouvons garder le prix de l'énergie en deça du coût social et obtenir les mêmes économies d'énergie que si le prix reflétait le coût social, peu importe la politique adoptée. Un programme sérieux d'économie d'énergie exige des prix énergétiques fondés sur les coûts sociaux, des programmes d'information touchant certains produits consommant beaucoup d'énergie et peut-être une taxe d'accise sur certains produits, et, dans certains cas, des normes d'efficacité minimales établies en collaboration et qui ne forcent pas la technologie. Il ne s'agit pas là d'une réglementation massive mais d'une mini-réglementation.

La réforme du barème des tarifs pour des énergies telles que le gaz naturel et l'électricité constitue un aspect particulièrement important de l'établissement des prix. Le profil traditionnel des prix a encouragé l'utilisation excessive d'énergie et a mené indirectement à la formation d'une base industrielle particulièrement vulnérable à la hausse du prix de l'énergie. La détermination des tarifs d'électricité est devenue une question importante pour les entreprises de service public et les commissions de réglementation nord américaines

> (...) en raison du dilemme énergétique, de l'inflation généralisée, de l'augmentation de la demande aux heures de pointe, des pénuries de capitaux et d'une préoccupation croissante face aux intérêts des consommateurs et à l'environnement (Malko, Ray et Hassig).

Dans leur texte sur le mode innovateur de détermination des prix dans le Midwest américain, *Robert Malko*, *Dennis Ray* et *Nancy Hassig* tentent

> (...) de fournir aux commissions de réglementation, aux compagnies d'électricité et aux autres groupes intéressés des renseignements concernant les dernières mesures législatives et les activités de mise en oeuvre touchant la facturation de l'électricité selon l'heure du jour. Au lieu de « réinventer la roue », les compagnies d'électricité et les commissions de réglementation devraient étudier sérieusement les programmes de facturation selon l'heure du jour, qui ont permis de progresser dans l'élaboration, l'application et l'analyse de ces structures tarifaires innovatrices et éventuellement utiles, et d'en tirer profit.

La question centrale de l'efficience économique qui se dégage de l'étude des niveaux du prix de l'énergie en général, et plus précisément des structures tarifaires, est également pertinente à la mise au point de mécanismes fiscaux et au contrôle de l'industrie réglementée. *Ernst Berndt, Jonathan Kesselman* et *Campbell Watkins* se penchent sur la « question de savoir si l'on devrait exiger que les entreprises de service public réglementées transfèrent à leurs clients les avantages fiscaux procurés par l'amortissement accéléré ». Les auteurs examinent cette question controversée à la lumière du critère de l'efficience économique. Ils insistent particulièrement sur la solution de rechange que constitue la « normalisation » en vertu de laquelle les taxes imposées à des fins de détermination de tarifs sont celles qui seraient exigibles si l'on exigeait un amortissement linéaire plutôt qu'accéléré ». Les auteurs concluent leur évaluation du concept de normalisation en affirmant que

(…) la normalisation fiscale correspond davantage aux normes concurrentielles que le transfert fiscal. En outre, la normalisation tendra à prévenir le comportement inefficient ailleurs dans l'économie.

Le gouvernement doit évidemment s'attaquer à un vaste éventail de problèmes sociaux, économiques et politiques. Afin de ménager des intérêts divers et divergents, il faut utiliser un ensemble de critères d'évaluation pour la résolution des principales questions relatives à la politique d'État. Bien que l'efficience économique comme telle ne représente qu'un des nombreux guides qui orientent l'activité des gouvernements, son importance relative-ment à l'élaboration et à la mise en oeuvre de la politique énergétique est primordiale, compte tenu du rôle décisif des approvisionnements en énergie dans notre régime économique et des exigences considérables de capital en vue de l'exploitation énergétique des prochaines décennies. C'est pourquoi il est particulièrement important que les gouvernements évitent de fausser, par inadvertance, l'établissement des prix de l'énergie.

MM. Averch et Johnson, dans leur article classique publié en 1962, avaient donné une première évaluation perspicace de l'effet de la réglementa-tion sur l'efficience des industries réglementées. Dans leur collaboration à ce volume, *Scott Atkinson* et *Robert Halvorsen* utilisent une approche théorique analogue afin de déterminer les effets des clauses de rajustement du prix des énergies sur le choix de ces entrées dans les entreprises de service public américaines. Les auteurs concluent que

(…) l'on peut s'attendre qu'une clause de rajustement du prix des énergies occasionne un recours à une quantité d'énergie supérieure à celle requise pour minimiser les coûts, eu égard au capital et à la main-d'oeuvre, en période d'augmentation des prix du combustible.

MM. Atkinson et Halvorsen ont identifié une déformation possiblement importante occasionnée par les mesures de réglementation du gouvernement en réaction aux récentes hausses substantielles du niveau des prix en général et du coût de l'énergie en particulier.

Il est de toute première importance que les mesures élaborées en réaction aux grands défis que posent la production et l'utilisation de l'énergie soient bien conçues, bien réfléchies, et libres des lacunes et des contradictions qui pourraient conduire à des échecs et donner des résultats contre-productifs. C'est surtout là que la communauté universitaire a un rôle décisif à jouer en aidant le gouvernement à élaborer et à mettre en oeuvre une politique énergétique rationnelle, efficace et efficiente. Ce rôle est particulièrement important lorsque les apports théoriques ont un rapport direct au processus de décision. *Peter Blair*, *Thomas Cassel* et *Robert Edelstein* fournissent un exemple de cette relation entre l'université et le gouvernement à l'occasion de leur examen de l'optimisation des décisions d'investissement sur plusieurs périodes en matière de génération d'électricité.

> La dimension théorique de l'exposé porte sur l'incorporation des diverses hypothèses ayant trait au monde, y compris l'incertitude, à la structure analytique de la méthode de solution de problèmes.

Le fait que les concepts élaborés par les auteurs « soient en voie d'être appliqués et concrétisés dans des recherches effectuées sous les auspices du United States Department of Energy » est d'importance.

Bien que de nombreux pays soient souvent aux prises avec des problèmes énergétiques semblables, d'autres problèmes relatifs à l'énergie proviennent des différences particulières dans la structure constitutionnelle des États. Deux articles de ce volume traitent d'une question qui revêt de plus en plus d'importance pour le Canada en tant qu'archétype de l'État fédéral, soit les aspects constitutionnels de la production de ressources énergétiques et la répartition des revenus qui en proviennent. *Leo Barry* effectue une étude juridique

> (...) des façons dont la structure constitutionnelle canadienne peut affecter les projets d'amélioration de la planification, de la génération et de la transmission électriques au Canada, sur les plans national et régional (...) comme dans bien d'autres domaines du droit constitutionnel canadien, le partage des compétences entre les gouvernements fédéral et provinciaux (...) constitue la principale question.

Selon les conclusions de l'auteur,

> (...) les gouvernements fédéral et provinciaux ont chacun un rôle à jouer, mais il existe des domaines constitutionnels où règnent l'incertitude et des possibilités de modifications radicales de l'équilibre des pouvoirs entre le fédéral et les provinces si les deux paliers n'arrivent pas à s'entendre pour collaborer.

Tandis que M. Barry identifie la possibilité d'un conflit futur au sein de la Confédération, *Leonard Waverman* se penche sur un problème qui a déjà occasionné passablement de frictions au Canada—la répartition des rentes tirées des ressources procurées par les industries pétrolière et gazière. La conclusion de M. Waverman est bien arrêtée et non équivoque :

(...) une participation du fédéral aux rentes tirées des ressources est bien justifiée. Les subventions antérieures du fédéral sont en partie responsables de l'actuelle prospérité de ce secteur. En outre, dans une fédération que préoccupe la répartition des revenus, il est inacceptable que les régions riches en ressources tentent d'échapper à la Confédération par des manoeuvres fiscales.

CONCLUSION

Il est évident, à la lecture des divers exposés de ce volume, que l'élaboration et la mise en oeuvre d'une politique énergétique constituent un processus d'une extrême complexité. Comme l'affirme *David Williams* dans sa description du contrôle de l'environnement et de la politique énergétique en Grande-Bretagne,

> (...) les problèmes sont d'ordre scientifique et technique, politique et administratif, social et économique, ainsi que juridique, tant au niveau national qu'international.

L'énergie se distingue des autres questions importantes et interdépendantes relatives à la politique d'État en ce qu'elle est une condition indispensable à la civilisation industrielle moderne. Le rôle prépondérant de l'énergie dans notre régime social, économique et politique exige que les débats de la société et de l'État contemporains accordent une priorité sans précédent à l'étude des choix dont nous disposons en cette matière.

BIBLIOGRAPHIE

Averch, H. et Johnson, L.L. 1962. « Behavior of the Firm Under Regulatory Constraint ». *American Economic Review*, 52 (décembre), p. 1052-1069.

Canada, ministère de l'Énergie, des Mines et des Ressources. 1973. *Politique canadienne de l'énergie, phase 1.* 2 vol. Ottawa, Information Canada.

Canada, ministère de l'Énergie, des Mines et des Ressources. 1976. *Une stratégie de l'énergie pour le Canada : politique d'autonomie.* Ottawa, Approvisionnements et Services Canada.

Canada, ministère de l'Énergie, des Mines et des Ressources. 1980. *Le programme énergétique national.* Ottawa, Approvisionnements et Services Canada.

Nemetz, Peter N., éd. 1979. *Energy Policy: The Global Challenge.* Montréal, l'Institut de recherches politiques.

Wildavsky, Aaron. 1980. « Richer Is Safer ». *The Public Interest*, 60 (été), p. 23-39.

Chapter One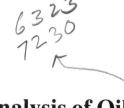

An Economic Analysis of Oil
Self-Sufficiency in Australia

by
Alistair Ulph and *Michael Folie**

SUMMARY

Although well endowed in other energy sources, Australia is currently only 65 per cent self-sufficient in crude oil, with the prospect that this will fall sharply in the 1980s as existing oil fields are depleted. This has led to suggestions that government policy should aim to ensure that a significant level of self-sufficiency in oil be maintained. This paper presents an economic analysis of various policy responses to such suggestions. It is shown first that under a wide range of scenarios, the fears raised are groundless—Australia may well become totally self-sufficient in oil. Nevertheless, under some scenarios, there will be a sharp fall in the level of self-sufficiency. There are two reasons why this could be of concern—the impact on the balance of payments and the threat of supply insecurity. Policies for dealing with both these problems are assessed and it is shown that, in both cases, attempts to maintain high levels of self-sufficiency are inappropriate. The final section considers broader aspects of the self-sufficiency debate, but these raise general issues of energy policy that would arise even in a closed economy.

INTRODUCTION

Australia is well endowed with energy resources (Australia, NEAC, 1978), having substantial reserves of black and brown coal and uranium, and significant reserves of natural gas. Unfortunately, Australia is not particularly well endowed with crude oil. Indigenous production accounts for about 65 per cent of domestic consumption, and the bulk of this production is from the offshore Gippsland Basin, in the Bass Strait between Victoria and Tasmania. Production from this source is forecast to decline sharply in the mid 1980s.

The combination of falling domestic production and rising import prices for crude oil has generated considerable public concern, especially about the impact on the balance of payments and security of supply. This concern has led to the belief expressed in some quarters that governments (federal and state) should pursue policies that would ensure a high level of self-sufficiency in crude oil for Australia for the rest of this century.[1]

The purpose of this paper is to present a simple economic analysis of some of the issues raised by the public debate on self-sufficiency. The first step is to analyse what might happen in the absence of any government policy (other than that already affecting the crude oil industry). By making a number of assumptions about the factors likely to affect the demand for and supply of crude oil (and its substitutes), it is possible to generate a number of scenarios about future levels of Australian self-sufficiency in crude oil. Under these assumptions, it transpires that the fears are not realized in most scenarios. There appears to be sufficient elasticity in the crude oil sector that the forecast high crude oil prices actually cause self-sufficiency to increase. There are a few scenarios, however, which show a marked decline in self-sufficiency from the mid-1980s, falling to under 20 per cent by the turn of the century.

The next step in the analysis is to ask what government policies might be appropriate should these pessimistic scenarios be realized. One reason for concern at the low level of self-sufficiency is the effect this could have on the balance of payments, and it is shown that there is indeed a sharp increase in Australia's oil import bill, both in absolute terms and as a share of total imports. However, it is argued that it is more sensible to spread the adjustment to this deficit throughout the export- and import-competing sectors of the economy, rather than concentrate the adjustment in one sector by pursuing policies of maintaining a high level of self-sufficiency.

A second reason for concern about low levels of self-sufficiency is the cost this would impose in the event of an embargo. Again, it is shown that these costs could indeed be high, but that a more efficient policy is to build stockpiles rather than aim for high levels of self-sufficiency.

The analysis presented in this paper does not support the case for policies designed to ensure that Australia maintains a high level of self-sufficiency in crude oil. Of course, it can be objected that the conclusion rests on the extremely simple nature of the analysis. Thus, no attempt is made to tackle the differences in energy resources confronting individual states; oil is treated as a homogeneous resource; concepts such as consumer surplus are used without qualification about their theoretical underpinnings; and so on. There are two justifications for this rather simple treatment.

First, most of the arguments presented are really arguments of principle and, as such, do not depend on the detailed quantitative analysis outlined here. The demand for self-sufficiency in oil is no different from that for protection of any domestic industry, and there are standard arguments why such protection should be avoided on efficiency grounds.

Second, the period of the study is 1980–2005, and over such a time period the major consideration is uncertainty, about future oil prices, costs of alternative technologies, and other factors. The paper attempts to capture the effects of the major elements of uncertainty, and it is inappropriate to attempt a very sophisticated analysis whose impact on the broad conclusions of this paper would be of second order. In this connection, it should be emphasized that the figures presented in this paper are not intended as forecasts. Rather, they are an attempt to establish broad orders of magnitude about the dimensions of the self-sufficiency debate.

ASSUMPTIONS AND SCENARIOS

The methodology adopted in this paper is as follows. It is assumed that Australia will follow a policy of setting crude oil prices equal to world prices; so the first step is to forecast world oil prices, and two possible paths are presented.[2] The assumption on prices allows domestic consumption and production of crude oil to be forecast independently, the difference being met by trade. Demand is forecast using a simple demand relationship incorporating price and income elasticities. With two possible income growth paths, this generates four possible demand scenarios. Supply is considered in terms of production from known fields, production from undiscovered fields, and a number of alternative technologies for producing oil. In each case, a range of assumptions are made about the costs and likely dates of availability of these technologies, and these are aggregated to produce three scenarios about supply. Both the demand and supply sides have the characteristics of being fairly price inelastic in the short term, but more elastic in the long term.

The assumptions about supply and demand are then combined to generate a range of scenarios for the level of imports of crude oil and hence for the degree of self-sufficiency in Australia. The detailed assumptions are now summarized.

Prices

There are two important assumptions about crude oil prices. The first is that Australia has no impact on the world price of crude oil. This is perfectly plausible, but it should be noted that for a country such as the United States, for which this assumption would not be plausible, this would provide an additional argument for increasing self-sufficiency—to alter the elasticity of demand for imported oil and hence the price set by OPEC (Newlon, 1977). The second assumption is that crude oil prices in Australia will be set equal to world prices. In the past this has not been the case, and the Australian government has controlled the domestic price of oil, initially to protect indigenous producers, but latterly to protect consumers. This policy is now being phased out, and it is assumed that, from 1980, world parity pricing will prevail.

There have been numerous attempts to forecast the prices that OPEC will set for oil in the future, a number of which have been summarized by Fischer *et al.* (1975), Hammoudeh (1979), and Gately (1979). Models based on treating OPEC as a rational, optimizing agent seemed agreed that prices would stay fairly constant (in real terms) through the late 1970s until the mid-1980s and then rise steadily to a backstop price.

Recent events cast doubt upon this forecast. However, as a lower price scenario, it will be assumed that the current price of oil is well above the long-term trend, and that the trend will be to keep real prices fairly steady until 1985 and then raise it gradually to reach a backstop price around US$30 by the turn of the century. The high-price scenario assumes that alternatives to crude oil are more expensive than envisaged, that oil reserves are lower than expected, and that OPEC will continue to raise the price of crude oil fairly sharply until 1985, and then more steadily until the end of the century towards a backstop price of US$50. Table 1 summarizes the two price paths used in this study (figures were calculated for every year but are only shown at five-year intervals to save space).

It should be emphasized that the price paths used in this paper are to be viewed as long-term trend paths. Actual prices could be expected to fluctuate significantly around these paths in response to short-term surpluses and shortages.

Demand

Demand is forecast using a simple demand relationship of the form

$$Q_t = AQ_{t-1}^{\alpha}P_t^{\beta}Y_t^{\gamma}$$

where Q_t is demand at time t, P_t is price at time t, and Y_t is GNP at time t. No

Table 1
FORECAST AUSTRALIAN CRUDE OIL PRICES
(A$ c.i.f. Westernport)

Year	Low Price	High Price
1980	20.00	20.00
1985	22.00	37.00
1990	27.00	42.00
1995	29.50	44.50
2000	32.00	47.00
2005	32.00	49.50

direct econometric evidence was available on the parameters of such a demand relationship, so elasticities were chosen that seem to be broadly in line with international results (see Taylor, 1977, for a survey). Specifically, it was assumed that the income elasticity is 1.0, the long-term price elasticity is $-.33$, and the short-term price elasticity is $-.05$. This generates parameters $\alpha = .85$, $\beta = -.05$, $Y = .15$. The parameter A was then estimated from past data as A = 1.297.

The assumptions about future prices have already been noted, so all that is required are assumptions about future GNP. Historically, real GNP in Australia has grown at fairly moderate rates by comparison with other OECD countries. It grew at an average annual rate of 4.7 per cent from 1950 to 1970, and 4.9 per cent from 1960 to 1975. In per capita terms, the corresponding rates were 2.6 and 2.2 per cent. More recently, growth has slowed to 3.4 per cent (1971−72 to 1976−77). Population projections suggest a slowing of population growth and, as a low-growth scenario, it will be assumed that the economy grows at 2.75 per cent to 1990 and 2.5 per cent thereafter, yielding an average growth rate of 2.57 per cent over the period 1980−2005. The high-growth scenario assumes that immediate prospects are rather gloomy, with growth of only 2 per cent till 1985, but that the economy reverts gradually to a pattern of high growth, with 3.25 per cent betweeen 1985 and 1990, 4 per cent between 1990 and 1995, and 4.5 per cent thereafter. These two paths appear to encompass the spectrum of possible growth patterns suggested for Australia over the next quarter century.

Combining the assumptions on prices, incomes, and the demand function yields the four possible demand scenarios shown in Table 2.

Under all scenarios, demand is forecast to fall until 1985, a reflection of the abolition of controlled prices for oil in Australia, whereby prices rise from an average of $9 a barrel in 1979 to $20 in 1980. Until the middle of the

Table 2
FORECAST DEMAND FOR CRUDE OIL IN AUSTRALIA 1980−2005
(Million Barrels)

Price Path:	High		Low	
Income Path:	High	Low	High	Low
Year				
1980	240.7	240.7	240.7	240.7
1985	215.8	218.8	232.7	235.9
1990	214.2	217.9	241.4	245.7
1995	232.9	230.1	265.6	262.4
2000	269.3	249.9	307.0	284.9
2005	321.4	274.9	368.2	315.0

1990s, the major factor affecting demand is the assumption about prices, since it is only at the end of the forecast period that income starts to rise rapidly on the high-growth path.

Supply of Conventional Crude Oil

Production from known oil fields is assumed to be both known and insensitive to price fluctuations. Current production is 160 million barrels a year, and this will fall to 110 million barrels by 1985, 55 million barrels by 1990, and 30 million barrels by 2005 (Folie and Ulph, 1979). The justification for the assumptions is that higher prices would have only a marginal effect on production; for example, enhanced recovery is not feasible for the Kingfisher and Halibut fields, which account for 88 per cent of current production, since the natural water drive is so strong that 60 per cent of *in situ* resources will be recovered, a very high recovery factor by world standards (McKay and White, 1978). Moreover, uncertainty about production from known fields is of second order relative to production from undiscovered fields.

The analysis of production from undiscovered fields is based on a study by the Bureau of Mineral Resources (Australia, BMR, 1978). Table 3 provides the estimate of the extent of such undiscovered reserves.

These figures for reserves are broadly in line with others available for Australia (Esso, 1978), though slightly more pessimistic due to assumptions about the extent to which fields discovered will be gas bearing rather than oil bearing. The major uncertainty concerns the probability of finding oil in Area

Table 3
UNDISCOVERED RESERVES OF OIL
(Million Barrels)

Zone	Description	Probability of Finding at Least the Stated Reserves		
		80%	Mean	20%
Area 1	Onshore & offshore in water depths less than 200 metres	700	986	1118
Area 2	Medium-depth water 200-500 metres	328	788	956
Area 3	Deep water below 500 metres	—	1074	1946
TOTAL		1028	2848	4020

Source: Australia, BMR (1978).

3, essentially the Exmouth Plateau. Preliminary drillings do not seem encouraging. To convert these figures to production data, assumptions were made about field size distribution, rate of discovery and order of discovery, lead time to commencing production, and other factors. It was not supposed that price variations would significantly alter these production rates, given the relatively high prices being forecast. The assumed production rates are given in Table 4.

Supply from Alternative Sources

There are a number of alternative sources for producing liquid fuels—L.P.G., methanol, oil shale, and conversion of coal to oil. Space precludes a detailed treatment of each technology (it can be found in Folie and Ulph, 1979). For each technology, assumptions were made about the likely costs of operating the technology, the earliest date at which that technology could become available, and the production rate at which the technology could operate. Of course, there is considerable uncertainty about these factors. For this paper, uncertainty is taken to relate only to the costs of the alternative technologies, and this is summarized in Table 5, where costs relate to the price A$ per barrel of crude oil that would make the technologies economic.

Table 6 shows the earliest dates at which the technologies are likely to be available and the scale of production on which they would operate.

It should be noted that in the case of shale oil and coal liquefaction, the dates refer to operation of the first plant. It is assumed that a second plant of the same size could be introduced five years after the introduction of the first plant. For L.P.G., the output of substitute includes that for supplement.

Table 4
PRODUCTION FROM UNDISCOVERED RESERVES
(Million Barrels/Year)

Year	Low Supply (80% Prob.)	Medium Supply (Mean)	High Supply (20% Prob.)
1980	0	0	0
1985	55	76	80
1990	85	168	185
1995	42	220	320
2000	9	69	129
2005	1	15	34

Source: Australia, BMR (1978).

Table 5
COSTS OF ALTERNATIVE TECHNOLOGIES
(A$/Barrel of Oil Equivalent)

	Low Success	Medium Success	High Success
Shale Oil	37	30	18
Coal Liquefaction			
• Pyrolysis	37	30	25
• Hydrogenation and Synthesis	45	37	27
Methanol	32	25	18
L.P.G.			
• Supplement	12	12	12
• Substitute	17	17	17

Table 6
AVAILABILITY AND SCALE OF ALTERNATIVE/ TECHNOLOGIES

Technology	Earliest Date of Availability	Scale of Operation
Shale Oil	1985	45 mbbls/year
Coal Liquefaction		
• Pyrolysis	1990	11 mbbls/year
• Hydrogenation and Synthesis	1995	36.5 mbbls/year
Methanol	1985	10% of Motor Spirit Demand
L.P.G.		
• Supplement	1980	2.5% of Motor Spirit Demand
• Substitute	1985	5 mbbls/year plus 7.5% of Motor Spirit Demand

Supply Curves

It is now possible to combine the assumptions made about conventional and unconventional supplies of liquid fuels to construct supply curves for each year under various assumptions about uncertainty. To characterize the range of possibilities, the pessimistic assumptions for each source of supply are combined, and similarly for the optimistic and medium assumptions. Alternative sources of supply are assumed to be introduced in the year in which price reaches the breakeven level or the earliest dates of availability, whichever is later. The possible range of supply for the two price scenarios outlined earlier is shown in Table 7.

The supply pattern displays the plausible characteristics that in the immediate future, supply is fairly certain and not much affected by variations

Table 7
SUPPLY PATTERNS
(Million Barrels/Year)

Price Path:	High			Low		
Supply Assumptions:	High	Medium	Low	High	Medium	Low
Year						
1980	144.2	144.2	144.2	144.2	144.2	144.2
1985	195.7	191.7	170.3	193.6	189.6	168.5
1990	377.8	310.4	222.3	374.4	251.3	152.5
1995	551.9	398.9	176.4	510.4	291.8	94.7
2000	387.1	274.2	208.3	384.2	199.4	67.5
2005	321.4	221.7	202.9	291.5	147.9	62.3

in price. However, over the long run, there is considerable uncertainty but also a significant supply elasticity, especially under the pessimistic supply scenario.

Self-Sufficiency Scenarios

The basic assumptions of the paper have been outlined, and it is now possible to analyse their implications for self-sufficiency. For each of the four possible demand scenarios outlined earlier, there are three possible supply responses, generating twelve scenarios for self-sufficiency. Space limitation prevents all scenarios being shown, but the eight most interesting ones for this study are shown in Table 8.

Table 8
SCENARIOS FOR SELF-SUFFICIENCY (%)

Scenario No.:	1	2	3	4	5	6	7	8
Price Path:	Low	Low	Low	Low	High	High	High	High
Income Path:	High	Low	Low	Low	Low	High	High	High
Supply Path:	Low	Low	Med	High	Low	Low	Med	High
Year								
1980	59	59	59	59	59	59	59	59
1985	72	71	80	82	77	78	88	90
1990	63	62	102	152	101	103	144	176
1995	35	36	111	194	76	75	171	236
2000	22	23	69	134	83	77	101	143
2005	16	19	46	92	73	63	68	91

A number of conclusions can be drawn from these results. First, under all scenarios, problems of a fall in self-sufficiency are unlikely to emerge until the 1990s. This is because the assumptions of low growth and high prices act to constrain demand in the 1980s, so that even with low supply assumptions, self-sufficiency actually increases.

Second, by comparing scenarios 1 and 2 and again scenarios 5 and 6, it can be seen that variations in the path of income growth are relatively unimportant for self-sufficiency. Variations in the uncertain factors affecting supply are more important, not surprisingly, as can be seen by comparing scenarios 2, 3, and 4 and again scenarios 6, 7, and 8. However, low supply forecasts are not by themselves sufficient to cause the level of self-sufficiency to fall, as can be seen by looking at scenarios 5 and 6. It is the combination of low supply and low prices, as in scenarios 1 and 2, that causes the level of self-sufficiency to fall markedly in the 1990s. While demand and supply individually may not be particularly price elastic, what matters for imports is the combined elasticity, and this is shown to be quite high.

One minor point should be noted. In those scenarios that do not assume a low supply response (3, 4, 7, 8), a significant level of *exporting* is shown in the mid 1990s. This is the result of the assumption that supplies of liquid fuels will be made available whenever conditions justify it. In practice, there would be a need to establish an export market before such supplies were introduced, and no attempt has been made in this study to assess whether such markets would be feasible. However, since this is not the area on which this paper is focused, the issue will be ignored; but this serves to reinforce the point that these figures should not be treated as forecasts.

In conclusion, this section has set out the assumptions made in this study and shown the implications for self-sufficiency. Under a wide range of scenarios, the fears of dwindling self-sufficiency appear groundless, and Australia will enjoy higher levels of self-sufficiency than it currently experiences. A major factor underlying this result is the effect of high prices and low growth in dampening demand and making other sources of liquid fuels economic. However, under some scenarios, the level of self-sufficiency will fall sharply, though not until the 1990s. The rest of this paper will assess the policy implications of these results.

TRADE EFFECTS

As noted in the Introduction, one reason why the government might be concerned about a fall in the level of self-sufficiency is the implications this would have for the balance of payments, and this section of the paper analyses these implications under the various scenarios and assesses a number of policy responses. In the late 1970s, Australia's oil bill has been approximately A$1 billion in 1980 dollars, and has accounted for 7 per cent of Australia's imports by value. It is straightforward to calculate what will happen to the value of oil imports under each scenario, but to assess the relative effect it is necessary to forecast what will happen to other imports. The simplest approach has been adopted in this study and that is that, other things being equal, non-oil imports will rise in line with GNP. Since it will be necessary to analyse the balance of trade later, it should be noted that, in the late 1970s, Australia was in approximate balance-of-trade equilibrium, and that for the future it will be assumed that the value of exports, in real terms, also grows in line with income, *ceteris paribus*.

Table 9 shows what might happen to the absolute and relative size of Australia's oil import bill, on the assumption that nothing else changes, for scenarios 1 and 2. It is only under these scenarios that there is a steady and significant deterioration of the position, although two points should be noted. Under all scenarios, the rapid rise in oil prices in 1979−80 leads to an immediate doubling of the oil bill, with the share of oil in imports rising to 11.4 per cent. This effect persists for two to three years in the other scenarios. Second, the combinations of low prices and medium supply or high prices and low supply lead to a deterioration in the oil import position, but not until the first five years of the next century, with oil's share of imports rising to 16.7 per cent by 2005 under scenario 3. Nonetheless, the severe cases are scenarios 1 and 2.

As can be seen from Table 9, Australia's oil bill could rise tenfold from its late 1970s' level (in real terms), and oil's share of total imports could rise three and a half times. There is not a substantial difference between scenarios 1 and 2, with 2 being slightly worse in relative terms; and so the analysis in the rest of the section will focus on scenario 2.

Two possible policy responses will be considered. The first policy is to aim to maintain a high level of self-sufficiency and thus cut the oil import bill. Table 10 summarizes the effect of maintaining the level of self-sufficiency above 50 per cent.

Table 9
AUSTRALIA'S OIL IMPORTS

Scenario No.:	1		2	
Year	A$b	%	A$b	%
1980	1.9	11.4	1.9	11.4
1985	1.4	7.6	1.4	7.7
1990	2.4	11.0	2.5	11.4
1995	5.0	19.0	4.9	19.8
2000	7.6	23.2	6.9	24.6
2005	9.7	23.8	8.0	25.3

Table 10
SCENARIO 2 WITH 50% SELF-SUFFICIENCY

Year	Old Price A$/bbl.	New Price A$/bbl.	Self-Sufficiency %	Oil Imports A$b.	%
1980	20.00	20	59	1.9	11.4
1985	22.00	22	71	1.4	7.7
1990	27.00	27	62	2.5	11.4
1995	29.50	37	68	2.3	9.5
2000	32.00	45	76	1.9	7.0
2005	32.00	45	70	2.7	8.6

Under a policy of maintaining self-sufficiency above 50 per cent, nothing changes until 1995, when self-sufficiency would have fallen below 50 per cent. Note that the lumpiness with which new technologies enter will usually lead to self-sufficiency being maintained well above 50 per cent. Given the assumptions made about non-oil imports and exports, a policy of maintaining self-sufficiency above 50 per cent would substantially mitigate, but not eliminate, the balance-of-trade problem. The second column in Table 10 shows the price that would need to be set for crude oil to secure the 50 per cent self-sufficiency policy. The difference between the new and old prices indicates the level of tariff on imported crude oil that would need to be imposed, a tariff being the most efficient way of achieving the self-sufficiency policy.

However, a policy of self-sufficiency is not an efficient response to any pressure on the balance of payments that may arise due to rising oil imports; for it concentrates all of the adjustment process on one sector of the economy—the oil sector. It will be more efficient, in general, to spread the adjustment throughout the traded-goods sector, encouraging the expansion of all export- and import-competing sectors. To achieve this, the government needs to implement a ''switching'' policy that will switch resources from the non-traded to the traded-goods sector. This paper will consider devaluation as representative of such a switching policy. It is not being argued that such a devaluation would work or that it should be employed, merely that it gives an insight into the order of magnitude that would be involved in implementing a broadly based trading adjustment policy.

The assumptions underlying the approach are based on traditional trade models, as outlined for example in Corden (1977), and are explained in greater detail in Folie and Ulph (1979). Briefly, the ''small country'' assumption is invoked that Australia can buy and sell as much as it wants at prevailing world prices, denoted in some foreign currency. A devaluation raises the prices of all traded goods, expressed in Australian dollars, and the effect this has on the balance of payments depends on the sum of the elasticity of demand for imports and the elasticity of supply of exports. Specifically, the devaluation required to achieve a balance-of-payments equilibrium is given by $(1 + \frac{D}{M})^{\frac{1}{\alpha+\beta}}$ where D is the deficit, M is the level of imports, α is the elasticity of demand for imports (in absolute value), and β is the elasticity of supply of exports. $\frac{D}{M}$ is simply the share of oil in total imports less 7 per cent. Estimates of elasticities were based on a study by Gregory (1976), in which he suggested that, for Australia, elasticities of demand for imports lay between $-.65$ and -1.66, while elasticities of supply for exports were between .7 and 3.0 for the rural sector, and 0.5 and 6.0 for mining. To illustrate the worst that might happen, a pessimistic estimate of $\alpha + \beta = .8$ has been used. The required devaluation is shown in Table 11. The figure for the devaluation is to be interpreted as the change in the exchange rate required relative to 1979, so that by 2005 the exchange rate will have deteriorated by less than 25 per cent over a twenty-five year period. Clearly this is not a very significant change, and suggests that the increase in oil imports, while large in absolute and even relative terms, can be readily accommodated by a policy of steadily switching resources into the traded-goods sector.

Of course, in seeking to raise the price of traded goods, one of the prices that will be increased by the devaluation is that for crude oil. Table 11 also shows the resulting prices for crude oil that would prevail under such a devaluation. These higher prices must increase the level of self-sufficiency, but by comparing them with the prices in Table 10, it can be seen that the new levels of self-sufficiency must fall well short of that obtained under the 50 per cent self-sufficiency policy.

To summarize, then, while a decline in the level of self-sufficiency could lead to substantial increases in both the absolute and relative size of the oil import bill, it is more efficient to respond to this by allowing a broad-based expansion in all traded-goods sectors, rather than concentrating

Table 11
DEVALUATION FOR SCENARIO 2

Year	Required Devaluation (1979 Exchange Rate = 1.0)	New Price A$/bbl.
1980	1.055	22.11
1985	1.000	22.00
1990	1.055	28.49
1995	1.162	34.28
2000	1.225	39.19
2005	1.234	39.48

the response on the oil sector by increasing the level of self-sufficiency. The changes required by the broad-based approach seem moderate, and will automatically cause an increase in the level of self-sufficiency in oil, though of a fairly modest amount. As a further perspective on the problems posed by the rising oil bill, it should be recalled that the share of mineral exports in total exports rose from 9 per cent in 1964/65 to 26 per cent in 1970/71, an increase of similar magnitude to that for oil, but in a period of six years rather than twenty-five. Of course, this is not meant to minimize the fact that there could be important structural adjustment problems, as there have been following the growth of mineral exports, but the longer time periods involved should make these more manageable.

It could be argued that the analysis of the required devaluation is too simplistic, in that it ignores capital flows and treats exports and non-oil imports in a mechanistic way. For example, it might be that higher oil prices will affect Australia's trading partners, especially Japan, more severely than Australia, making it difficult for Australia to maintain export growth. While these are valid comments, they really miss the point. The above analysis is not intended to be a forecast of the actual balance-of-payments situation in the future or of any devaluation that might occur. The analysis shows a substantial increase in the absolute size of Australia's oil import bill, under one scenario, but this has to be seen relative not to the *current* trading position of Australia but to an expanded trading position. The analysis shows what would occur if nothing else happens and trade grows in line with income, and the devaluation calculated is that required to restore the *relative* position of Australia's oil imports. Of course, developments in the capital side of the account or in other areas of the trading balance may lead to other balance-of-payments problems. But these other developments will happen anyway, even if Australia is 100 per cent self-sufficient in oil. The analysis shows what adjustments are required to deal solely with the problem of a rising import bill, and any other developments in the balance of payments will require their own adjustments. It should be noted that rising oil prices will make alternative fuels more attractive, which would be to Australia's advantage. Trying to predict the future trading position of Australia is clearly difficult, but the analysis has not sought to do this.

Finally, it should be noted that there is another argument related to trade that could support the case for a higher level of self-sufficiency—that the existing trade pattern is significantly distorted by various policies on tariffs and quotas, which act to the disadvantage of exporting or non-protected, import-competing sectors. This case for tariff compensation has been widely debated in the context of agricultural exports (Harris, 1975) and will not be pursued here. Where possible, governments should seek to remove the original trade distortion rather than introduce further distortions in the oil sector.

Figure 1
COSTS OF AN EMBARGO

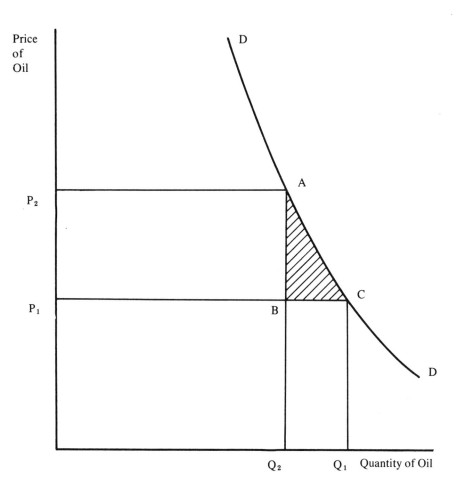

SECURITY OF SUPPLY

The second reason given for concern at low levels of self-sufficiency is that it leaves the economy vulnerable to short-term supply interruptions. This presupposes that international sources of supply are more likely to be disrupted than domestic ones, and it is not clear that this would always be the case. However, accepting the underlying premise, the question arises how best to protect against such supply instability. Two policies will be considered—increasing self-sufficiency and stockpiling—and it will be shown that stockpiling is considerably cheaper. To illustrate this, attention will again be given to scenario 2 presented earlier.

Before assessing the costs of various policies, it will be useful to consider the benefits, which are the avoidance of the costs imposed by a sudden reduction of supply. Such costs are measured by a loss of consumer surplus, as shown in Figure 1. DD is the short-run demand curve for oil, and the original price is P_1, at which amount Q_1 is purchased. An embargo reduces the amount available to Q_2, causing prices to rise to P_2, and the cost of this is a loss of consumers' surplus measured by the shaded area ABC. It is clear from the diagram that supply interruptions could also take the form of a sudden increase in the price at which supply is available, from P_1 to P_2.

Given the specification of the short-run demand curve, it is straightforward to calculate the costs of embargoes of various extents. Table 12 presents the costs each year of embargoes amounting to reductions of 10 and 20 per cent of annual imports for scenario 2, together with the associated prices that would need to be charged to ration the available supply.

Clearly, as the level of self-sufficiency falls, the cost of embargoes rises, and it rises more steeply the longer the embargo. Doubling the length of an embargo causes the losses to rise by a factor between 5 and 10. It is not suggested that governments would let the price rise to the levels shown in Table 12, but the cost to consumers must be at least that shown in the table,

Table 12
COSTS OF EMBARGOES FOR SCENARIO 2

Year	Self-Sufficiency	Old Price A$/barrel	10% Embargo		20% Embargo	
			Annual Cost A$b.	New Price A$/barrel	Annual Cost A$b.	New Price A$/barrel
1980	59	20.00	.10	45.30	.60	106.27
1985	71	22.00	.05	39.27	.27	71.33
1990	62	27.00	.13	58.48	.71	130.66
1995	36	29.50	.53	110.48	4.08	454.31
2000	23	32.00	.99	156.44	9.28	877.02
2005	19	32.00	1.26	170.33	12.55	1056.16

no matter how the government reacts to the embargo. It should be noted, however, that embargoes of the extent shown in Table 12 are large by historical standards. The OPEC embargo of 1973−74 was equivalent to a reduction of annual OPEC production of less than 5 per cent, and the reduction in annual world trade in oil caused by the Iranian difficulties in 1979 was also relatively small.

Given the low, short-run demand elasticities, the costs of embargoes are high when self-sufficiency is reduced, and the government would clearly wish to pursue policies to reduce such costs. To analyse the costs of various policies, a number of unrealistic assumptions will be made to simplify the exposition. It will be obvious, however, that the restrictions can be relaxed without changing the qualitative results to be presented.

It will be assumed that the government believes that, in any one year, there is a maximum possible duration of an embargo—say, a 20 per cent reduction in annual imports, which, as has been noted, is well above the historical experience. Then holding a stockpile equal to 20 per cent of annual imports will be sufficient to protect against any embargo. It will be assumed that this stockpile is to be in addition to stocks held by the oil industry anyway, and that the government buys at the beginning of each year enough oil to cover 20 per cent of the *increase* in imports for the coming year (but does not sell oil if imports fall). The costs of this policy then are the costs of buying and storing this extra oil. Purchase costs are given by the assumptions about world oil prices. Storage costs are more difficult to assess. A recent report (Australia, NEAC, 1979) quotes a figure of A\$8 per barrel, so, to be conservative, it is assumed that storage will add A\$10 per barrel to the price of oil. Given these assumptions, Table 13 shows the annual costs of a policy of holding stocks equal to 20 per cent of imports under scenario 2, together with the present value of such costs using a real discount rate of 10 per cent. Note that there is a very large initial expenditure as stocks are raised from zero to 20 per cent of 1980 imports, and no costs are then incurred until 1991, since imports in the 1980s never exceed the 1980 level under this scenario.

The second policy considered is for the government to ensure that at least 50 per cent self-sufficiency is maintained. The levels of self-sufficiency actually achieved by such a policy, and the price levels required to sustain it, were set out in Table 10. The costs involved in such a policy are illustrated in Figure 2. DD is the demand curve for oil, and SS is the indigenous supply curve. At the prevailing world price, P_1, demand is OQ_1, indigenous supply is OR_1, and self-sufficiency is OR_1/OQ_1. To raise self-sufficiency, the government imposes a tariff P_2P_1 on imported crude oil, causing demand to fall to OQ_2 and indigenous supply to rise to OR_2, so that self-sufficiency is now increased to OR_2/OQ_2, say, 50 per cent. The costs of such a policy are the loss of consumers' surplus, shown as area ABC, and the increased cost of domestic production, relative to importing, given by area EFG.

Figure 2
COSTS OF SELF-SUFFICIENCY

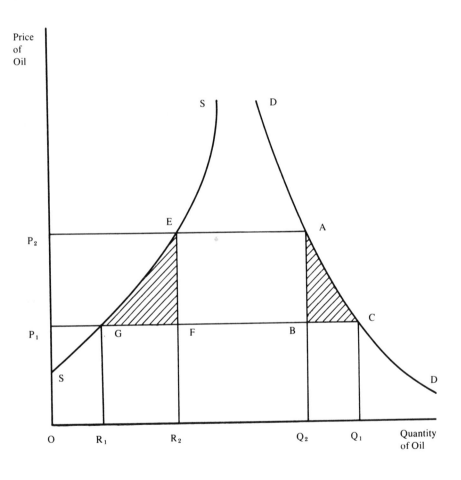

Given the information used in this study, it is possible to calculate the welfare costs of maintaining a high level of self-sufficiency, and these costs are shown in Table 13. The costs do not start until 1994, which is when the level of self-sufficiency would have dropped below 50 per cent. The annual costs are quite substantial relative to stockpiling costs, but because the costs are incurred later, the present value of the self-sufficiency strategy is only double that of stockpiling. Of course, given the assumptions made, stockpiling will ensure complete protection against embargoes, while increased self-sufficiency provides only limited protection, at twice the cost.

The above analysis suggests that stockpiling will be a more efficient policy than trying to increase the level of self-sufficiency, but it does not suggest what the optimal stockpiling policy will be. In general, the optimal policy will be to provide less than complete protection against embargoes—at some low level of embargo, it will be cheaper to bear the expected embargo cost than to build up stockpiles.

Another question to be considered is whether the government needs to provide the stockpile or whether private firms or individuals will have sufficient incentive to do so. In principle, with a full set of competitive future and contingent markets, the market mechanism will provide an efficient level of stockpiling. In practice, markets may fail to do so for a number of reasons, which will be listed briefly. In the absence of price signals to stockpile, agents will have to guess the prices that would prevail under various levels of embargo. This is difficult to do, especially since prices will depend on the level of stock other agents decide to hold. If agents think that no one else will stockpile, they have an incentive to provide large stockpiles, and if all agents behave this way, there will be excessive stockpiling. On the other hand, if agents believe that others will carry large stockpiles, they will under-stockpile. Absence of contingent markets leads to insufficient spreading of risks in private markets, and since stockpiling is risky, this may lead to underprovision of stockpiling. Perhaps most important, stockpiling will only

Table 13
COSTS OF PROTECTING AGAINST EMBARGOES (A$m)

Year	20% Stockpile	50% Self-Sufficiency
1980	579.6	—
1985	—	—
1990	—	—
1995	173.0	545.7
2000	70.6	1442.5
2005	51.2	1589.7
PV@10%	839.1	1695.0

be undertaken if stockpilers can sell at a price sufficient to cover costs of holding stocks when the embargo occurs. However, this often appears as if the stockpilers are benefiting from the consumers' misfortune and, for distributional reasons, governments may prefer to institute a price freeze and ration the restricted supply. Clearly, this would eliminate any incentive for private stockpiling.

On balance, therefore, it is likely that private markets would undersupply stockpiles so that some government provision of stockpiling would be warranted.

CONCLUSIONS

The analysis presented in this paper has shown that the fears expressed of a dwindling level of self-sufficiency in liquid fuels may not materialize under a wide range of scenarios, and if it does, it will not be until the 1990s. It is often suggested that the government should take action to prevent a reduction in self-sufficiency for both balance-of-payments and security-of-supply reasons. This paper has shown that more sensible policies will consist of reacting to any balance-of-payments problems by switching resources into all traded-goods sectors (for example, through devaluation) and not just the oil import-competing sector, and of stockpiling to protect against any supply instability.

The paper concludes by noting a number of broader considerations that arise in connection with the policy recommendations, but most of these would arise even if Australia were a closed economy. The fact that Australia's resources of crude oil arc finite means that as these resources run out, Australians will become worse off than they would have been if crude oil reserves were limitless. This may sound obvious, but it has the important implication that there is nothing the government can do to prevent these losses; the aim of government policy rather is to minimize these losses.

Now the way in which the losses arise is that more resources must be devoted either to producing alternative (more expensive) sources of liquid fuels or to producing exports to pay for increased imports of crude oil. These are resources that could otherwise be deployed in raising living standards. Given the relatively small share of energy in total costs, and the likely orders of magnitude for costs of alternative technologies, the loss of increased living standards involved should be fairly slight, amounting to a slowing down of economic growth of about half a per cent a year. The fact that Australia's oil production is likely to peak before the rest of the world's means that there will be a phase when it will pay Australia to increase its level of oil imports, rather than adopt some extremely expensive methods of producing liquid fuels. In this light, then, a fall in self-sufficiency has some merits, for it means that Australia is substituting relatively cheap imported oil for the much more expensive supply technologies it would otherwise have to adopt (or abandon some uses of energy that yield considerable benefits). This will cause adjustment problems as more exports will be needed to pay for the imports, but adjustment problems will occur anyway due to the rising cost of liquid fuels, and importing more oil is a means of reducing the increase in costs of liquid fuels.

If the costs imposed by dwindling oil reserves are to be minimized, it is of crucial importance that users of liquid fuels and potential suppliers of alternative sources of liquid fuels be confronted with the higher cost of oil. Thus, the fact that under many scenarios, self-sufficiency stayed high is due to the impact of higher prices in moderating demand and making alternative sources of liquid fuels economic. Of course, this must make the consumers of oil worse off, and will confer windfall gains on some sectors of the economy, such as owners of cheap sources of oil. As already noted, some reduction in welfare is inevitable, but it is right for governments to be concerned about the distribution of welfare losses. Blanket subsidies to energy consumers is not a desirable weapon to achieve such a balancing of welfare losses, and other instruments of policy such as windfall profit taxes and increases in pensions are available to meet these distributional objectives.

It could be argued that higher prices by themselves will not be sufficient to ensure that the required levels of conservation and supply expansion take place. Certainly, the absence of a complete set of future and contingent markets destroys any assumption that markets will necessarily work smoothly. But the problems this introduces for government policy are rather subtle, and these are discussed at greater length in Ulph and Folie (1980). Governments may well have to undertake policies to ensure that alternative technologies are brought to the market at the appropriate time and that conservation measures are undertaken that are warranted by the prevailing energy prices. However, such policies would need to be undertaken anyway, even if the issue of self-sufficiency did not arise, and so this is not the place to elaborate on such policies. Failures to undertake such policies will result in lower levels of self-sufficiency than indicated in this study, but such failures would manifest themselves in different ways even in a closed economy.

These final comments suggest that to concentrate attention on self-sufficiency is to misdirect the debate on energy policy. The specific problems that arise in connection with self-sufficiency—balance-of-payments difficulties and security of supply—can be handled in a fairly straightforward manner, though not by trying to maintain high levels of self-sufficiency. This can lead into issues concerning the equitable sharing of welfare losses, and the ability of markets to provide adequate signals for risky activities with long lead times. However, these latter issues would be important even if there were no international trade, and it is on the latter set of questions that policy makers and advisers should concentrate their efforts.

NOTES

* This paper is a substantially condensed and updated version of an earlier report, Folie and Ulph (1979). In compiling that report, we received valuable assistance from the Department of National Development, the Joint Coal Board, the Bureau of Mineral Resources, and other public and private organizations. We would also like to thank John Revesz and Thomas Mozina for research assistance, and our colleagues Stuart Harris, John Hewson, Greg McColl, Hugh Saddler, and Ben Smith for comments and discussions on various parts of the report.

¹ For example, one politician's view is given by Baume (1980): "without a reasonable degree of self-sufficiency, Australia would be susceptible to large balance of payments difficulties and possible supply interruptions which would have an adverse impact on the economy and jeopardise our defence capability. A basic energy policy objective for Australia is therefore to maintain the maximum possible degree of self-sufficiency in liquid fuels."

² All prices in this paper are in real 1980 dollars.

REFERENCES

Australia, Bureau of Mineral Resources (BMR). (1978) *A Rapid Assessment of Production from Australia's Undiscovered Resources*. Canberra: The Bureau.

Australia, National Energy Advisory Committee (NEAC). (1978) *Australia's Energy Resources: An Assessment*. Canberra: Australian Government Printing Service.

Australia, National Energy Advisory Committee (NEAC). (1979) *Liquid Fuels Self Sufficiency in Australia*. Canberra: Australian Government Printing Service.

Baume, M. (1980) "Comments." In *Oil and Australia's Future*, edited by T. von Dugteren, pp. 187-91. Sydney: Hodder and Stoughton.

Corden, W.M. (1977) *Inflation, Exchange Rates and the World Economy*. Oxford: Clarendon Press.

Esso Australia Ltd. (1978) *Australian Energy Outlook*. Sydney: Esso.

Fischer, D.; Gately, D.; and Kyle, J.F. (1975) "The Prospects for OPEC: A Critical Survey of Models of the World Oil Market." *Journal of Development Economics* 2 (December): 363-86.

Folie, G.M. and Ulph, A.M. (1979) *Self-Sufficiency in Oil: An Economic Perspective on Possible Australian Policies*, C.A.E.R. Paper No. 6. Sydney: University of New South Wales.

Gately, D. (1979) "The Prospects for OPEC Five Years After 1973/74." *European Economic Review* 12 (October): 369-79.

Gregory, R. (1976) "Some Implications of the Growth of the Mineral Sector." *Australian Journal of Agricultural Economics* 18 (August): 71-91.

Hammoudeh, S. (1979) ''The Future Price Behaviour of OPEC and Saudi Arabia; A Survey of Optimisation Models.'' *Energy Economics* 1 (July): 156-66.

Harris, S. (1975) ''Tariff Compensation: Sufficient Justification for Assistance to Australian Agriculture?'' *Australian Journal of Agricultural Economics* 19 (December): 131-45.

McKay, B.A. and White, J.A.W. (1978) *Enhanced Recovery of Petroleum—Applications to Australia*. Report 1978/43. Canberra: Bureau of Mineral Resources.

Newlon, D.H. (1977) ''The Demand for Energy Imports and Energy Independence.'' In *International Studies of the Demand for Energy*, edited by W.D. Nordhaus. Amsterdam: North-Holland.

Taylor, L.D. (1977) ''The Demand for Energy: A Survey of Price and Income Elasticities.'' In *International Studies of the Demand for Energy*, edited by W.D. Nordhaus. Amsterdam: North-Holland.

Ulph, A. and Folie, M. (1980) ''Energy Policy for Australia.'' In *Industrial Economics: An Australian Study*, edited by R. Webb and R. Allen. Sydney: George Allen and Unwin.

Chapter Two

Missing Elements in Canadian Energy Policy

by
*N. Berkowitz**

72 30

I

Much of what now passes for energy policy in Canada centres on the urgent need to secure long-term oil supplies. It recognizes that deliveries from petroleum reservoirs of the western Canadian basin—which met the greater part of the country's oil needs for almost three decades, and for some years made Canada a net exporter of oil—will fall increasingly short of demand (see Figure 1). And it reflects recognition of the unacceptable consequences of continued dependence on imports. Without internal closure of the supply-demand gap, the annual outflow of capital on oil accounts would reach some $10 billion (in constant 1980 dollars) by 1985 and exceed $14 billion by 1990. Security of supply would at best be uncertain—and often would depend upon whether Canada's conduct of foreign affairs met the *supplier's policy objectives* (for example, the Jerusalem embassy issue and recent events in Iran). As global oil supplies become tighter, Canada could even find itself unable to compete for adequate contracted supplies, and would then suffer the additional heavy penalties associated with spot purchases.

Current energy policy therefore proclaims itself as designed to recapture self-sufficiency in oil by 1990, and expects to achieve this objective by promoting conservation, fuel substitution, and development of alternative indigenous petroleum sources. Appropriate action in these areas is to be encouraged by a variety of incentives, the most important of which envisages substantial, progressive increases in the price of Canadian oil.

Unfortunately, mere *statements* of goals do not attain goals; and what practical steps have been taken towards them cannot possibly reach self-sufficiency in oil by 1990. Without some reordering of priorities and rethinking of public policies respecting energy resource development in general, self-sufficiency is, in fact, not even likely to be reached by the end of the century.

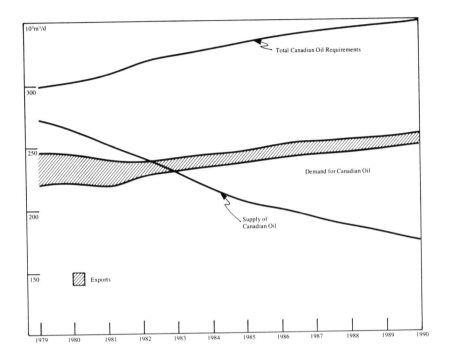

Figure 1
PROJECTED CANADIAN OIL SUPPLY AND DEMAND, 1979—1990

Note: "Demand for Canadian Oil" relates to that fraction of total Canadian oil requirements that cannot be met by presently expected levels of imports.
For the source and discussion of Figure 1, see Energy Resources Conservation Board (1979), "In the Matter of an Application of Esso Resources Canada Limited under Section 43 of The Oil and Gas Conservation Act, and Section 27 of The Coal Conservation Act," October.

II

In the short term, to the mid-1980s, there is now little that can be done to reduce our dependence on foreign oil. Voluntary conservation—for example, voluntary curtailment of personal gasoline consumption—is unlikely to impact significantly on total oil requirements.[1] Long lead times—commonly five to six years, and up to ten to twelve years for development of remote "frontier" resources—rule out substantial reliance on new indigenous sources. And savings from substitution of natural gas and/or coal for imported fuel oil in central and eastern Canada, which could be significant, will not accrue quickly even if an effective programme, still only talked about, were to be initiated immediately. Some new supplies could conceivably become available before 1985 from such wells as Mobil's Hibernia P-15 off Newfoundland. Overall, however, this will have little effect; and in the immediate future, we have in fact no choice but to pay the price for our failure to act in the wake of the 1973 Arab oil embargo, which foreshadowed OPEC's stranglehold on us.

In these circumstances, it is more important to focus attention on the ten to fifteen years after 1985, to assess what course of action could bring us to oil self-sufficiency over that period—and to observe that we seem to be misdirecting our efforts over this longer term.

Of the three "pillars" of current energy policy, the most ardently pursued by government and the mass media is *conservation*—which, in so far as it bears on oil, is likely to yield the smallest gains in the battle for self-sufficiency. That conservation can and will effect some economies is, of course, indisputable. However, since savings in oil would for the most part come about from "retrofitting" (e.g., of older oil-heated homes with better insulation) and from replacement of existing housing and automobile stocks, they will accumulate slowly and tend to be offset by population growth, by population shifts into more personal-transport-intensive western Canada and, above all, by deeply entrenched social policies that require continued growth of the GNP through industrial expansion. Unless promoted by statutory restrictions on the use of motor fuels and heating oils, conservation *per se* can therefore not be expected to do more than slow the growth of demand for oil and oil products. While it may result in declining per capita personal demands for liquid fuels, it would at best leave overall requirements at present levels—or, more realistically, restrain growth of oil demand to 1.3−2 per cent per year (National Energy Board, 1978).

Nor can much hope be reposed in currently fashionable adjuncts of conservation philosophy—in "soft energy" paths that not only demand considerable life-style changes but also place quite unwarranted trust in renewable resources.[2] Of the latter, only biomasses (e.g., wood wastes or vegetation specifically raised for energy production) could be converted into gaseous and/or liquid fuels. Solar energy or wind power could obviously do no more than provide some of the processing energy for such conversion. However, a need to synthesize liquid fuels (e.g., alcohols for use as ''gasohol'') could be far more efficiently and economically met from coal, or from natural gas (if gas proved sufficiently cheap for the purpose).

The more significant gains might therefore be expected to accrue from carefully planned *direct fuel substitution*—that is, from, wherever possible, displacing oil by natural gas or coal. What makes this so attractive is the abundance of these fuels. Recent appraisals (National Energy Board, 1979) set Canada's established recoverable gas reserves[3] at 2×10^{12} m^3, mostly (~87%) in Alberta; and the available surplus—that is, the excess over contracted deliveries and the volumes retained in Alberta under that province's protection policy—runs to over 87×10^9 m^3. Thermal coal, far less well explored, is even more ubiquitous. Federal agencies (Canada, 1979) report at least 4×10^9 tonnes immediately recoverable with current technology and at current market prices, while other authoritative sources (Energy Resources Conservation Board, 1979*a*) consider that in the Alberta plains region alone, over 5.5×10^9 tonnes, all recoverable by surface-mining, are now ''proved.''

But how far fuel substitution can take us toward oil self-sufficiency, and how quickly it could do so, cannot now even be guessed at. As matters stand, we still do not know how fast the price of Canadian oil will be allowed to rise, how it will eventually be pegged to the ''world'' crude oil price, and how natural gas will track oil; and until these questions are resolved, one can only speculate about future patterns of interfuel competition. Probabilities suggest, however, that current thinking about substitution, with its almost exclusive emphasis on natural gas as a replacement for oil,[4] may turn out to be counter-productive. In view of the present western Canadian gas surplus, and the need to provide cash flows that would encourage further exploration (as well as keep the smaller independent producers solvent), this emphasis is not entirely misplaced. But in the longer run, the neglect of coal as an alternative substitute is bound to prove to the disadvantage of consumers. The price of gas is certain to escalate faster than that of coal—and seductive price offers now will especially not stand the test of time if new gas demands, stimulated by substitution, require deliveries from less favourable (tight, deep, or ''frontier'') reservoirs. An effective fuel substitution programme requires, therefore, not only that coal be gradually phased in, but also that a national energy policy actively promote moves in this direction and ensure timely provision of the infrastructure for expanded coal production and transportation. This becomes all the more important when it is borne in mind that coal use is not confined to electric utilities and heavy industry—that contemporary equipment and practices equally allow its deployment in light industry and as a heating fuel in large buildings.[5]

III

And this, by a somewhat circuitous mental route, leads me to see the most serious shortcomings of current Canadian energy policy in its third pillar, in its approaches to the development of alternative sources of petroleum.

If, as I suggest, "soft" energy paths are inherently unrealistic in the foreseeable future, and even direct displacement of oil by other fossil fuels can only make limited inroads into the total Canadian demand for liquid fuels, oil self-sufficiency can only be attained if new, commercially producible, "giant" petroleum pools are discovered or, failing that, if greater reliance is placed on "synthetic" crude oils from "unconventional" sources (which, technically, include coal as well as heavy oils and oil sands).

Of these two alternatives, the first, *if pursued as more than a secondary route*, is prima facie the more hazardous because it rests on long-range expectations and is, in fact, a gamble. Over the past ten years, additions to western Canadian conventional petroleum reserves—all accruing from step-outs and adoption of previously too costly enhanced recovery methods rather than from new strikes—have only averaged 24×10^6 m^3 per year, or less than a third of the average annual rate of withdrawal from these reserves; and while exploration in frontier areas, most notably in the Beaufort Sea and off the Newfoundland coast, has yielded some encouraging finds, the extent and commercial potential of these discoveries remain uncertain.

Further exploration activity may quite conceivably resolve these uncertainties in our favour, and then make very significant contributions to Canada's inventory of producible liquid hydrocarbons. So, of course, will enhanced recovery operations in older pools, which will gain momentum as oil prices rise. But what is already quite evident is that "giant" pools, if found at all, will only be found in remote areas and will be expensive to produce; and in these circumstances, prudence dictates higher priority for the second alternative—for production of "synthetic" crude oils ("syncrudes") from unconventional sources with which this country is, in fact, singularly well endowed.

In-place, heavy oil resources, mostly straddling the Alberta-Saskatchewan interprovincial boundary in the Lloydminster region, have been conservatively estimated at 1.1×10^9 m^3 and could, even if only recovered with 20 per cent efficiency, ultimately provide at least 0.2×10^9 m^3 of an (upgraded) marketable syncrude. Surface-mineable oil sands in Northern Alberta contain some 6×10^9 m^3 of proved recoverable bitumen from which about 3.6×10^9 m^3 syncrude can be produced. More deeply buried oil sands, from which bitumen will have to be extracted by *in situ* methods, are expected to yield an additional 24×10^9 m^3. And ultimately recoverable thermal coal in the three western provinces—which, if converted into syncrude by contemporary technology, would yield an average of approximately 0.5 m³ per tonne—offers a reservoir that exceeds the combined potential of all oil sands and heavy oils.

Against this almost embarrassing wealth of resources stands, of course, at present a distinct poverty of technology. Practical methods for exploiting these enormous energy reservoirs as sources of oil are still in their infancy, lack commercial performance records, and are therefore perceived to be risky.

In the case of heavy oils and deeply buried oil sands, the most immediate difficulties relate to extraction. Primary recovery of heavy oils by conventional petroleum recovery methods yields on average less than 7 per cent of the oil in place; and the still more viscous oil sands bitumen cannot be produced at all by such means. Reasonably efficient extraction, which is as much dictated by economics as by resource conservation policies, depends therefore on application of innovative techniques (e.g., steam stimulation, partial *in situ* combustion, etc.), which lower oil viscosity by raising reservoir temperatures; and the operability of these techniques is critically affected by the specific geological features of the reservoir.[6]

Where oil sands can be surface mined, recovery is less of a problem—although mining itself is sometimes difficult because of northern Alberta's harsh winters. But separation of bitumen from the sands by the Clark process (Energy Resources Conservation Board, 1979*b*), still the only commercially proven method, is environmentally hazardous. It consumes very large amounts of water and requires disposal of spent sands, associated toxic sludges, and chemically contaminated waters in large tailings ponds. Alternative processes that would minimize these hazards are under study, but they are still in early stages of development or too costly.

In contemplating production of oil from coal, we find ourselves seriously hampered by unfamiliarity with engineering and performance details, which, until recently, made "indirect" coal liquefaction (via gasification and subsequent Fischer-Tropsch synthesis) appear inefficient as well as uneconomic,[7] and restrained by the fact that "direct" liquefaction processes are barely beyond testing in pilot plants.

Yet, notwithstanding their real or perceived shortcomings, the technologies at hand are *workable*; and on the basis of half a loaf being better than none, they are *adequate*. The fact, bluntly put, is that we cannot continue to sidestep questions of future oil supply without escalating our present difficulties into a serious emergency. And in this situation, the abundance of resources at our disposal allows us to downgrade concerns over technical efficiency and/or resource conservation; to use available means for syncrude production wherever these means suffice; and to gain time for development of better technology *without* remaining dependent on foreign oil.

What could be achieved can be inferred from recent moves toward further development of oil sands, and from a more realistic appreciation of coal conversion than is currently accorded it.

The producibility of marketable syncrudes from oil sands by means at hand is attested by commencement of operations at the 20,000 m³/d Syncrude plant near Fort McMurray, Alberta; by current development of the similar, slightly larger, Alsands plant; and even more significantly, by the Esso Cold Lake project, which differs from the other two (and the previously commissioned Suncor plant) in being a first commercial *in situ* scheme. All three developments have received wide public attention and need no further comment here beyond the observation that their very initiation implies confidence in the effectiveness of available technology. Like Suncor (previously, Great Canadian Oil Sands), each of these projects will undoubtedly encounter technical difficulties from the unprecedented scale of operations. Syncrude has, in fact, already experienced such problems with its fluid coker units, reportedly the largest so far built. But these set-backs—although costly and time consuming—are evidently not seen as more than "teething troubles" that accompany the commissioning of any large, complex industrial installation.[8]

From a broader Canadian viewpoint, the question is therefore not *whether* development of oil sands as sources of liquid fuels is practical, but rather *how* such development can be accelerated and supplemented in order to close the forecast supply-demand gap as quickly as possible.

Though less directly supported by Canadian experience, much the same bottom-line question presents itself with respect to heavy oil and coal.

Effective utilization of heavy oil depends primarily on more efficient recovery. Subsequent upgrading to a lighter crude, without which it cannot easily be transported or further refined, involves much the same processing sequence as upgrading of oil sands bitumen and does not have to wait upon development of new technology. But in engineering terms, heavy oil is merely a somewhat lighter hydrocarbon mix than oil sands bitumen, lies typically in reservoirs comparable to Esso Resources' Cold Lake lease, and—as has, in fact, already been demonstrated by field pilot tests—is recoverable by the same methods. That it has not been aggressively developed as a source of syncrude is therefore more due to self-inflicted lethargy than to lack of know-how.

As to coal: here, paradoxically, we have potentially a quite unique technical freedom because its conversion to a marketable syncrude can now be accomplished by any one of three quite different techniques.

The first, to which I have earlier alluded as "indirect" liquefaction, entails gasifying the coal with oxygen and steam to produce a syngas mostly composed of carbon monoxide and hydrogen, "shifting" this gas to the required $CO:H_2$ ratio, and then reacting it over catalysts at elevated temperatures and pressures. All these steps are commercially proved by more than fifty years' industrial performance[9] and, more particularly, by South Africa's Sasol plant (which began operations in 1955 and has now been replicated in the almost three times larger Sasol II facility).

Rooted in German coal-based Fischer-Tropsch technology—which attained commercial status in the late 1920s, but fell into disuse after World War II when low-cost oil and natural gas became abundantly available in Western Europe—"indirect" liquefaction holds particular interest because of its inherent flexibility. Since reaction products depend on operating conditions—that is, on the specific catalyst-temperature-pressure combinations—a single plant can be operated in several different modes (see Table 1) and thus respond to periodic demand changes.

The second route of synthetic liquid fuels from coal involves *complete direct* liquefaction. A further development of Bergius hydrogenation, which was also commercially used in Europe during the 1930s and 1940s for production of gasoline, diesel fuel, and heating oils, this is less well established than the indirect technique in the sense that contemporary processes do not yet have industrial track records. However, extensive pilot plant testing of such technologies as Gulf Oil's SRC and Hydrocarbon Research Inc.'s H-Coal has demonstrated their basic reliability and also provided design data for scale-up to commercial sizes; and if expedited, full-size plants could be brought into operation in the late 1980s. What is required is, primarily, a test programme that would identify the most suitable Canadian coals and establish the optimum processing conditions for them.

The third option for producing liquid fuels from coal presents itself in *partial conversion*—that is, in stripping the coal of its most valuable hydrocarbon fractions before using it as a steam-raising fuel or gasification feedstock. Technology for this option is offered by Occidental Petroleum's (Garrett) Flash Pyrolysis, by the similar Lurgi-Ruhrgas process, and by FMC's COED (or Cogas) process, each of which can yield some 0.20 m^3 of "primary" coal liquids per tonne for upgrading to a marketable syncrude and

Table 1
OPERATIONAL FLEXIBILITY OF INDIRECT COAL LIQUEFACTION

Product	Operational Mode*			
	(a)	*(b)*	*(c)*	*(d)*
Gasoline	75%	49%	38%	25%
Diesel/Jet Fuel	20%	46%	35%	22%
Chemicals	5%	5%	27%	3%
SNG	—	—	—	50%

* Set for maximum production of (*a*) gasoline; (*b*) diesel fuel; (*c*) chemicals; (*d*) substitute natural gas (SNG). Yields are approximate.

leaves a charred solid fuel for use under utility boilers. All three processes are new and lack a convincing history of commercial performance in North America.[10] But because of their inherent simplicity, they pose little risk and could be operated as an integral function of any large, base-load, coal-fired power station. Subject to confirmation of European and U.S. combustion tests, which showed that the residual solid fuel can be burned as easily as raw coal, partial conversion could, in fact, be gradually phased in on a station-by-station basis.

The potential of such a programme, which could be implemented within a relatively very short time-frame, is illustrated by what would accrue in Alberta, where annual coal consumption for electricity generation is now projected to climb from 9.2 million tonnes in 1979 to approximately 16 million tonnes in 1985–86 and approximately 34 million tonnes by the end of the century (Energy Resources Conservation Board, 1979c). Processed to recover liquid hydrocarbons before being dispatched to boilers, these tonnages could provide some 3 million m³ syncrude/year (roughly equal to the annual output of the Suncor oil sands plant) by mid-decade, and approximately 6.5 million m³ (not much less than the design capacity of the Syncrude plant) by century's end.

IV

Except for indirect liquefaction, for which detailed capital and operating costs can be developed from Sasol data, coal liquefaction is at this time still beset by considerable economic uncertainties. Published analyses often vary widely—partly because they relate to different liquefaction processes, assume different financing methods, and sometimes make contingency provisions that do not reflect the actual status of the technology. There is now, however, broad consensus that the spectrum of operated or operable liquefaction techniques contains several that would be viable against OPEC oil;[11] and on that basis, they offer practical means for augmenting syncrude production from oil sands and heavy oils.

What must therefore be addressed is, first, whether oil self-sufficiency by the end of the century could be achieved solely by accelerated exploitation of oil sands and/or heavy oils; and secondly, if this were possible, whether it *should* be so accelerated or whether greater advantages might lie in fostering a wider resource base for syncrude production.

With regard to the first question: while forecasts of indigenous oil supplies published in the early 1970s envisaged, and were in fact predicated on, bringing new oil sands plants into operation at regular, two-to-three-year intervals, no such sequence actually materialized. Even if one additional large plant beyond those now on stream or scheduled to commence operations in 1985–86 could still be commissioned in this decade, the combined syncrude output from oil sands—then totalling some $30-32 \times 10^6$ m³ per year—will not suffice to offset declining deliveries from traditional western Canadian pools and, even in such best of circumstances, will not begin to reduce our present levels of oil imports. In the following decade, more rapid development, sufficient to compensate for further decline of western production *and* substantially eliminate the need for imports, is also quite unlikely: we would need to build a minimum of six (or, more realistically, seven) additional 23,000 m³/d plants during those ten years. Such pressure on oil sands can, of course, be relieved by concurrent development of (Lloydminster-type) heavy oil, and will undoubtedly also be reduced by contributions to marketable oil supplies from new "conventional" frontier or offshore reservoirs.[12] But bear in mind where oil sands and heavy oils occur (see Figure 2); unless heavy oil development is shifted into Saskatchewan, the overwhelming bulk of syncrude would have to come from production facilities in the sparsely settled northeast quarter of Alberta and, even with significantly improved technology, impose quite unacceptable social, economic, and environmental burdens that would extend far beyond this region.[13]

That, as much as grave uncertainty over how much oil sands and heavy oil development programmes can in fact be speeded up, makes it in my view imperative to base a significant part of future syncrude production on coal. The environmental impact of a coal liquefaction plant would not be less than that of a similarly sized oil sands operation; and since the most easily accessible (lowest cost) coals often lie under shallow cover in farming regions, mining itself may give rise to some land-use conflicts.[14] But the ubiquity of such coals in western Canada would afford us greater freedom in selecting plant sites and also provide significant opportunities for syncrude production elsewhere than in Alberta and Saskatchewan.[15]

Within the twenty-year time-frame in which oil self-sufficiency could be attained, coal will undoubtedly only be a secondary source of syncrude; and dispersal of syncrude capacity through reliance on coal would therefore be quite limited. But even so, a synfuels development programme that now begins to encompass coal as well as oil sands and heavy oil offers some compelling advantages. It would make for greater flexibility in promoting additional capacity, and through a much wider range of potential development sites, go some way towards easing anticipated manpower shortages as well as lowering needs for new infrastructure. It would reduce the technical hazards of syncrude production by spreading them over a wider spectrum of methods. It would enhance the beneficial spin-offs from application of innovative technology.[16] And it would coincidentally, though not unimportantly, gain a more truly *national* dimension than our present endeavours possess.

But aside from need for a wider resource base for synfuels production than Canada's energy policy (and energy companies) now contemplate, the

Figure 2
DISTRIBUTION OF OIL SANDS AND HEAVY OIL DEPOSITS IN ALBERTA
(All existing, tentatively approved, and expected new oil sands plants are, or will be, located in the Athabasca and Cold Lake Deposits)

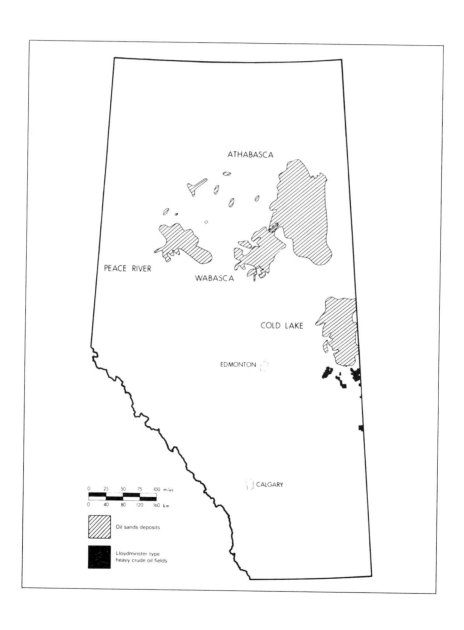

required pace of development appears to demand other actions that the policy does not adequately, if at all, address.

There can be no argument about the need for a much higher price than Canada is presently setting on its indigenous oil. Without that, production of syncrude (from whatever source) and development of frontier petroleum are clearly impossible. Technically, however, this does not necessarily mean going *indiscriminately* to, or close to, a ''world'' crude oil price; and it can, in fact, be argued that such a move would impede rather than accelerate development of new syncrude capacity. Different technologies for producing syncrude do not always possess cost parity even when producing from the same resource; and since developers will naturally seek to maximize their returns on investment, a substantially uniform plant-gate price offer is virtually bound to slow development by removing incentives for the deployment of production methods that are financially less rewarding, but essential, from a public standpoint. More is likely to be gained from recognizing different production costs and thereby encouraging a more diversified capability that can meet declared needs for oil subject to an economically acceptable average price for the total ''pool.'' As to the mechanics of such a policy, I note that we are already implementing something quite similar when differentiating between ''old'' and ''new'' natural gas, and that we come close to it wherever government imposes royalties and taxes in a manner that is designed to prevent an individual development making ''windfall profits'' *or* going under.[17]

However, if we are really serious about wanting to regain oil self-sufficiency within the shortest, practically possible time, appropriate oil pricing is only one imperative. Equally essential are lead roles by government that are not yet exercised. Specifically, I refer to the need for the following:

a. Clearly enunciated, firm ground rules for development of syncrude capacity that would replace the current adjudication of each proposed development ''on its own merits''

b. A timetable that sets out unequivocally where Canada intends to be with respect to oil supplies at different points between now and the end of the century

c. A form of collaboration between government and the energy industry that can give effect to this timetable.

Because oil is so obviously essential to us, the public interest makes a greater degree of governmental involvement in programming of syncrude development imperative. And what I see as necessary is perhaps, in the final analysis, not much different from what Alberta has long found essential for ensuring orderly and timely development of additional capacity for electric energy generation. In that province, the Electric Utilities Planning Council, composed of privately and publicly owned utility companies and observers from other directly concerned public agencies, maps long-range development and development options; utility companies in effect compete for authorization to install the required new capacity at the appropriate times; and rates of return to the companies are set by the Public Utilities Board. It does not seem unreasonable to see in this system a model for a more coherent approach towards ensuring that self-sufficiency goals set by national policy will, in fact, be met.

NOTES

* I am indebted to F.J. Mink, P.Eng., and R.A. Funk, respectively Department Manager and Economist, Economics Department, Energy Resources Conservation Board, for some helpful discussions and for validating certain of the reserves data.

[1] Evidence from Western Europe as well as North America suggests that consumers tend to accommodate rising gasoline prices by reallocating discretionary income rather than curtailing consumption, and Canadian policy may well encourage such a response: *gradual price increases* over an extended period of time have little effect on reducing consumption (Willenborg and Pitts, 1977).

[2] Incredibly, a leading Canadian politician proposed, as recently as February 1980, alleviating future oil shortages by manufacturing ethyl alcohol from potatoes. The only excuse for this proposal can be that it was made in Prince Edward Island, a potato-growing province, in the heat of a federal election campaign.

[3] The following metric equivalents have been used in this paper:
a. For liquid hydrocarbons: $1 m^3 \sim 6$ barrels.
b. For gas: $1 m^3 \sim 35$ scf (standard cubic feet).

[4] Only Nova Scotia and New Brunswick are now taking active steps to replace oil by coal, and even in these provinces, attention is confined to electric utilities that burn oil. Ontario Hydro, facing somewhat lower than expected demand for electric energy, has simply shut down its oil-burning Lennox power plant.

[5] Oddly enough, while there is general appreciation of the forward strides that gas and oil technologies have made over the past thirty to forty years, there is almost no acknowledgement of the fact that in jurisdictions more dependent on coal than North America, coal technology has similarly advanced. The popular comparison—often fostered by the mass media—is usually between oil and gas *as now perceived*, and coal *as it was one or two generations ago*. In this situation, the Canadian coal industry faces educational tasks that it has not yet addressed.

[6] This site-specificity is one of the reasons for the many (seemingly duplicative) experimental schemes that have been and are presently being undertaken in various oil sands deposits.

[7] Recent publications (Fluor, 1979) and proposed new energy-related action programmes in the United States indicate a profoundly changed view of this technology.

[8] Suncor (as GCOS) brought the world's first commercial oil sands plant, designed for 7150 m^3/d, on stream in 1968 and reportedly accumulated a loss of nearly \$100 million before beginning to record offsetting profits ten years later. However, practical experience gained in the 1970s should ensure that Syncrude and Alsands will not be similarly penalized.

[9] It might be pointed out here that except for the initial coal gasification, these steps are also replicated in Canadian operations in which "reformed" natural gas is used for manufacturing, for example, methanol and ammonia.

[10] A sizeable Lurgi-Ruhrgas facility has, however, been operated since 1963 at Lukavac, Yugoslavia.

[11] For example: a Sasol-type plant, taken to construction now, would yield attractively priced motor fuels when commissioned (in 1985), even if "world" crude oil prices escalated only at an average 12 to 15 per cent per year (as compared with very nearly 35 per cent per year since 1973) (Fluor, 1979). A Lurgi-Ruhrgas facility in Alberta, producing 2360 m^3/d of 28° API gravity syncrude from coal destined for electricity generation, would require a capital investment of $362 million (in 1980 dollars) and, for a 21.2 per cent DCF rate of return, yield syncrude at $145.25 per m^3 ($23.10/bbl) (Herring and Tollefson, 1979).

[12] It is important, however, to observe that it would take some seven wells, each as prolific as Mobil's Hibernia P-15 is expected to be, to yield the equivalent of one syncrude plant of the scale here envisaged.

[13] Serious concern over such burdens—mostly springing from heavy in-migration of (partly transient) labour forces, inadequate social infrastructure, massive water requirements, process wastes, and cumulative impacts on ambient air and water quality—was widely expressed at public hearings of Esso Resources' Cold Lake project and Alsands' Fort McMurray project. And there is now some worry over availability of labour for these two ventures.

[14] In view of contemporary mandatory land reclamation requirements, such conflicts would, however, be temporary. As a rule, lands disturbed by mining can be returned to their previous productivity within five to seven years.

[15] Such opportunities would exist, in particular, in British Columbia. In the coal-producing Maritimes and in coal-consuming Ontario, where coal costs are likely to remain too high for economically viable complete liquefaction, they would be confined to partial liquefaction of coal required for generation of electric energy.

[16] The history of the pre-World War II German synthetic fuels industry, and the resultant pre-eminence that its present-day successors still enjoy in many synfuels-related matters, is a case in point.

[17] To the extent that additional syncrude capacity generates new employment, it also offers means for offsetting production costs that are modestly higher than immediately acceptable levels through return of part of the revenues accruing to government from personal and corporate income taxes. If the equivalent amount of oil were to be purchased abroad, public treasuries would gain nothing; and I can see no reason for not similarly operating on a break-even principle where the goal of oil self-sufficiency might make it expedient. We would still reap major benefits from the new employment *per se* and from our independence from foreign oil sources.

REFERENCES

Canada, Department of Energy, Mines and Resources. (1979) *Coal Resources and Reserves of Canada*. Report ER 79-9. Ottawa: The Department.

Energy Resources Conservation Board. (1979*a*) "Reserves of Coal, Province of Alberta—Dec. 31, 1978." Calgary: The Board.

Energy Resources Conservation Board. (1979*b*) "In the Matter of an Application by the Alsands Project Group Under Section 43 of The Oil and Gas Conservation Act." December.

Energy Resources Conservation Board. (1979*c*) "Energy Requirements in Alberta: 1977–2006." ERCB 78-1, pp. 5-6. Calgary: The Board.

Fluor. (1979) "Liquid Fuels from Coal." Fluor Engineers and Constructors Inc., December.

Herring, Ian W. and Tollefson, E.L. (1979) "The Economics of Coal Liquefaction in Alberta by Low Temperature Flash Pyrolysis as a Precursor to Thermal Electric Power Generation." Paper presented to the Canadian Gas Processors Association, Calgary, 23 November. See also "Synthetic Crude from Coal Economic." *Oilweek* (3 December 1979): 71-73.

National Energy Board. (1978) *Canadian Oil: Supply and Requirements*. Ottawa: Minister of Supply and Services Canada.

National Energy Board. (1979) *Canadian Reserves of Natural Gas*. Ottawa: Minister of Supply and Services Canada.

Willenborg, J.F. and Pitts, R.E. (1977) "Gasoline Prices: Their Effect on Consumer Behavior and Attitudes." *Journal of Marketing* 41 (January): 24-31.

Chapter Three

6323
7230
Canada

The Geopolitics of Oil

by
F.H. Knelman

The history of the world exhibited a major discontinuity at the beginning of this decade. By the end of 1973, the global jolt was felt throughout the economically developed world. The Middle East oil embargo was a serious shock, a revelation of things to come.

The euphoria of the late 1960s with its vision of unlimited growth has been shaken. We are witnessing a dress rehearsal for living with permanent shortages. The end of the oil age is in view on the horizon and, in the telescoping of time through accelerated depletion, the world faces increasing political and economic stress for at least the next three decades. The politics and economics of oil may well produce some of the most intense strains on the present precarious balance of power. The Green Revolution, which has been considered the salvation for global hunger, is a voracious consumer of petroleum and petroleum products. The ecological perspective is frightening, for food and fibre connect all humanity, and fuel (petroleum) is the source of both.

Modern industrial societies are basically oil economies. Oil represents some 50 per cent of their primary energy consumption. But only when one examines the contribution to economic activity of the automobile, diesel engine, farm chemicals, plastics, and synthetics, all of which use oil or are derived from oil, can we fully comprehend what is meant by an oil economy.

Given the inventory of existing technologies, the combined demand of transportation, agriculture, and petrochemical feedstocks is largely directed towards a single resource—petroleum. The single unifying concept of power in its physical (the rate of energy use) and political sense provides the key to comprehending the multiple interactions between the global mal-distribution of oil and the mal-distribution of health, wealth, and justice, the intrinsic conflict of sovereignty and ecology and, ultimately, the major source of global conflicts. For these reasons alone, the geopolitics of oil is a central but neglected research area in the study of peace and war. It is the hope of this author that this modest contribution might catalyse a serious expansion of interest.

THE GLOBAL PERSPECTIVE

The rich petroleum reserves of the Middle East, for years the special domain of the multinationals—the seven sisters: Exxon, Mobil, Texaco, Standard, Sunoco, Shell, and BP—are now centre-stage in the drama of international conflict. The price of oil has risen dramatically in eight years, about a 1400 per cent increase. Iraq and other countries made sales at spot prices of U.S. $40 per barrel early in 1980. There seems little doubt that the world price will move up steadily to even higher levels within the next few years. This would mean gas rationing in the United States where coupons have already been printed and the plan is ready. It could also mean rationing in Quebec and eastern Canada.

Moreover, it is likely that OPEC will reduce its exports despite or because of a peace treaty between Egypt and Israel. The profound impact of Islam on the geopolitics of oil is still not comprehended. While Iran's total contribution to OPEC exports is marginal, this very marginality is extremely critical. Total imports from OPEC, including the 5+ million barrels per day (mb/d) from Iran, were about 30 mb/d in 1978 out of a rated capacity of only 32 mb/d (OECD, 1977). One can see how the withdrawal of 5 mb/d squeeezes the precarious supply/demand equation. Even before its war with Iraq, the new Iranian regime indicated that it would probably reduce production, an entirely reasonable policy in view of the high rate of valuation of oil left in the ground. The political vulnerability of Saudi Arabia is as real and imminent as that of Iran and for the same reasons.

It is strange that the arithmetic of oil, so stark in its simplicity, should elude the understanding of so many. It may be myopia or technological euphoria that has led us to such a state of blindness. Current global consumption is some 65 mb/d and is growing at least 3 per cent per year (the Western nations at 2.5 per cent per year). This represents a current annual consumption in excess of 23 billion barrels. But the demand in the Third World, which now consumes only 20 per cent of the total, is growing at 6 per cent per year. By the year 2000, this demand will more than double, and the Third World will command an increasing percentage of the total reserves. Total demand by the year 2000 could be as high as 40 billion barrels. This quantity threatens the depletion of all present reserves in a relatively short time. Competition for rapidly depleting reserves will become much more intense. The Warsaw bloc will be directly competing for the same reserves as NATO countries and in the same regions of supply. Both will be competing with the economically developing world. To ignore the significance of this arithmetic of oil would be folly.

In a speech to a U.S. special Senate committee, Admiral Turner, former head of the CIA, made an identical assessment. Now, for the first time, independent experts, the multinationals, governments, and the CIA are in close agreement on this ominous situation.

Saudi Arabia's oil minister, Sheik Ahmed Yamani, is the West's staunchest supporter in that he recognizes the advantages, if not the necessity, of mutual support for the preservation of the *status quo* in Saudi Arabia. This is a precarious hope at best. Sheik Yamani has been quoted as saying about his U.S. relations, ''Sometimes I feel like a Catholic who can't divorce a wife he isn't happy with'' (Kettle, 1978, p. 46).

Sheik Yamani has no illusions concerning reserves or the consequences of running out of oil. In Paris on 25 March 1980, he warned the world to actively seek alternatives or face the destruction of civilization. The danger comes not from the suppliers but the ''conventional wisdom'' that guides the policies of the West, particularly the United States and Canada. With Ontario turning to nuclear power and Quebec to hydro cum nuclear, there is the illusion of creating and imposing an ''electrical economy'' when the real demand for electricity simply does not exist. In Quebec, a preoccupation with electricity generation is fused with nationalism.

It is not at all clear that OPEC will increase the present capacity to 35 mb/d to meet OECD demand in 1985 (OECD, 1978). This uncertainty is as real as the fact that OECD dependency on oil will not change until 1985, representing a roughly constant demand of about 50 per cent of total energy since 1974. Neither nuclear power nor energy projected from other sources of some 5.5 mb/d by 1985 will reduce this dependency or provide the required security. At the assumed 4 per cent increase in annual growth by the importing countries, their demand will rise to approximately 45 mb/d by 1990. What will be the consequences of this threat to American security, prestige, and power or to the security of the Middle East, and what is the significance of oil to a viable Middle East peace? The relative power of the major actors in the Middle East theatre has now changed. The recent war between North and South Yemen, one of the mini-series of surrogate wars between great powers, is ominous for the security of Saudi Arabia, the last of the United States' friendly oil suppliers in the Middle East. There is a common delusion that new discoveries or more exploration will solve the oil supply problem. This is the most dangerous barrier to a sane policy.

It is the earth science of plate tectonics rather than the imagination or short-run profit maximization of oil exploration companies that determines the remaining reserves in the world. Professor Ken North has rightfully identified the Canadian ''Cornucopia Syndrome,'' that is, our faith that our resource base is infinite. While oil and gas fields are widespread in the earth's crust or the ocean bed, *giant ones are not*. There are relatively few, significantly large sedimentary basins. Moreover, where oil is found is critical in determining its economic viability. Finding a few million barrels close to markets is one thing, but to exploit oil in the Beaufort, Baffin Bay, or even the North Sea requires very large oil fields (over 300 million barrels). Moreover, the key is not the size of these potential reserves but the size of world demand. So far, the promise of "large" discoveries off eastern Canada has had a great influence on the stock market but not on the Canadian supply picture. The largest visible oil "find" to date has been the spill by the tanker *Arrow*.

In our self-defeating style of responding only to the most current crisis, we tend to forget that there are numerous crises facing the world. Perhaps energy is the most immediate and pressing, particularly for the global poor and dispossessed. But the reality is that we face multiple and converging crises of energy, environment, economics, and equity. War, of course, in its nuclear form is the crisis of crises and the inevitable product of our inability to solve our multiple problems.

One would be insensitive as well as imperceptive if one viewed the geopolitics of oil and food merely as causes rather than weapons in the struggle for an emerging new international economic order (Bergsten, 1975). There are profound shifts taking place in the dominant world order. They began, not in 1973 with the Middle East oil embargo, but earlier with the late war in Viet Nam, the retirement of Lyndon Johnson, and the failure of the Nixon/Kissinger policies for salvaging American economic and political pre-eminence by reluctantly exchanging mutual security of a bi-polar world for a tripartite American-European-Japanese alliance (Brezinsky, 1970). While we can no longer argue with Richard Falk (1975, p. 80) on the fundamental inadequacy of "a world order system constituted primarily by sovereign states of unequal size and wealth," the problem is which way the world will turn—to the lifeboat or the spaceship model, and who will be thrown to the "sharks," if the former. As a matter of reality, the whole notion of "lifeboat ethics" will not float. The majority left in the water will surely sink the lifeboat and all will drown together. The lesson of Iran is, in part, that a policy of global triage does not guarantee who will turn out to be dispensable.

American foreign policy under Kissinger was perhaps more subtle—a blend of carrots and sticks. There was a direct appeal to the oil-consuming nations to respond with solidarity against OPEC. Sensible tactics like the 1974 OECD agreement to curb member countries' energy consumption through conservation worked well for almost every nation except the United States and Canada. There was even some suspicion that the United States' somewhat reluctant response to OPEC was, as the London *Economist* observed, because "they saw increased oil prices as a quick and easy way of slowing down the Japanese economy" (Barraclough, 1975). But both sticks and carrots were nevertheless present in a variety of forms as military, technology, and food aid. Viewed from the conservative wing of OPEC, represented by Iran and Saudi Arabia, and now only the latter, military aid was one of the carrots. Both countries required this aid to maintain their own regimes against internal and external Arab revolutionary forces. But certainly

Kissinger viewed the accumulation of balance-of-payment surpluses by OPEC as leading to financial and economic chaos in the West. The mutual interest of the United States and the conservative wing of OPEC appeared prepared, if necessary, to give the Arabs some larger stake in a modified world order. Massive military and technology trade could offset and recycle petro-dollars as it successfully did in Iran. But perhaps the most powerful stick was what has been called "the brutal facts of the world food market" (Schertz, 1975, p. 180), the most significant being U.S. pre-eminence in food production. Assistant U.S. Secretary of State, Thomas O. Enders, put it bluntly at the November 1974 World Food Conference: "the food producer's monopoly exceeds the oil producer's monopoly," and "food will give us affluence . . . and would enable the United States to formulate the rules of the game" (NYT, 1974). "Food is a tool in the kit of American diplomacy," as Earl Butz said (Barraclough, 1975, p. 24), but it is a tool that must be used delicately. Agripower and technipower have not yet directly confronted petro-power. The United States would prefer carrots to sticks as long as the new world economic order does not displace the position of U.S. multinationals. The years 1974–75 witnessed a powerful debate concerning Third World aid, and OPEC achieved representation on the boards of the International Monetary Fund and the World Bank. Although Kissinger was forced to abandon hawkish rhetoric by the early summer of 1975, reconciliation with OPEC was not secured.

President Carter's policy on fulfilling the full goals of the peace treaty between Egypt and Israel was given urgency by the collapse of the Shah's regime and the ensuing conflict with Iraq. The current focus of the Reagan government is to stabilize Saudi Arabia and to trade military and technological aid for Egyptian political support, perhaps to place an American military presence in the Middle East. News of oil discoveries in Egypt will reinforce this policy. It seems clear that the United States is now prepared to resume to some degree a military presence in areas where its interests are threatened. The Soviet invasion of Afghanistan has provided the excuse as well as a cause for this type of action. It is also clear that the revolution in Iran places new pressures on Israel because there has been a real shift of power in the Middle East theatre.

THE CANADIAN SITUATION

The oil embargo of December 1973 by OPEC traumatized Japan and the oil-importing Western nations. This included Canada because of the crucial decision that had been made in the Diefenbaker years to formally partition the country into two oil-marketing areas divided by the Ottawa Valley line. Areas east of the line were to depend on oil imports predominantly from OPEC (about 70 per cent for Quebec), while Alberta would supply the western areas. Then and now, our policies are influenced by the multinationals who were supplying eastern Canada with off-shore cheap crude, cheaper than a truly national distribution system could supply. When the Eisenhower quotas on imported oil were lifted at the end of 1970, Canadian producers scrambled to satisfy the new export market, reaching 1 mb/d by 1973.

In 1970, Jean-Luc Pepin, then minister of industry, urged the Canadian energy industry to sell off its oil. "It would be crazy to sit on it. In maybe 25 or 50 years we'll be heating ourselves from the sun and we'd kick ourselves in the pants for not capitalizing on what we had when oil and gas was a current commodity" (Kettle, 1978, p. 39). Oil and gas represent 80 per cent of Canada's energy consumption. Surely the export of either fuel then or now is folly.

Another critical interaction between oil and politics was the Prudhoe Bay discovery by Atlantic Richfield on the Arctic slope of Alaska, the largest oilfield and the second largest gas field in North America. This increased pressures on the Canadian government for energy continentalism, the Nixon policy for securing stable access to Canadian oil and gas resources.

The period between 1968 and 1971 was the equivalent of a diplomatic resource war between the United States and Canada. The issues, as expected, were sovereignty and economics. Canada wanted the pipelines that were to carry oil and gas from Alaska to pass through Canada. Our federal government also wished to sell more oil and gas to the United States. In addition, the issue of territorial and coastal waters in the Arctic was raised by the voyage of the *Manhattan*. Canada responded with territorial claims. This was in part the birth of the struggle over the oceans and their continental shelves. We shall return to this theme later, but for now it is important to establish the fact that we initiated policies that led to the selling off of our proven economic resources only to force ourselves to seek these same resources farther into the northern hinterland, where both the society and the environment are far more fragile and the costs are much higher.

One digression is important at this point. For many of the basic oil-consuming activities described above, neither natural gas nor coal are viable substitutes, the former because it too is a relatively limited resource and the latter because of mismatching between producing and consuming regions, a microcosm of the global predicament of geopolitics. Conversion of our large western coal reserves to synthetic fuels, gaseous and liquid, could be very significant, but a combination of factors—economic, political, and environmental—presents serious barriers to this development.

If we turn to nuclear power as our salvation, this is equally problematic. For one thing, nuclear power provides baseload electricity only and, thus, substitution for oil is extremely limited. In addition, nuclear requires fossil fuel back-up plants in the absence of hydro. The domestic nuclear programme in Canada has suffered the same cut-backs as in the United States, perhaps not so drastic, but significant. The year 1978 was the first year since the beginning of commercial nuclear power that no new orders were on the "books" in the United States. Canada's earlier projection in 1976—some one hundred reactors by 2000—had faded. There may be twenty reactors (mainly in Ontario), but this is only the most recent, scaled-down projection. Nevertheless, the Canadian and Ontario governments—the real decision makers—are dedicated to a nuclear option by default, default of real analysis or a viable energy policy. The policy direction of the major vendors of reactors in the world is no longer to domestic programmes but rather to economically developing regions. Canada has made sales recently to Rumania and is courting Mexico. All the nine vendors in the London Group are eyeing the huge potential China market. The prospect for proliferation will increase rapidly as a result of these technology transfers.

The recent announcement about an international fusion research centre that might be sited in Canada—even in Quebec—merely adds to the confusion. The Tokamak fusion experimental reactor is a large net user of energy. While there is a vague promise of energy in the long term, there is not as yet any certainty of technical feasibility, and certainly not of commercial feasibility. Contrary to its public image, there are serious environmental and economic problems inherent in the fusion process. The mass of the containment unit is greater by an order of magnitude than that in fission reactors. Very high rates of neutron activation and tritium emissions constitute serious radiation problems. Net energy considerations are totally unresolved. The fusion promise, like the earlier fission promise, should be viewed with caution as any new so-called salvation technology. Technological optimism over fusion is dangerous. Solar power is more advanced yet neglected because it does not conform to the current distribution of political or economic power.

Still another significant fact that seems to be particularly difficult to comprehend is that the United States is both the largest producer (or has been until recently) and the largest consumer of petroleum. And since consumption has continued to exceed production, the United States is also the largest importer. California alone—still a major producer—imports some half million barrels (mb/d) of oil per day from Indonesia. The United States as a whole consumes almost 18 mb/d and must rely on imports for some 8 mb/d. This is more than Saudi Arabia produces. Mexico cannot be expected to supply more than 1 mb/d and 1 trillion cubic feet (tcf) of gas per year by the 1990s, neither being significant in terms of U.S. consumption. The United States consumes some 26 tcf of natural gas per year.

In terms of global geopolitics of oil, the Middle East has the largest reserves followed by the Soviet Union. The Soviet Union is the leading single producer in the world, that is, about 11 mb/d, or more than U.S. peak production. The Soviet Union consumes 8 and exports 3 mb/d, and its production will probably peak before 1984. China's export potential is small in terms of global or even U.S. demand. North Sea oil and gas face huge Western European demand and will peak before 1990, after which Britain, the major owner, will again be dependent on imports. The incontrovertible fact is that the oil economies of the world are doomed and doomsday will occur before 2000; that is, if business as usual cum conventional wisdom continues to guide their policies. So far, we have had mere tokenism towards alternative energy sources and conservation, despite some promising signs. Marginal or replacement pricing of oil would surely indicate economic feasibility of a broad range of conservation measures as well as several renewable energy possibilities. The Canadian energy system is the most wasteful among the Western industrialized countries, even when correcting for geography, climate, and the energy content of exports. We are the most voracious consumers, exceeding even the United States on a per capita basis.

THE FUTURE IS NOT WHAT IT USED TO BE

If we examine the consumption of petroleum in our energy system, the minimal or threshold amount below which we cannot satisfy present obligatory demands, that is, by transportation, agriculture, and petrochemical feedstocks, is estimated to be between 25 and 35 per cent of our total energy demand (Kettle, 1978). By 1990, we will barely be able to satisfy this demand and will not be able to sustain that level for more than a couple of years. The grim fact is that Canada, like the United States, is now well towards becoming a major importer of petroleum in this century. Already, Canada had lost an Exxon commitment to Imperial Oil of 100,000 barrels per day from Venezuela, a country producing at full capacity. This erupted into a crisis over Petro-Canada. New discoveries, extremely rare, like West Pembina or Hibernia, can only help delay the dilemma for a matter of a few years at the most. We are not prepared for a major nuclear commitment or a serious commitment to coal. Our major fossil fuel reserve will be natural gas, and we will need all we have and all we can find to last through the end of the century.

This analysis bears repeating. The demand for thermal energy, heat as in space heating or process heat, can be met by electricity or coal, that is, by nuclear, or by coal directly or indirectly as synthetic fuels. But the demand for motive power or mechanization in transportation and agriculture, with the latter's additional demand in the form of fertilizers and pesticides, and the demand for petrochemical feedstocks for most synthetics from plastics to food additives cannot be satisfied by electricity; that is, nuclear cannot meet this irreducible and growing demand. Theoretically, coal could, but the lead time and the problems make it unlikely if not impossible in the medium time range. There are constraints of land and water that require close analysis. The facts are that there are no surpluses of oil or gas in Canada and any policy to export natural gas, as recently recommended by the National Energy Board (on the basis once again of purely temporary excess capacity), is myopic and clearly not in the national interest. If we examine the United States' appetite for oil and gas and their enormous shortfall, any contribution Canada could make is almost meaningless (a matter of a few months' relief) and only feeds an appetite that requires serious dieting.

The technological optimists still continue to argue that there are no real shortages and no economic shortages. In a perfectly free market, a particularly popular fiction, price alone will determine the amount of energy available; perfect substitutability and the perfection of the technological fix being part of the paradigm.

Canada's energy policy has been correctly described as a "bizarre case of bungling" (Crane, 1976); but even this is generous. One should recall that our National Energy Board was created by Mr. Diefenbaker's Cabinet, which was reasonably nationalist in composition. Nevertheless, it was susceptible to the Canadian (multinational) oil lobby, so that sections of the energy act dealing with oil have never been proclaimed in the House. The oil companies have managed to keep Canada at once a major exporter and importer of oil; this device being a technique for maximizing returns from the tax-adjustment haven that is Canada.

Very few people realize the largely hidden and diplomatic struggle between Canada and the United States that took place between the Prudhoe Bay discovery of 1968 and Canada's plea to the United States to buy our oil from 1970 to 1972. Arctic sovereignty was at stake, as we recall from the challenge of the *Manhattan*. In Feburary 1970, Nixon announced his policy of energy/resource continentalism. Canada was encouraged to respond, hoping that the United States would move its Alaskan oil through a pipeline built in Canada (and later a gas pipeline as well). In September 1970, the Canadian government approved the largest single export of natural gas to the United States in its history (6.3 tcf), a major mistake yet to receive its proper accounting.

At this point, it is important to introduce the question of how we determine our reserves of non-renewable resources. Space does not permit an exhaustive analysis. Nevertheless, certain broad issues may be analysed. Methodological questions introduce serious uncertainties into estimates of reserves; so much so that wide variations can be selectively legitimated. One of the methodological problems is associated with the use of geological analogy, which can combine politics and geology. The use of the words "potential," "reasonably assured," "inferred," "expected additional," "prognosticated," and so forth, are so susceptible to value judgements as to render them readily amenable to manipulation. The fact that industry forecasts, either directly or through trade associations or institutes, have consistently overestimated oil reserves is in itself suspicious but possibly amenable to different interpretations. The classic example was the speech by former energy minister Joe Greene on 2 June 1971 to the Canadian Institute of Mining and Metallurgy, where he stated that "Canada's total oil reserves were 469 billion barrels at the end of 1970 while total natural gas reserves were 725 trillion cubic feet. At 1970 rates of production, these reserves represent 923 years' supply of oil and 392 years for gas" (Crane, 1976, p. 10). The ultimate source of these figures was the Canadian Petroleum Association, although the National Energy Board presented them to Mr. Green as the product of their own research.

Is this the result of irresistible optimism or other motives? Since the multinationals play it both ways, that is, claim shortages for some purposes and surpluses for others (whether they are seeking depletion allowances or export licences), a certain degree of incredulity is justified.

It has been revealed that the leading oil companies deliberately manipulated the market, possibly in collusion with OPEC, in order to create a false shortfall of supplies during the 1979 gasoline shortage (Kimche, 1980, p. 14). Nor could the Iranian disruption be blamed. Such manipulation (Mobil has now been officially charged by the U.S. government) is intrinsic to the nature of the multinational oligarchy. Occasional declines of spot prices fuel this confusion. Using a similar technique, the oil companies seek deregulation of prices and beg for tax incentives or demand other social subsidies to stimulate exploration, thus enlisting and delisting the free market at the same time. One thing is clear—regardless of whether deception or self-deception is involved, the multinationals have a monopoly on information, which governments cannot match. This hoarding of information would be reason enough to support a more powerful Petro-Canada with the full range of capacities of a multinational. In particular, former energy minister Alastair Gillespie's decision last March to allow Petro-Canada to negotiate directly for Venezuelan imports is a positive development.

Canada currently imports over 350,000 barrels a day from foreign sources at an average price of approximately $32/bbl. The balance-of-payments deficit on the 350,000 bbls/day imported to the eastern provinces is about $2.2 billion per year. It is unfortunate that we do not have technology to refine the heavy oil sludge remaining after the production of gasoline and heating oils from conventional light crude. We simply export this to the United States, which has such capacity. If Montreal refineries, for example, could acquire this additional capacity, they could reduce our import burden by 75,000 bbls/day in a few years. Even this could hardly keep up with increased dependency on imports, which could be 500,000 bbls/day by 1982. Consequently, we should be applying refining capacity for sludge on a much larger scale to reduce our dependency. A true cost comparison of such a development with tar sands and frontier sources would now be highly advisable. Conventional wisdom, unfortunately, is tied to conventional sources. Bad energy, like bad money, drives out the good. Moreover, bad energy is both costly and uncertain. The billions of dollars required for tar sands, pipelines, and frontier and offshore sources will surely cut into the money we desperately require for necessary social services, let alone for conservation. Thus bad energy drives out both good energy and good money.

Examining the current forecasts of oil supplies in Canada and the United States, one immediately notes that we are dealing with ranges of pessimism (Hayes, 1979; CONAES, 1978; U.S., 1977). The unrelenting optimists have lost credibility but unfortunately not power. Even when the facts on conventional crude become inescapable, non-conventional sources are introduced to continue to paint a picture of ultimate plenty. Environment and social and economic constraints somehow disappear from the arithmetic.

It is interesting that Wilbur Hopper, chairman of Petro-Canada, unlike the former head, Maurice Strong, is a technical optimist. He has gone so far as to state that a rapid construction of tar sands plants could make Canada energy independent by 1995, and he includes both the present surface-mining method and the still undeveloped *in situ* technique. Syncrude has proven so inefficient that it has sued the engineering firm for design defects. And this is a so-called "mature technology." As far as *in situ* technology is concerned, Dr. C. W. Bowman, chairman of the Alberta Oil Sands Technology and Research Authority (AOSTRA), suggests that the challenge to develop the technology is even greater than was CANDU technology because "we've got both the basic research and the engineering design to do" (Calamai, 1979).

To add to the monumental engineering and scientific task of tapping the deep deposits (under more than 50 metres of overburden), the static net energy on the present surface-mining Syncrude process is seven to one, that is, we require 18,000 bbls of oil equivalent to produce 129,000 bbls/day (Calamai, 1979).

This ratio is highly questionable. If we assume that a Syncrude plant has no better power ratios (i.e., $\frac{P_o}{P_i}$) than the best of nuclear power plants, that the life of the plant is twenty-five years and the time to build one is five years, then at a doubling rate of building new plants of about three years, the energy pay-back time will be delayed many years. The doubling time of three years is chosen because there would still be plants under construction by 1991. Under this particular set of conditions, there would be no net energy output until well into the 1990s. Moreover, this programme would face impossible physical, fiscal, and environmental barriers. We would have made such a large energy investment to meet the target date of twenty producing Syncrudes by 1995 that we would be imposing huge shortages on ourselves between now and then. Even using Peter Calamai's 18,000 bbls/day, each plant consumes 32.85 million barrels per year in construction (Price, 1975). The "heavy oils" involve a similar set of problems and uncertainties that are possibly even greater.

The tar sands issue is yet another example where the illusion of a perfect market purportedly creates the necessary technology for absolute and timely substitutability.

It is instructive to provide some idea of the extent of synthetic crude in Canadian oil sands and other heavy oil deposits. There are over 1 trillion barrels, the vast proportion being in Athabasca. This is a total figure and less than 50 per cent will ever be recoverable. This also gives us some insight into the enormous waste problems. Recovery of this bitumen oil would require about 5 trillion barrels of water. The amount of sulphur associated with these deposits is significant (about 5 per cent by weight). Sulphur dioxide is already a contentious issue for Syncrude, and costs do not include proper emission controls. The quantity of solid wastes in the surface-mining process is also enormous and environmentally significant. And finally, there is no *in situ* technology that is economically and technically feasible at present. Lead time will be at least ten years for the first prototype plant, and both capital requirements and net energy considerations could constrain the rate of exploration. The concept supported by Atomic Energy of Canada Limited that we should use nuclear power for *in situ* mining (Gander and Belaire, 1979) is environmentally unacceptable.

A critical indicator for oil-producing countries is the ratio of their reserves to their production, or the R/P ratio. For oil, a minimum desirable ratio is ten. This is not an arbitrary number. When an R/P ratio falls to ten, it means current annual production is 10 per cent of our remaining known stock; that is, there is only ten years' supply. Due to the nature of oil field mechanics, if a country's oil comes from fields that have been worked for some time at the maximum efficiency rate, the remaining recoverable oil (about 50 per cent) can usually be tapped at about 10 per cent per year. With no significant new discoveries, or only discoveries in costly frontier areas, there is a long lead time required to bring such fields into production, that is, at least ten years. Even worse, these new discoveries are for the wrong market, as was the case in Canada in 1973 and is again now. Thus, as Professor North (1979) has so aptly put it, "Canada is in double jeopardy."

For natural gas, our R/P ratio is sixteen and will decline to about twelve by 1985, another peril point. Under the National Energy Board's old formula, a gas R/P ratio of twenty-five was considered the minimum desirable level.

One should remember that OPEC pricing is in U.S. dollars. At U.S. $30 per barrel, the Canadian price would be $35. Before 1990, Canada could be importing as much as 1.6 mb/d and the price could be Can $50 /bbl, or $29 billion per year of expenditure (Kettle, 1978). This is close to the federal government's total current annual expenditure. By 1990, if we have not radically reduced our dependence on oil, we will face a combined energy and economic disaster.

Canadian oil demand by sector (excluding losses) is as follows: transportation (50 per cent), domestic and farm (20 per cent), industrial (18 per cent), commercial (7 per cent), and energy industry (5 per cent). In 1978, oil provided 54.8 per cent of our total energy. The proponents of electric power would have us believe that we can transform our technological structures in time to convert to an electrical economy. This is unlikely and is a policy with unacceptable risk.

Thus, the present and medium-term future realities are bleak. In 1974, the United States produced 10.4 mb/d of oil and imported 5.8 mb/d. They produced 20.9 tcf per year of natural gas and imported 0.9 tcf per year. For 1980, and the 1985 reference case (U.S., 1977), U.S. import dependency will increase for oil to 9.3 mb/d and 9.7 mb/d respectively, and for gas to 1.3 and 1.7 tcf—a definite and significant increase in dependence on imports. It should be noted that all of Canada's proven oil reserves (6.5 billion barrels) would only provide total U.S. needs for a matter of twelve months, and all Canada's proven gas reserves (about 60 tcf) for a matter of two to three years. The reference case also indicates that Canadian oil imports will rise from −0.2 (a net export) to 1.1 mb/d in 1985. These numbers should be taken with caution, since they may be unduly optimistic. Only a massive national commitment to reduced demand through elimination of waste, renewable substitutions, and demand management can avert crises in this century.

Professor Ken North (1979), petroleum geologist of Carleton University, and King Hubbert (1974), of the U.S. Geological Survey, are the two most prophetic oil reserve analysts in the Western world. For well over a decade, both have been making the most consistent and correct forecasts. One cannot credit them to highly.

One can only conclude that the oil age is ending. By the year 2000, if we have not made the necessary adjustments, and time is now also a scarce resource, we will face truly massive discontinuities in our economy and our life-style. Due to the direction of the threat, such as the special vulnerability of the United States, international pressures over energy resources will intensify rapidly in the 1980s. If the United States cannot find a new salvation technology and rapidly bring it to commercial fruition, then Canada among other countries will find itself affected by serious spillover effects from the American energy problem.

GEOPOLITICS IN CANADA

The politics of oil not only exists between countries but also within them. Thus, geopolitics in Canada and the United States mirror the regional, jurisdictional, and political conflicts in the global scene. Texas, Oklahoma, and Alberta express, in political terms, the mentality of OPEC countries. In the United States, the political struggle over deregulation and ''windfall'' profits is strictly divided along lines between producers and consumers.

The Canadian election of 1980 found energy at the top of the agenda of issues. Many Canadians were concerned about the price hikes for gasoline announced by the Conservatives. It is strange that no Conservative candidate quoted Prime Minister Pierre Trudeau at the First Minister's Conference of April 1975 when he remarked:

> We cannot expect those who search for oil—whether they be Canadians or others—to look for it and develop it in Canada if our prices are far below other countries. We cannot go on year after year being extravagant in our use of oil far beyond what almost every other country in the world consumes—mainly because it is being sold cheaply in Canada, a lot cheaper than elsewhere and a lot cheaper than our future supplies will cost. We cannot expect Alberta and Saskatchewan to go on year after year selling their oil to Canadians at a price which is far below what they could get by exporting it.
>
> So my colleagues in the government and I have come reluctantly to believe that the price of oil in Canada must go up—up towards the world price. It need not go all the way up. We should watch what happens to the world price and decide from year to year what we should do (Canada, 1977, p. 21).

Alberta has and will continue to take steps towards more independent decision-making power for oil exports and prices. The province has already legislated in a way that creates a direct constitutional challenge to the new Liberal federal government. This throws into question interprovincial trade, pricing, and international export rights. Alberta may be able to make a strong case on constitutional grounds for much greater power in trade and pricing policies. It is no longer totally out of the question to envisage a time in the future when there is a direct confrontation between Alberta and the federal government, a confrontation so serious in its implications as to bring pressure on the federal government to use its emergency powers.

The east-west polarization in Canada has already witnessed a resurgence of western separatism. Ontario will find itself in a separatist squeeze from Quebec and Alberta. In many respects, except for lip service to federalism, Alberta behaves like Saudi Arabia. In Saskatchewan, the government has adopted similar attitudes with respect to uranium and potash. Political differences between these two provincial governments do not determine any real difference in their resource policy. Thus, in Canada, we have the equivalent of the OPEC crisis of the Western industrialized oil-importing nations.

Resource/energy nationalism is even more rampant than this. Ontario is placing new emphasis on nuclear power in its latest energy policy; in part undoubtedly due to the fear of disruptions and substantial price increases in oil. The fact that this plan to expand nuclear power is politically, economically, and environmentally risky does not deter this undertaking. That Ontario finds itself in 1980 with an excess of 42 per cent in electrical generating capacity has apparently taught the provincial government little. The real limits on substitution and the need for fossil-fuel back-up for nuclear power are somehow glossed over or misunderstood.

British Columbia, on the other hand, has decided on a moratorium on uranium mining, a somewhat surprising decision, but very likely due more to the government's precarious political situation than any other consideration. Ontario seems to have underestimated the depth of opposition to nuclear power or is merely less sensitive to the signals. Or possibly, it is Ontario Hydro that makes the real energy decisions, much as Hydro-Québec does in Quebec.

In Quebec, current energy planning embraces the notion of a dominantly electric economy, again in apparent ignorance of the limits of substitution. Electricity is being resurrected by the Parti Québécois government as its answer to oil dependency and as an export item to make up the balance-of-payments deficit for oil imports.

Current Canadian energy policy conflicts are multiple—between the federal government and the provinces of Alberta, Quebec, and Newfoundland; between Ontario and Alberta; and between Quebec and Newfoundland (over the price agreement on Churchill electricity). Alberta threatens a "slow-down" on tar sands development. The federal government has implied that it does not need the tar sands and has suggested that it might not even need Alberta's oil, if the promise of significant Maritime offshore discoveries are realized. Ontario chooses to go to nuclear power and coal in a major way to offset oil dependency. Quebec decides on a heavy commitment to electricity. The project to bring liquefied natural gas to Quebec and extend the gas pipeline to the Maritimes is a direct threat to the oil refinery capacity in these regions while at the same time allowing for oil substitution by gas.

Petro-Canada has been rejuvenated under a Liberal administration and has been involved in negotiating the purchase of the large refinery complex at Come-By-Chance in Newfoundland, providing it is found viable. The intent of this arrangement is to process crude from the offshore Hibernia field (164 miles off the east coast). The size of this field has been variously estimated from 250 million to 1.5 billion barrels.

As a direct federal challenge to Alberta, it was announced that Ottawa has revoked its 1976 guarantee of the world oil price for Syncrude Canada Limited (*The Gazette*, 1980*a*). This takes advantage of a *force majeure* clause in the original agreement due to the escalation of world oil prices from $13 in 1976 to $36 in 1980. This decision will also apply to the other oil sands plant, known as Great Canadian Oil Sands.

Alberta seems intent on challenging federal authority on a fundamental constitutional question. This author has had private discussions with one former western provincial attorney-general who believes that Alberta could win the right to ship its oil to whomever it wished at the price of Alberta's choosing. There is, apparently, a constitutional precedent. The oil of Alberta, retained as the property of Alberta until the point of transfer, at Chicago for example, might be a provincial right, rendering unconstitutional current control by the federal National Energy Board. A more subtle form of pressure is restricting production. As we have discussed earlier, the reserves-to-production ratio is now at the "peril point" of ten. Due to the physical restrictions of oilfield mechanics, Alberta's proven reserves have reached the maximum efficiency rate of recovery (MER) of 10 per cent per year (North, 1979), with only ten years of reserves left.

Alberta's recent cut-back in oil production is a direct challenge to the federal assumption of rights in interprovincial trade. The constitutional question hangs upon the balance between the federal power to regulate trade and commerce (under section 91 of the *BNA Act*, 1867) and provincial ownership of resources (under section 109). There is also a question about the ability of the federal government to tax the property of a province. There was a Supreme Court ruling on such an issue on 10 October 1922, when the Province of British Columbia purchased one case of Johnny Walker Black Label in Glasgow and refused to pay custom's duties. The Collector of Customs, on behalf of the Dominion of Canada, then refused delivery. The Supreme Court eventually ruled in favour of the Dominion, choosing to interpret custom's duties as *not being a form of taxation*. What Alberta may have in mind, however, is the opposite; that is, exporting a commodity constitutionally belonging to them where such ownership is not relinquished in crossing the international boundary. This is an interesting constitutional question that is currently being put to the test before the courts.

The controversial "Peace, Order and good Government" clause of the *BNA Act* used in recent times for a "perceived insurrection" and for the control of wages and prices could be used against Alberta. It is of interest that Claude Ryan's "beige paper" would keep this clause in the constitution but have each declaration "ratified" by a parliament of provinces. We can envisage in this scenario Saskatchewan, Quebec, British Columbia, and Newfoundland supporting Alberta, while Tory Ontario would support Liberal Ottawa. The geopolitics of oil would have become nationalized. The result could be the Balkanization of Canada into three, four, or more "states," that is, the Oil Republic of Alberta, the Uranium Republic of Saskatchewan, the Electric Republic of Quebec, and the Oil Republic of Newfoundland.

In a speech to the Canadian Manufacturer's Association in Ottawa on 28 May 1980, Marc Lalonde stated that "raising oil prices without altering the revenue system will only make matters worse." He was referring to the financing of a national energy policy. The division of oil revenues at that time was 50 per cent to the petroleum industry, 28 per cent to Alberta, about 10.5 per cent to the federal government, and the remainder to Saskatchewan and British Columbia. Lalonde went on to say that "We're not trying to halt the westward shift of industry to Alberta. We're not trying to wrest control of natural resources from the provinces. We're simply emphasizing that the nation is in a bind and that province-building shouldn't be at the expense of nation-building" (*The Gazette*, 1980b).

Since the summer of 1980 and the unsuccessful oil-pricing negotiations between Ottawa and Alberta, the most significant change in the geopolitical situation has been the federal *National Energy Program* of 28 October 1980. Under this programme, Ottawa has taken major new initiatives to raise the levels of Canadian ownership in the petroleum industry, to allocate a larger share of resource revenues to the federal government, and to lessen the national reliance on uncertain supplies of foreign petroleum. The national energy debate is, nevertheless, sure to continue as both Alberta and the multinationals have voiced objections to numerous provisions in the energy programme. In particular, Alberta has already initiated unilateral cuts in provincial crude oil production in an attempt to force larger and faster increases in oil prices than those proposed by the *National Energy Program*.

ENERGY AND PEACE

The new global agenda is dominated by four critical problems: economics, energy, environment, and equity. Economics comprise a cluster of issues concerning growth, development, employment, inflation, trade, and aid. Energy is the fundamental force that provides the goods and services that we need or desire. Environment comprises Nature's gifts of essential environmental goods and services as well as the physical, biological, and built environments in which we live, produce, consume, and generate waste. Environment is both a gift and a constraint on human activities and establishes the limits within which we must live. Equity is the ultimate historical human value. It is the driving force of political action within and between nations.

These four issues are often posited as either positively or negatively correlated when taken in pairs. Thus, energy consumption is assumed to be highly correlated with gross national product (GNP). Economic growth and the increased production of energy are assumed to be the principal source of equity. Even environmental protection is thought to be dependent on energy and economics. Using the popular notion of necessary trade-offs, we are told we must trade away environment for energy, growth, and equity. This group of views constitutes a mythology of progress, a dominant social paradigm, which is suffering a series of shocks through an increasing number of significant anomalies. While there is certainly a threshold of energy availability that societies require for survival and fulfilment, there is also an upper level beyond which environmental, social, and psychological costs are unacceptably large and represent a variety of limits. Unfortunately, these limits and costs are frequently hidden by deferral, subsidization, and other economic and political factors.

The growth/progress paradigm, which assumes that GNP is a direct linear correlate with energy consumption and that both are the exclusive source of equity, is becoming increasingly mythic. The key ratio of Total

Primary Energy (TPE) to GNP for Canada was 1.87 in 1976 and is still about 1.8. For Norway, however, this ratio declined in the same time period from 1.4 to 1.24, for Sweden, from 1.32 to 1.19, and even for the United States, from 1.5 to 1.41. Canada's energy effectiveness ratio is the poorest in the Western industrialized world, as is its TPE per capita of 9.7. Sweden, which only consumes 6.4 primary energy units per capita, leads Canada in GNP per capita and is superior in literacy, infant mortality, and life expectancy (Sivard, 1977). In the last three social indicator rankings, Sweden is number one in the world. Canada, on the other hand, has had the most dismal record in real energy conservation, certainly the worst among OECD countries. Canada has merely slowed down growth rather than eliminated waste.

What is of particular interest is that every serious and responsible study on conservation potential (Stobaugh and Yergin, 1979; Knelman, 1975; Mauss and Ullman, 1979; Landsberg *et al.*, 1979; EFE, 1977; Grossman *et al.*, 1979) indicates that a significant commitment to efficiency and conservation, without a radical change in American or Canadian life-style or social order, can be achieved at about 50 per cent of present growth projections to 2000. One might ask, if such a policy is so promising, why have Canada and the United States failed to implement it. What we have is a sound policy in search of a government. The reasons for this failure are profound and relate to the entire market and price systems, to the distortions of short-term planning, and to the power of existing private and public institutions whose short-term interests are not in accord with conservation.

President Carter stated that the energy crisis is like a state of war. But he was not able to muster his forces on that issue. The fearful problem is that real war might become the extension of energy politics by another name. As surmised earlier, resulting threats to our economic and social order may well occur in the 1980s.

The Canadian policy of self-reliance is commendable but tragically distorted in direction, focusing on supply rather than the reduction of waste. Even if we were to achieve self-reliance, no country is an island, particularly our country, whose proximity to the United States will continue to shape our destiny, perhaps more divisively than ever before.

It is obvious that Canada is plagued by a complex of economic, energy, environment, and equity problems. In fact, if we could find a Canadian solution, it would provide a model for the planet, since Canada, in many ways, is a microcosm of the world.

Regardless of our possible involvement in a new wave of resource continentalism, we should attend to the major energy policy task facing us nationally. This is to convince our government to implement a massive programme of conservation and efficiency and a genuine commitment to renewable energy sources, particularly biomass applications for the production of liquid fuels. There is no reason why we cannot have an acceptable level of economic growth of over 3 per cent per year at an energy growth of less than 2 per cent per year. Renewables could also reasonably account for 20 per cent of our total energy budget by the year 2000, particularly at our reduced energy growth (Von Hippel and Williams, 1975; Stobaugh and Yergin, 1979; "Soft Energy Paths," 1979, 1980). Such a policy will not only be productive for our country but also will assist us in all our international commitments. The transfer of appropriate technology and the contribution of appropriate development to the poor and hungry will be assisted, not restricted, by our adoption of this proposed energy policy. A massive conservation policy can make a contribution to the resolution of the four critical problems of equity, environment, energy, and economics. In this way, we can achieve "self-reliance" (Knelman, 1979) and independence while promoting global equity. Thus, energy policy can become the extension of peace by the best means.

REFERENCES

Barraclough, G. (1975) "Wealth and Power: The Politics of Food and Oil." *New York Times Review* (2 August): 23-30.

Bergsten, C. Fred. (1975) *Toward a New International Economic Order: Selected Papers of C. Fred Bergsten, 1972-1974*. Lexington, Mass.: Lexington Books.

Brezinsky, Z. (1970) *Between Two Ages: America's Role in the Technotronic Era*. New York: Viking.

Calamai, P. (1979) "The Tar Sands Challenge." *Science Forum* (March-April): 12-17.

Canada, Department of Energy, Mines and Resources. (1977) *An Energy Strategy for Canada: Policies for Self-Reliance*. Ottawa: Minister of Supply and Services Canada.

Committee on Nuclear and Alternative Energy Systems (CONAES), Demand and Conservation Panel, National Academy of Sciences. (1978) "U.S. Energy Demand: Some Low Energy Futures." *Science* 200 (14 April): 142-52.

Crane, D. (1976) "Canada's Energy Crisis: A Bizarre Case of Bungling." *Northern Perspectives* 4: 1-12.

Environmentalists for Full Employment (EFE). (1977) "Jobs and Energy." Washington, D.C.

Falk, R. A. (1975) *A Study of Future Worlds*. New York: Free Press.

Gander, J. E. and Belaire, F. W. (1978) *Energy Futures for Canadians*. Report EP 78-1. Ottawa: Dept. of Energy, Mines and Resources.

The Gazette. (1980a) 29 March, p. 1.

The Gazette. (1980b) 29 May, p. 1.

Grossman, R. *et al.* (1979) *Energy, Jobs and the Economy*, chapters 3 and 4. Boston: Alyson Publications.

Hayes, E. T. (1979) "Energy Resources Available to the United States, 1985−2000." *Science* 203 (19 January): 233-39.

Hubbert, M. K. (1974) "U.S. Energy Resources: A Review as of 1972." A National Fuels and Energy Policy Study, 93rd Cong., 2d sess., Senate Committee on Interior and Insular Affairs, Serial No. 93-40 (92-73). Washington, D.C.: Government Printing Office.

Kettle, J. (1978) "Energy: The Bewildering Uncertainty." *Executive Magazine* 20 (December): 35-48.

Kimche, J. (1980) "The Oil Crisis That Never Was." *The Gazette* (15 March), p. 14.

Knelman, F. H. (1975) *Energy Conservation*. Science Council of Canada Background Study No. 33. Ottawa: Information Canada.

Knelman, F. W. (1979) *Anti-Nation: Transition to Sustainability*. Oakville, Ont.: Mosaic Press.

Landsberg, H. *et al.*, eds. (1979) *Energy: The Next 20 Years*. Cambridge, Mass.: Ballinger for Resources for the Future.

Mauss, E. A. and Ullman, J. E., eds. (1979) *Conservation of Energy Resources*. New York: Annals of the New York Academy of Sciences.

The New York Times. (1974) 3 November, p. 1.

North, F. K. (1979) "Canada's Oil and Gas: Surplus or Shortage?" In *Energy Policy: The Global Challenge*, edited by P. N. Nemetz. Montreal: The Institute for Research on Public Policy.

Organisation for Economic Co-operation and Development. (1977) *World Energy Outlook*. Paris: OECD.

Organisation for Economic Co-operation and Development. (1978) *Energy Policies and Programmes of the IEA Countries, 1977 Review*. Paris: OECD.

Price, J. H. (1975) "Energy Evaluation of Energy Conversion Processes." In *Non-Nuclear Futures: The Case for an Ethical Energy Strategy*, edited by A. B. Lovins and J. H. Price. Cambridge, Mass.: Ballinger.

Schertz, L. P. (1975) "World Food Prices and the Poor." In *The World Economic Crisis*, edited by W. P. Bundy. New York: Norton.

Sivard, R. L. (1977) *World Military and Social Expenditures*. Dundas, Ont.: Peace Research Institute.

"Soft Energy Paths, Part 1." (1979) *Alternatives* (Summer and Fall).

"Soft Energy Paths, Part 2." (1980) *Alternatives* (Winter).

Stobaugh, R. and Yergin, D., eds. (1979) *Energy Future* (New York: Random House).

United States, Project Independence. (1977) *U.S. and World Energy Outlook Through 1990*. Washington, D.C.: Government Printing Office.

Von Hippel, F. and Williams, R.H. (1975) "Solar Technologies." *Bulletin of the Atomic Scientists* 31 (November): 25-31.

Chapter Four

1323
7230

Energy: The Future Has Come

by
J.G. Melvin

> There is a tide in the affairs of men,
> Which taken at the flood, leads on to fortune;
> Omitted, all the voyage of their life
> Is bound in shallows and in miseries.
> Shakespeare
> *The Tragedy of Julius Caesar*
> Act IV, Scene III

The transformation of Canada which accompanies the new energy era might be as dramatic as that introduced by the railroads.

Gander and Belaire (1978)

SUMMARY

The approaching scarcity of conventional petroleum presents a severe challenge to mankind. Canada is exceptionally well endowed with energy resources that can provide the time and the means for transition to a post-oil energy economy. Both economics and technology point to a major role for electricity, at least in the Canadian context. Canadians, by seizing the opportunities offered by their unique endowment of hydrocarbon resources and low-cost electricity, can build a prosperous economy extending into the indefinite future. The essential first step is to complete the substitution of electricity for imported oil during the next ten to fifteen years. This is shown to be technologically feasible and economically beneficial.

A PROBLEM OF OIL

The oil age, as a tide in the affairs of men, began to flow at the start of this century. It has profoundly influenced every aspect of life has provided unprecedented freedom from natural afflictions, and has fuelled the growth of man's material and cultural wealth. In a period when the world's population trebled, oil has helped to support a vast number of those people—more than the total of those who lived at any one time prior to

61

1850—at a level of prosperity and health beyond the imagination of earlier generations.

Figure 1 illustrates our history and prospects in terms of population and energy. The "oil" curve represents conventional petroleum; most authoritative projections (Flower, 1978; Hubbert, 1971; WEC, 1978) fall within the range shown. Thus, the oil tide is near its crest and will soon begin to ebb. "The supply of oil will fail to meet increasing demand before the year 2000" (Flower, 1978, p. 42); that is, before a child born today reaches maturity. Conventional oil, which now provides half the world's energy, will decline, slowly in absolute terms, but rapidly relative to the needs and aspirations of an expanding human family.

Figure 1
WORLD POPULATION AND ENERGY

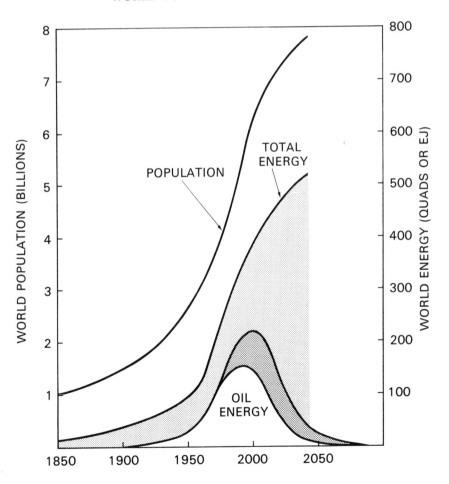

This is the "energy crisis" in its starkest form and it leads to two important observations: it is not a crisis of energy but of oil, and the time available for corrective action is short. In this global context, new oil discoveries will be of little help because of the magnitude of the problem. Another Persian Gulf, for example, would buy only a few years' respite.

The Canadian outlook is similar though less drastic. Again, oil provides about half the total energy. Projected oil demand (Figure 2) tends to level off due to declining population growth and the assumed effectiveness of conservation measures. The supply of conventional oil continues to decline but is eventually offset by expanding production from oil sands and heavy oil deposits, the non-conventional supplies that represent a trump card in Canada's energy hand. Even so, we face a growing demand for oil imports that we can ill afford; a doubling of our annual balance-of-payments deficit is a dismal prospect, and constant vulnerability to disruption in supply is even worse. This leads to a second observation: oil is the problem, not the solution.

CHANGING PERSPECTIVES

What then is the solution? There is no single answer, no quick fix, no panacea, but the necessary approach can be described in two words: conservation and substitution. To quote the LEAP (Gander and Belaire, 1978) report, we must "reduce and replace," and the thing to be reduced and replaced is oil, not energy in general, at least not in the Canadian context.

The problem is not as severe for Canada as for most industrial countries because we are more self-sufficient in oil (Table 1). Moreover, the resources on which to build a solution are more diverse and more plentiful in Canada than in most countries. In addition to non-conventional oil, we hold other trump cards such as natural gas, hydroelectricity, coal, and uranium. Our potential for long-term energy security might well be envied even by Saudi Arabia.

Table 1
OIL SELF-SUFFICIENCY

Country	Imported Fraction (1978)
Canada	0.14
United Kingdom	0.43
United States	0.44
West Germany	0.97
France	0.99
Japan	1.00

Source: BP (1979).

Figure 2
OIL SUPPLY AND DEMAND, CANADA

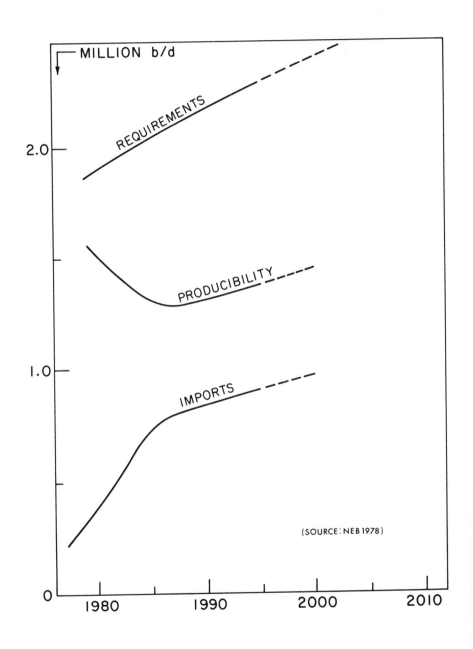

Given this potential, it is difficult to support a case for extreme conservation, retrenchment, and economic slow-down, and yet this sort of energy death-wish is widely touted.

To adopt such a policy would be to refuse the challenge presented by our resources and thereby to reject the opportunities for economic growth and social well-being. But precisely this course is advocated by at least two groups, those who see virtue in austerity and those who make a "dismal science" of economics. The former are beyond the purview of this discussion, but the difficulties facing the latter deserve recognition. Econometric modelling, of necessity, uses historical data and relationships to project the future. One difficulty is that these relationships change under the influence of social and technological factors, which can seldom be identified, let alone quantified, in advance. A more serious difficulty, in the present context, is the discontinuity injected into the global economy by the actions of the OPEC cartel. The world changed abruptly in 1973; prior econometric relationships are largely irrelevant, and models based on them are rear-view mirrors rather than windows on the future. Quite simply, today's prices and relationships are beyond the range of past experience; it is a new world.

The new energy world is an astonishing place to those of us whose perceptions are firmly rooted in the old, for things are no longer as they used to be. We are equipped with knowledge that is no longer true. We are burdened with policies and attitudes that are inconsistent with the new, post-1973 realities. In consequence, we confront a problem not of energy, nor even of oil, but of mind-set.

As evidence, consider the following statement: it is economically and technologically feasible to displace most of Canada's imported oil with electricity. Intuition, based on three quarters of a century of history, tells us that the statement is untrue, because electricity, though clean and convenient, is expensive. Most policy analysts share this view, as do their political masters. The facts, however, are otherwise.

A ROLE FOR ELECTRICITY

Present technology would allow electricity to substitute for a portion of our oil sufficient to achieve almost total displacement of imported oil. Nearly half of Canada's oil is consumed in transportation and cannot soon be displaced by electricity. We lack the technology for electric propulsion of personal vehicles, and we are not yet prepared to build, let alone to use, large-scale electrified public transportation. That leaves the other half of our oil. Some of this is necessary for portable or remote power purposes, such as earth moving and well drilling, and some is used for non-energy products such as asphalt, lubricants, solvents, and chemicals. The balance, about one third of our total oil, is burned to produce heat in domestic furnaces, industrial and commercial boilers, and a wide variety of industrial processes. The technology to substitute electricity for this portion of our oil is available on normal commercial terms in the appropriate forms—furnaces, heaters, boilers, and so forth. In a very real sense, we are importing oil at $32 per barrel merely to incinerate it.

The magnitude of this level of substitution, relative to current and projected imports, is illustrated in Figure 3. With some penetration of electricity into the transportation sector later in the century, by a combination of electric cars, mass transit, and some diversion of freight and passenger traffic to electrified railways, Canada could become self-sufficient in oil.

In purely technological terms, therefore, we need not be importers of oil; with existing technology, electricity is a ready substitute.

Figure 3
SCOPE FOR ELECTRICITY/OIL SUBSTITUTION

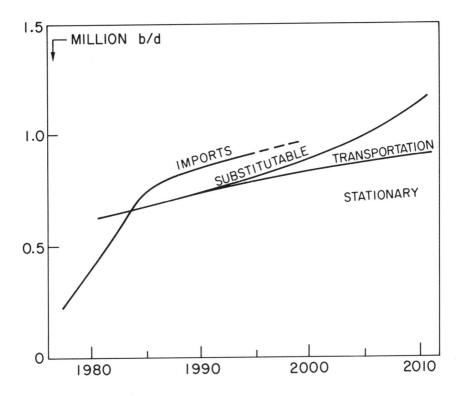

ELECTRICITY TO FILL THE ROLE

But how much electrical capacity would be needed and how long would it take to put it into place? Intuition suggests that the required capacity would be enormous, but again intuition plays us false. Figure 4 presents the same information as the previous chart, with the addition of a curve showing the rate at which electricity might be made available to displace oil. The curve of course results from some assumptions:

a. Electrical capacity is expanded at an annual rate of 7 per cent. This rate had been sustained for many decades in Canada prior to the post-OPEC economic slump and conservation boom. Rates between 6 and 7 per cent continue even now in several regions, for example, Quebec, Alberta, and British Columbia.

b. Electricity demand, for normal, non-substituting purposes, grows at an annual rate of 4 per cent. This approximates the currently projected growth of gross national product and exceeds the electricity demand projections in some regions.

c. The excess electricity resulting from the 3 per cent difference in growth rate between supply and normal demand is available for oil substitution.

d. The efficiency with which oil is used, relative to that of electricity, is 67 per cent on the average, so that one unit of electrical energy displaces one and a half units of oil energy.

Figure 4
POTENTIALLY AVAILABLE ELECTRICITY

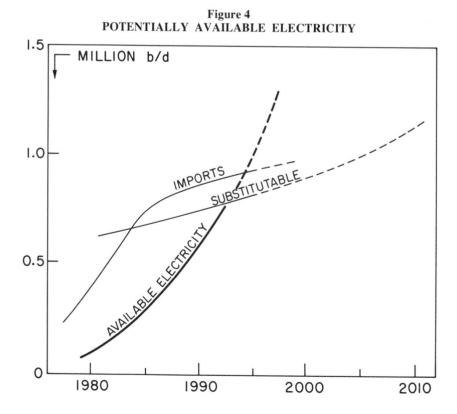

The picture that follows from these modest assumptions (Figure 4) is, at first sight, astonishing; oil imports could be virtually eliminated by one more decade of electrical expansion at the traditional rate. Upon reflection, this result is reasonable, because oil imports, though growing rapidly, currently represent less than 8 per cent of our energy while electricity contributes almost twice this amount, and because of its higher end-use efficiency, is equivalent to a proportionately larger quantity of oil. The fact that Canada's dependence on imported oil is relatively low places us within striking distance of self-sufficiency. In this respect, as in several others, our problem is more tractable than that of most countries.

It is important for Canadians to assess the energy problem and its solution in the context of Canada's circumstances. This is not as easy as it might seem, because we are constantly exposed, through our news and entertainment media, to foreign views and perspectives. It takes some effort to realize that the U.S. problem is not our problem and that a U.S. solution, if there is one, would probably be inappropriate, and perhaps impossible, for Canada. (As an extreme example, the United States, lacking our per capita wealth in oil sands and hydroelectricity, can contemplate a migration to "sun-belt" states (CONAES, 1978) as one response to energy shortage.) The theme "think Canadian" will surface again later, not in the cause of economic nationalism but of practical necessity.

AFFORDABLE ELECTRICITY TODAY

Reviewing the argument to this point, we have found that existing technology allows the substitution of electricity for most of our imported oil and that the required electrical capacity could be installed in about a decade while also accommodating normal growth in electricity demand. It thus appears that electricity offers a way of reducing, almost eliminating, oil imports. But we are not desperate, not yet, so it is necessary to ask about the bottom line: could Canadians individually and collectively afford the electrical solution? The answer is that if we can afford oil today, we can equally afford electricity, and tomorrow we can afford electricity but probably not oil.

Right now, electricity is marginally competitive with oil in Canada at the going prices. That is to say that in regions where electricity rates are low and for applications in which oil is not used with high efficiency, the two are competitive. This is illustrated in Figure 5, which contains no assumptions and allows the reader to compare oil and electricity costs in his own situation. To illustrate, consider domestic space heating. Point A represents this task using oil at a price of 16.5 cents per litre (75 cents per gallon) and an efficiency of 60 per cent. The chart shows that electricity at 2.5 cents per kilowatt hour would do the job at the same cost; a lower electricity price, a higher oil price, or a lower oil-heating efficiency would make electric heating the cheaper alternative. Retail prices for oil and electricity in some regions of Canada are (1980 February) in the vicinity of Point A. As one specific example, the author has the choice of paying 3 cents per kilowatt hour for electricity or 16.5 cents per litre for oil; in fact, he heats his home with electricity because for some time the cost differential has been, in his view, a low premium for the other advantages gained.

Figure 5
ENERGY COSTS, ELECTRICITY VERSUS OIL

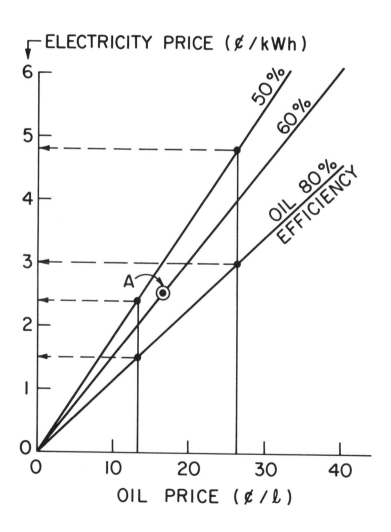

A commercial or industrial customer who uses large quantities of energy obtains oil or electricity at bulk or wholesale rates that are lower than those associated with Point A. Heavy fuel oil (Bunker C) is available at 13.2 cents per litre (60 cents per gallon), represented by a vertical line on the chart. If this oil is burned efficiently to provide heat directly to a building or process, an efficiency of 80 per cent might be achieved. If the boiler or heater is less efficient or if the heat must be distributed, say as steam, to dispersed points of application, the losses will be greater and the final efficiency correspondingly lower. Taking the range from 80 down to 50 per cent as representative of industrial oil use efficiencies indicates a competitive electricity cost from 1.5 to 2.4 cents per kilowatt hour. Bulk electricity rates are available at least in the upper part of this range, so that electricity competes at least marginally with oil at current domestic prices.

Domestic oil prices are not the relevant criterion, however, because the real target for substitution is imported oil, for which Canadians collectively pay the world price even though the individual consumer is subsidized. The world price is now $32 per barrel of crude. Allowing for oil consumed and lost in refining, and for processing and distribution costs, this implies a cost of at least 26.4 cents per litre ($1.20 per gallon) of product—with no margin for profits or tax. The chart shows that electricity in the range 3.0 cents to 4.8 cents per kilowatt hour is competitive, which is to say that electricity is a better buy than imported oil for most Canadians.

A BASIS FOR COMPARISON

But what of the future? We know only too well that the cost of electricity has been rising rapidly along with that of oil. What is less obvious is that the prices have been rising for different reasons and that their future prospects are quite different. Electricity costs have been responding to domestic conditions including a rapid escalation of construction costs, while the price of imported oil is set by an international cartel in a climate of scarcity. While some might hope that OPEC will collapse, the fundamental driving force is scarcity, which over the long term can only become more severe.

To assess the prospects for oil and electricity costs, it is helpful to remove the effects of general inflation, which increase the costs of energy and of all goods and services more or less equally. This can be done by dealing in constant-dollar costs, referred to as "real" costs.

As the basis for comparison, the present price of oil to a Canadian consumer (based on a crude oil price of $15 per barrel) is taken as unity (i.e., 1.0). On the same scale, the present world price is about 2.

Electricity, as shown earlier, is at best competitive with oil and, at the other end of the range, perhaps 40 per cent more costly. Thus, electricity ranges between 1 and 1.4 on the oil-cost scale.

Any prediction of future costs is speculative, but we are not completely without information on which to base decisions. In what follows, an attempt will be made to identify reasonable upper and lower bounds to the future costs of oil and electricity in order to see where we might be heading.

ELECTRICITY: A STABLE COST

First, electricity. A utility with a solid base of inflation-proof hydraulic capacity has built-in inertia to resist increases in real cost. If the utility has access to further economic hydraulic potential, or if it is large enough to accept nuclear units, the cost of electricity can remain constant in the long term. Ontario Hydro, for example, is able to project constant, or even declining, rates to the end of the century. This is made possible by the economic characteristics of nuclear plants, which, once built, produce energy throughout their lives at steadily decreasing real cost. A coal-based utility faces constant real costs, assuming, optimistically, that the cost of coal will not rise more rapidly than general inflation. An oil-based utility faces escalating oil costs from which it can escape only by plant replacement or electricity purchases, both of which are costly in the near term.

The lower bound of future real costs (Figure 6) is a constant level of 1.0, meaning that those utilities having low costs now will have the same real costs in the future. An arbitrary upper bound rises to a real cost 30 per cent higher than the present level of 1.4, as a liberal allowance for the cost of financing rapid expansion, and then levels off. In the longer term, when oil substitution has been completed and the expansion phase ends, both bounds should decline.

It is important to recognize that this prospect of low-cost electricity stems from our existing base of hydraulic power, which still provides more than half our electricity and is almost unique to Canada. The United States generates 30 per cent of its electricity from oil and gas; France produces 25 per cent from oil, and Japan even more. These countries cannot substitute electricity for oil until they complete the expensive process of replacing their oil-based electricity with other means of generation. Canada is right now where others might hope to be in two or three decades.

OIL: RISING COSTS

Prospective oil prices are something else (Figure 7). The upper bound represents the world price, currently 2.0 and assumed to rise at a real rate of 4 per cent per year. In reality, the price is likely to move erratically rather than follow a smooth curve, and the average rate might well be much higher. All indications are that by the end of the century, oil will be in short supply but still a vital necessity, resulting in competitive bidding in a sellers' market. How high might the price go? One indicator is the fact that Europeans

Figure 6
RELATIVE ELECTRICITY COST PROJECTION

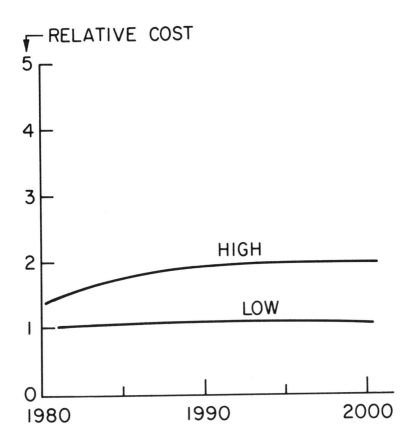

willingly pay $2.50 per gallon for gasoline, which suggests that crude oil prices might move into the range of $50.00 to $100.00 per barrel before consumers are forced to reduce their demands drastically. This range corresponds approximately to 4 to 6 on the scale of Figure 7, well above the upper bound shown.

The lower bound originates at 1.0, the present price of oil in Canada. The curve represents cost (not necessarily price) as a weighted average of conventional domestic oil at 1.0, oil sands product at 2.0, and imported oil at the cost represented by the upper bound.

Many of the assumptions are arbitrary, but it seems highly unlikely that the cost will fall below the lower bound, possible that it will exceed the upper, but likely that it will remain between the two for some time.

Figure 7
RELATIVE OIL COST PROJECTION

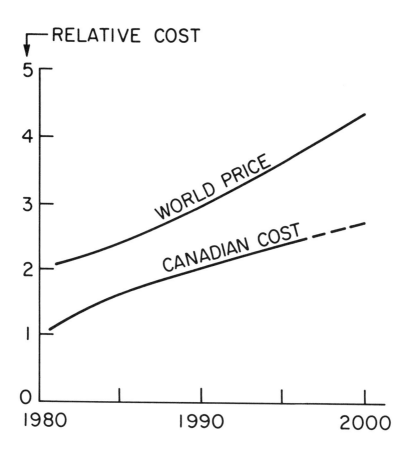

A WIDENING MARGIN

The projected costs of oil and electricity are shown together in Figure 8, which shows a growing advantage for electricity. The advantage is even more pronounced when it is realized that electricity is likely to approach the low end of its range, and that in the context of substitution for imported oil, the upper bound of the oil range is the relevant criterion.

It is hard to escape the conclusion that on this side of the 1973 watershed and in this fortunate country, oil has become a poor second choice. We are burning oil that we do not need at a price that we need not pay; we are paying a premium to incinerate oil.

Figure 8
COMPARATIVE COSTS, OIL AND ELECTRICITY

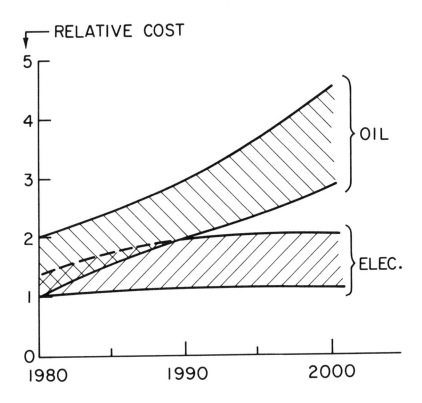

THE RESPONSE OF THE MARKET

If Canada chose to substitute electricity for imported oil and implemented policies to achieve that goal, the rate of substitution would be
limited either by supply or by demand. The limit imposed by the supply of
electricity has been estimated and is shown in Figure 4. The rate of demand
growth would be determined by the decisions of thousands of individual
consumers and is more difficult to estimate.

Facilities that use substitutable oil fall into two categories, new and
existing, with vastly different characteristics. A new facility, whether it be a
home, a commercial building, or an industrial process, can be designed to use
electricity or oil; the choice generally has little impact on the size of the
initial investment. Most people would now prefer to avoid a commitment to
oil, so that substitution of electricity or some other fuel is virtually assured

for new facilities henceforward. It appears, therefore, that the rate-limiting process is the retirement or conversion to electricity of existing facilities. The rate of this process will be influenced by many factors such as energy pricing policies, perceived reliability of oil supplies, performance of the economy and its impacts on expansion, obsolescence, and the turnover of capital stock. Any estimate of the rate must therefore be speculative, but history offers some guidance.

The displacement of oil by electricity can be regarded as the penetration of a market by a competing technology. A market-penetration model is available that closely matches the experience of seventeen different processes and products from steel making to detergents (Fisher and Pry, 1971). The penetration rate is characterized by a "take-over time," which ranged from about eight to sixty years for the cases studied. An unrelated analysis (Mansfield, 1968) of the diffusion of twelve new technologies through the relevant industries reported elapsed times between ten and twenty years from introduction to complete diffusion. On this evidence, a market penetration model with a fifteen-year take-over time was chosen as a reasonable basis for estimation, intermediate between extreme urgency and business-as-usual. The resulting penetration rate of electricity into the technologically feasible oil "market" (Figure 9) is slower than the previously estimated rate at which electricity supply might be increased. The difference between the two might be regarded as a cushion to allow for shortfalls in planned electrical capacity or for accelerated substitution triggered by instability of oil supplies or by rising natural gas prices. It has the great virtue of a smoothly declining rate of growth rather than the boom-and-bust of an energy bonanza.

On the basis of these estimates, it appears that market response to foreseeable conditions of supply and price could result in the substitution of electricity for most of Canada's projected oil imports in ten to fifteen years. This schedule for oil displacement is challenging but feasible; it is probably a minimum target for prudent planning and might need to be accelerated as the energy universe unfolds.

PRICING AND FINANCING

There are questions of pricing and financing that cannot be treated adequately in the space available.

Electricity is said by some to be underpriced because the utilities rely primarily on debt financing, because the rates do not include taxes and profits, or because of perceived virtue in something called marginal cost pricing. Regardless of technical arguments, the best pricing policy is that which is most consonant with public goals. It is not argued here that electricity ought to be subsidized in order to displace imported oil; that would be unnecessary. Given the oil problem, however, it is difficult to see what goals would be served by the opposite course, discouraging substitution

Figure 9
MARKET PENETRATION BY ELECTRICITY

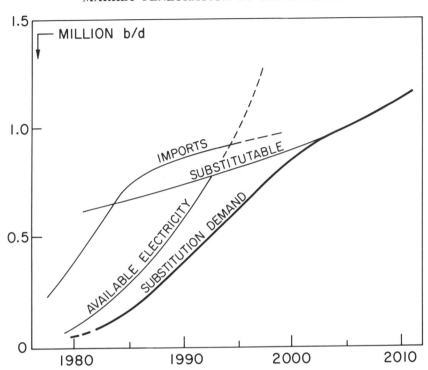

through an arbitrary increase in the price of electricity. In fact, if we did not already have cost-based pricing, now would be the time to institute it.

The ability of Canada to finance electrical expansion is a related question and two responses are offered. First, the judgement of authorities, including chartered banks (Sultan, 1978) and the federal government (Canada, 1977) appears to be that, though difficult, it is feasible. Second, Canada can scarcely afford not to displace imported oil. The total quantity that might be displaced by electricity up to the year 2000 can be estimated from Figure 9 at about three billion barrels, representing an expenditure, at today's price, of $96 billion. The generating capacity to displace this oil would be about 60,000 megawatts, requiring a total investment of some $90 billion, roughly equal to the import savings. Moreover, the investment would produce jobs and economic stimulus in Canada while providing long-term energy security rather than day-to-day satisfaction of an appetite for oil. It is better to invest in an electrical future than to remain hostage to oil while paying the equivalent in ransom.

THE PROBLEM OF LOAD FACTOR

It is argued by some that much of the oil to be displaced is used for space heating, which, because of its seasonal nature, is not a suitable job for electricity; the generating capacity to serve the heating load would be an idle burden during summer. It is true that a large portion of the substitution load is seasonal and that this would impose a penalty, but three ways of responding to the problem are available: accept the penalty, reduce the winter load peaks, or fill in the valleys.

The penalty is estimated to represent a cost increase of 10 to 15 per cent; it is not desirable, but it is not prohibitive and could be accepted. A way of reducing the peaks has been proposed by Clayton *et al.* (1980). This hybrid electric system would use electric heat at a constant, relatively low rate throughout the seven- to eight-month season and turn on a non-electric furnace for the additional heat needed during the colder periods. In this way, electricity would provide an estimated 50 per cent of the total heat requirement while presenting a remarkably flat load to the utility. An oil/electric hybrid system would reduce oil consumption by about half, while a gas/electric system would stretch out our gas supplies, thus providing an indirect substitute for oil.

In the longer term, any valleys in electrical demand can be seen as opportunities to put low-cost electricity to productive use in processes that will give a competitive advantage to Canadian industry. Canada is already well established in electricity-intensive industries such as non-ferrous metals. In the new, post-1973 energy world, a whole range of electrical technologies is ripe for exploitation. The transformation of existing processes and the creation of new electricity-based industries will present opportunities to Canada on a scale that has only been foreshadowed by aluminum. Such industries, with their high annual load factors, would improve the seasonal balance and perhaps offer scope for further improvement by load management.

Thus, while it is possible to regard the peaks as a problem, it is more useful to see the valleys as an opportunity.

MULTIPLE SOURCES

By demonstrating that electricity substitution is possible, we have not, of course, shown it to be necessary. After all, there are other possibilities and other resources.

In a brief article it is not possible to cover the energy "waterfront" in detail, but the situation is basically this: while there are indeed other hopes and possibilities, there is no other approach that might be adopted with confidence in both short-term feasibility and long-term potential. At the same time, any other energy sources that are able to pass the tests of availability, feasibility, and economy can make welcome contributions to energy self-sufficiency and thus ease the transition.

If, for example, the scope for energy conservation without economic damage is found to be greater than assumed, or if the contribution from decentralized renewable energy sources should grow with unexpected rapidity, the pressure on our energy systems would be reduced accordingly.

In this context it is significant that electricity is not an energy source, but a carrier, or vector. The principal sources of electricity are water, coal, and uranium. While these are adequate for the foreseeable future, electricity might well be the vector of choice for other energy sources, such as solar, tidal, wind, and biomass. Even minor or intermittent contributions from these sources could offer local benefits and enhance the resiliency of the electrical grid. Conversely, the existence of a grid capable of accepting energy when it is available from local sources and providing energy at other times would provide the climate in which local initiative and alternative energy technologies might evolve.

THE ROLE OF NATURAL GAS

Natural gas is the wild card in the energy game. The official projection (NEB, 1979) based on known reserves indicates that demand will overtake supply in the early 1990s. The picture might change drastically, however, if frontier gas is discovered and brought to market on the scale envisioned by some authorities. In this case, domestic supplies would be secure for some decades beyond 1990 and large quantities would be available for export.

The impact of expanded natural gas supplies on the electrical system would be to allow a lower rate of expansion; this should be welcomed. At present, gas provides about 27 per cent of Canada's secondary energy, effectively displacing the equivalent quantity of oil. Without this contribution, the substitution of electricity for imported oil would be a far longer and more difficult task. Ultimately, however, natural gas will decline, and in the absence of some unforeseen development, it too will have to be replaced by electricity.

Fortunately, it appears that the two jobs can be tackled in sequence, first imported oil subsitution, then natural gas. If new supplies of gas are available by the 1990s, the gas substitution process can be slower than otherwise, driven by relative price rather than by urgent necessity. Failing such new supplies, rapid expansion of electricity would have to continue into the first decade of the twenty-first century.

THE LONGER TERM

Expansion of Canada's electrical capacity to the extent that would be necessary to make it our major energy supply might seem an impractical goal. Electricity today represents about 16 per cent of the secondary energy supply. To provide a major fraction, say, 60 per cent of a future energy supply that might be two or three times as large as today's, electricity production would have to be multiplied by a factor approaching ten. Such an expansion might not be feasible within a few decades, and the "electricity fix" was rejected by Clayton *et al*. (1980) for this reason. It might, however, be possible to achieve the goal eventually with a smaller quantity of electricity.

The essential point is that energy is consumed in order to do work or provide heat; it is the useful output, not the energy input that is of concern. In energy terms, electricity is a finished product while fuels, such as oil and gas, are raw materials; a quantity of electrical energy can generally produce a greater output, often much greater, than the same quantity of fuel energy. A conservative factor of 1.5 was used in estimating Figure 4, but increasingly larger factors are likely to prevail in the more distant future. A heat pump with a coefficient of performance of 2.0, for example, is three times as effective as a fuel-burning system having an efficiency of 67 per cent. Equally large factors are possible in industrial processes, for cooking, for metallurgical operations, for paper making, and for many other operations that can be performed by microwave, electric induction, or a range of mechanical or electrochemical technologies. In the transportation sector, electrical energy should eventually replace many times its equivalent in liquid fuel. The energy required in an electrical future is therefore likely to be only a fraction of that projected on the basis of present-day fuel consumption.

Large potential for energy conservation is inherent in the use of electricity.

OPPORTUNITY KNOCKS

History and natural resources have combined to create in Canada the potential for a prosperous future based on indigenous supplies of relatively low-cost energy. We can look forward to long-term self-sufficiency in essential liquid fuels from the oil sands and to low-cost electricity for most of our other needs. Our remaining supplies of conventional oil and natural gas provide the most essential element, time, for conversion. The time is short, but adequate; the challenge is severe, but manageable.

There are two main energy systems to be built and they are complementary, both technically and geographically. Production of hydrocarbon fuels can provide more than sufficient economic stimulus in those regions where the resources exist. The challenge for other regions is to produce the equipment and technology for electricity supply and, equally important, to use low-cost electricity in the creation of energy-efficient facilities and internationally competitive industries. Oil is where you find it, but electricity is where you choose to put it.

The 1950s and 1960s were a period of rapid expansion, which provided an opportunity for Canada to build a strong, self-regenerating economy. The opportunity was missed and the 1970s found Canada in a state of economic decline measured by unemployment, inflation, and a growing balance-of-payments deficit.

The 1980s and 1990s offer the prospect of a second chance based on a virtual revolution in the supply of energy and its application throughout the world economy. Canada enters this period of revolution better equipped than perhaps any other country.

The Canadian energy problem is not one of resources or technology; it is a problem of decision and resolve. If we arrive at the year 2000 with limited and costly energy, a depressed economy, and a dispirited population, it will be because in the early 1980s, we looked at opportunities but saw only problems.

REFERENCES

British Petroleum Co. Ltd. (BP). (1979) ''BP Statistical Review of the World Oil Industry 1978.'' London.

Canada, Department of Energy, Mines and Resources. (1977) *Financing Energy Self-Reliance*. Report EP 77-8. Ottawa: Information Canada.

Clayton, R.; Lafkas, C.; Kreps, G.; and Miller, R. (1980) *Canadian Energy: The Next 20 Years and Beyond*. Working Paper No. 5. Montreal: The Institute for Research on Public Policy.

Committee on Nuclear and Alternative Energy Systems (CONAES), Demand and Conservation Panel. (1978) ''U.S. Energy Demand: Some Low Energy Futures.'' *Science* 200 (14 April): 142-52.

Fisher, John C. and Pry, Robert H. (1971) ''A Simple Substitution Model for Technological Change.'' In *Industrial Applications of Technological Forecasting*, edited by Marvin J. Cetron and Christine A. Ralph, pp. 290-307. New York: Wiley Interscience.

Flower, Andrew R. (1978) ''World Oil Production.'' *Scientific American* 238 (March): 42-49.

Gander, James E. and Belaire, Fred W. (1978) *Energy Futures for Canadians*. Report EP 78-1. Ottawa: Dept. of Energy, Mines and Resources.

Hubbert, M. King. (1971) ''Energy Resources for Power Production.'' In *Environmental Aspects of Nuclear Power Stations*, pp. 13-43. Paper No. IAEA-SM-146/1. Vienna: International Atomic Energy Agency.

Mansfield, Edwin. (1968) *Industrial Research and Technological Innovation*. New York: Norton.

National Energy Board. (1978) *Canadian Oil Supply and Requirements*. Ottawa: Minister of Supply and Services Canada.

National Energy Board. (1979) *Canadian Natural Gas Supply and Requirements*. Ottawa: Minister of Supply and Services Canada.

Sultan, Ralph G.M. (1978) ''Can Canada Finance the New Electrical Era?'' Address delivered to the Canadian Electrical Association, Jasper, Alberta, 2-4 July. Montreal: Royal Bank of Canada.

World Energy Conference (WEC). (1978) *World Energy Resources 1985−2020*. Guildford: IPC Science and Technology Press for the World Energy Conference.

Chapter Five

Uranium and Coal as Low-Cost Energy Sources—The Safety Issue

by
Ernest Siddall

The advanced societies in the world are concerned today that the steady rise in their standard of living, which they have enjoyed throughout the lifetime of their oldest members, will be arrested or even reversed by the declining availability and rapidly rising real price of oil and natural gas. The world has immense energy resources in the form of uranium and coal, which are potentially available at relatively low cost in the form of electricity from large generating stations. The widespread distribution and use of such electricity is well established, convenient, and economical. Expansion of its use can at least partly replace consumption of oil and natural gas. It would therefore appear reasonable for such replacement to go ahead quickly and for the use of electrical energy to be promoted as it has been almost since the technology was developed.

Unfortunately, the situation has become complicated by the issue of nuclear safety, which has caught the attention of the news media and the public and has raised concern to the point of bringing about a form of paralysis in energy matters in several countries. It now invokes questions of what constitutes a good life, how cities, provinces, and nations are to be governed, and even whether reason can prevail over emotion.

At the same time, there is a consensus of informed opinion, based on much evidence, that the generation of electricity from coal has caused and will cause considerably greater mortality and ill-health than the generation of electricity by nuclear methods.

When all these facts are considered together, it is clear that there are some important problems to be resolved. Is it the case that neither nuclear nor coal is safe enough? Is one safe enough but the other not? Or are both safe enough?

In the past, it has appeared that a balance must be sought between the material benefits arising from a low-cost energy source and the risks associated with it. It is now becoming clear, however, that our societies operate in such a way that part of the benefit may be an increase in safety in the society; if this nullifies or exceeds the direct risk, safety is eliminated as an issue or may even become a factor in its favour. This article attempts to establish the framework for consideration of the problem and to suggest what some of the numbers might be.

The impairment of safety resulting from the generation of electricity from uranium and coal can be divided into two elements—injury resulting from normal operation of the plant and injury resulting from accidents. The potential victims can be divided into the public and those employed in the industry, including those engaged in mining, fuel processing, and transport. All four cross-combinations can only be estimates; the routine effects depend on measurements of small effects in the presence of large components resulting from other causes; the accident severity and frequency are estimated from past experience of frequent events, which are components of more serious but less probable combinations. Nevertheless, such estimates can and must be made if rational decisions are to follow. There are alternatives, such as simply doing without low-cost energy or making arbitrary decisions based on legal semantics. The former is a choice towards freezing in the dark; both are really the negation of using our brains to live better.

Nuclear reactor technology is now about thirty-seven years old. It comprises many major variations, of which the use of ordinary water, heavy water, carbon dioxide, helium, and liquid metal as coolants are the most prominent. The variations in detail are so great that when the complex chains of events that characterize most accidents are considered, every plant may in one sense be viewed as a separate entity. Nevertheless, the principles that have been followed have had a certain basic uniformity and there has been much exchange of information and intercomparison of designs. All the major reactor types incorporate a high quality of design and inspection and are based on extensive testing of components and subsystems. Redundancy is extensively employed, and this permits many errors to be tolerated and corrected; in other words, it permits learning from experience without disastrous results. On the whole, there is no obvious concentration of merit or weakness in respect to safety between different reactor systems and different countries. The very few reactor accidents that have occurred have tended to be attributable to direct human error, but their consequences have been mitigated by very general defensive measures. The overall safety record in respect to accidents to electricity-producing plants is almost perfect. Routine releases of radioactivity from most nuclear plants are very low, and their consequences can be estimated with reasonable accuracy. As will be seen, a single risk figure is proposed for all nuclear plants; overall decisions do not appear to be particularly sensitive to the actual value; it is more a matter of deciding whether the figure is even roughly right. The occupational risk is estimated to be considerably higher than the risk to the public, but both are estimated to be very small.

In the case of coal, the worst effects are considered to result from the dispersal of stack effluents containing substances that are directly or indirectly harmful to the public, of which sulphur dioxide is the most notable. Actual figures depend on the quality of the coal, the stack gas treatment, and the density of population up to great distances.

The evidence of harmful effects from the principal methods of generating electricity has been reviewed by the Council of Scientific Affairs of the American Medical Association, which may perhaps be expected to combine impartiality in this issue with objectivity and with access to extensive specialized knowledge of factors harmful to people. Their report was endorsed by the House of Delegates of the AMA on 21 June 1978 (AMA, 1978). Table 1 is reproduced from that report. For the present purpose, the single figures listed in Table 2 are used. They are taken from Hamilton (1979), whose work forms part of the evidence reviewed in the AMA report. The last column of Table 2 is explained below.

The accident at Three Mile Island occurred after the AMA report was written. The injury to the public is assessed at about 3300 man-rem, which may be taken as roughly equivalent to one fatality. This has occurred in approximately 1500 reactor-years of nuclear power generation, so that it amounts to about .0007 fatalities per reactor-year, which may be compared

Table 1
COMPARISON OF HEALTH EFFECTS FOR FUEL CYCLES*

Effect	Coal	Oil	Nat. Gas	Nuclear
Occupational deaths	0.54- 8.0	0.14- 1.3	0.06- 0.28	0.035- 0.945
Non-occupational deaths	1.62-306.0	1.0 -100.0		0.01 - 0.16
Total deaths	2.16-314.0	1.1 -101.0	0.06- 0.28	0.045- 1.1
Occupational impairments	26.0 -156.0	12.0 - 94.0	4.0 -24.0	4.0 -13.0

* All figures are per 1000 MW station per year.

Table 2
MORTALITY ESTIMATES FOR NUCLEAR AND COAL*

			Failure to Increase GNP
Effect	Nuclear	Coal	Nothing
Public mortality	0.1	7	26
Occupational mortality	0.6	1	

* All figures are per 1000 MW station per year.
Note: Rounded from Hamilton (1979).

with the figure of .1 fatalities per reactor-year among the public shown in Table 2. The California Medical Association endorsed the AMA report in February 1980, that is, after the Three Mile Island accident, reversing a previous anti-nuclear stand (*Energy Daily*, 1980).

The figures used for coal in Table 2 refer to modern plants with scrubbers and are therefore at the low end of those shown in Table 1, which covered the whole range of plant types and ages.

As an illustration of what these figures mean, Table 3 shows how the premature (age less than 65) and total mortality in Ontario would appear if all its 1979 electrical consumption had been hypothetically supplied from coal or from nuclear sources.

Table 3 also forms an introduction to the concept that important benefits in respect to safety may result indirectly from the development of low-cost energy. This community incorporates, for example, hospitals, medical laboratories, and fire brigades. It has water supplies and sewers, with their

Table 3
MORTALITY IN ONTARIO[a]

Cause	Premature (Age 65)	Other	All Canada 1950 *Pro Rata*
Heart disease, strokes, etc.	6,951	23,553	
Cancer	5,036	6,932	
Other disease	5,408	6,242	
Road accidents	1,583	233	
Other accidents & suicide	2,769	680	
Hypothetical, all nuclear electricity[b]			
Public	~ 2	(−416[c])	
Occupational	~ 10		
Hypothetical, all coal electricity[b]			
Public	~112	(−416[c])	
Occupational	~ 16		
Sub Totals	~22,342	~37,534	
Total	~59,876[f]	(−11,920[d])	71,796[e]

Notes: a. Except where otherwise noted, all numbers are prorated for Ontario from the age distribution for Canada, 1973 (Statistics Canada, 1974).
 b. Sixteen stations of 1000 MW capacity, using figures of Table 2.
 c. Lives saved per year through association with GPP increase—see text.
 d. Reduced mortality because of lower rate in Ontario in 1973 compared with all Canada in 1950.
 e. All Canada 1950 rate (Statistics Canada, 1956), applied to same population as Ontario in 1973 (Statistics Canada, 1974).
 f. ~ indicates approximately. Totals are precise but categories do not add because of omission of some minor categories.

treatment plants. Its members include competent and well-paid surgeons, other physicians, medical technicians, and nurses. Much money is spent on better roads and street lighting, at least partly for safety reasons. In short, the community includes many activities whose main or sole purpose is to save life and to cure or prevent illness or injury. These life-saving activities are only possible because of the availability of surplus or disposable wealth, that is, money, in the community. This money must be raised in taxes or paid directly by members of the community in return for the benefits received or as an insurance premium in some form. The level at which these life-saving activities can be sustained is clearly a function of the general level of wealth in the community, which depends largely on industrial activity, which will consequently tend to depend on the availability of the low-cost electricity that makes its wheels turn.

As a preliminary attempt to quantify this indirect positive contribution to safety, the following figures, discussed in Appendix 1, may be considered:

a. Some 6 per cent of the gross national (or provincial) product of an advanced community is spent on life-saving activities.

b. An extra kilowatt hour of electricity produced and sold adds, or is essential to, a net increase of the GPP equal to its selling price, say, 2 cents.

c. Many thousands of lives are available to be saved in North America at a cost of about $300,000 per life.

From these figures, a generating plant of 1000 megawatts capacity running at 75 per cent capacity factor could, under rational safety policies, save

$$\frac{1000 \times 1000 \times 8766 \times .75 \times .02 \times .06}{300,000} = 26 \text{ lives per year.}$$

If, therefore, a failure to expand the availability of low-cost energy results in or is a result of a failure to expand the GNP, this will, in effect, be a loss of life, as shown in the last column of Table 2.

These figures are also shown in Table 3 as applying to sixteen generating stations of 1000 megawatts size. Also shown on the bottom line of Table 3 is the total mortality in Ontario in 1973 compared with what it would have been if the rate experienced in all Canada in 1950 had applied. It will be seen that there is a difference of 11,920 lives, which could well be attributed to the different GPP per head of population in the way outlined above. The suggested benefit of 416 lives saved per year from electricity generation could readily be visualized as one component of this total. Note that this is the safety benefit only, arising from 6 per cent of the GPP increase. The remaining 94 per cent would go towards better living and more leisure, which may be found to be very highly valued by the populace, particularly if it is denied to them by restrictive policies. There are some secondary environmental problems: acid rain and carbon dioxide emission with coal and nuclear waste disposal. They will be dealt with to the extent necessary if the will is there and this will simply depend on resolving the main safety issue.

Unfortunately, there is at present a gulf between the kind of reasoning exemplified by this article and the making of decisions in most of our societies (the OECD countries). The sovereignty of a majority of the people is unquestioned. However, the people are both led and followed by the news media, which must sell news to survive. The uneventful operation of a power station or a lucid discussion of risk statistics cannot hope to compete as news with sensations such as Three Mile Island and sinister suggestions of wrongdoing by arrogant technocrats. As a result, not only is the public disinclined to spend the time to understand the real issues, but they are also misled in many ways by being exposed to news instead of to a reasonably balanced diet of information. The possibility therefore exists that utterly wrong decisions may be made within a system that will be staunchly defended as the best (or the least bad) that man has been able to devise. This paradox can be resolved by the exercise of professional responsibility, by the development of the consensus of informed opinion, and by the delegation of responsibility by the public to their elected leaders. None of these steps is excluded or discouraged in any way by the democratic ideal.

In the present case, those professionally concerned with electrical energy and human safety should first get their own ideas clear about the issues, hopefully along the lines suggested in this article. They should then seek to develop a consensus among the general scientific and professional community, and from this base to persuade the public, the media, and political leaders that what they advocate is right and that they can be trusted with the detailed implementation of it. There can then be a reasonable expectation of favourable decisions from political leaders.

If a majority of the public rejects the advice of those members of the public who have specialized and professional knowledge and experience, there would be no redress except to go back to the beginning and try again. However, in the long run, there need be no despondency about this possibility; you cannot fool all the people all the time. The solid undercurrent of support for nuclear energy, at least by a substantial minority of people, throughout the period of panic following the Three Mile Island accident illustrates this point. (These words were written before the referendum in Sweden on 23 March 1980, which revealed a substantial majority in favour rather than a minority.)

In conclusion, this article attempts to discuss an important issue in a rational way. It presents what are believed to be facts and also estimates and hypotheses, and reaches conclusions based upon them. The whole is open to discussion. The first requirement for solutions to our low-cost energy problems is to ensure that reason prevails.

APPENDIX 1. RELATIONSHIP BETWEEN LOW-COST ENERGY AND LIFE-SAVING ACTIVITIES

1. Fraction of GNP spent on life-saving activities (*World Almanac*, 1978)

U.S. hospitals in 1976; cost for 1,433,515 beds	$55.7 billion
U.S. nursing homes 1973: 1,107,358 beds at, say, 30% of cost of hospital beds	$12.9 billion
Practising doctors: 345,659 in 1974, say, × $60,000 in 1976	$20.7 billion
Medical: ambulances, fire brigades; road, aircraft, railway shipping, food & drug technology—regulation & administration; plus miscellaneous—guess	$20.0 billion
Total	$109.3 billion
U.S. GNP in 1976	$1,706 billion

Say 6% of GNP

2. Relationship between electricity generation & GNP. Historical relationship in Ontario, 1% increase in GPP associates with 1.3% increase in electricity consumption.

   Ontario GPP in 1979 = $70 billion
   Electricity consumption in Ontario 1979 was 12,000 MW average = 105 billion kWh
   Ratio is 51¢ GPP per kWh increase. Cost of 1 kWh approx. 2¢.

   Therefore, it does not appear to be an overstatement that electricity contributes at least its own cost to the GPP — *say 2 ¢.*

3. Cost of saving an extra statistical life (CSX)

   A very wide range of present rates is listed in Siddall (1979). It would appear that many lives are presently available to be saved at around $100,000. An eventual figure of $300,000 is used to allow for the fact that lives that could be saved at low CSX would be saved anyway under a rational safety policy by money transferred from activities of high CSX as explained in Siddall (1979).

REFERENCES

American Medical Association Council on Scientific Affairs. (1978) ''Health Evaluation of Energy-Generating Sources.'' *Journal of the American Medical Association* 240, No. 10 (10 November 1978), p. 2193.

''California's Doctors Perform About-Face on Nuclear Power, at Meeting of the California Medical Association in San Diego Recently.'' *Energy Daily* (28 February 1980), p. 4.

Hamilton, L.D. (1979) ''Health Effects of Electricity Generation.'' Paper presented at Conference on Health Effects of Energy Production, Chalk River, 12-14 September. (Proceedings to be published.) Upton, New York: Brookhaven National Laboratory Associated Universities Inc., National Center for Analysis of Energy Systems.

Siddall, E. (1979) ''A Rational Approach to Public Safety—An Interim Report.'' Paper presented at the Conference on Health Effects of Energy Production, Chalk River, 12-14 September. (Proceedings to be published.)

Statistics Canada. (1956) *Mortality by Detailed Cause of Death, 1950—55*. Cat. No. 84-502. Reference Paper No. 65. Ottawa: Queen's Printer.

Statistics Canada. (1974) *Causes of Death, 1973*. Cat. No. 84-203. Ottawa: Information Canada.

The World Almanac. (1978) New York: Newspaper Enterprise Association, Inc.

Chapter Six

7230
6352 US

Electric Utility Investments: *Excelsior* or Confetti?*

by
Amory B. Lovins

Congressman Mo Udall recently described a billboard he had seen in Southern California. In bold letters it gave the Biblical warning:
YE SHALL PAY FOR YOUR SINS.
Underneath, someone had added the fine print:
Ye who have already paid, please disregard this notice.
Electric utilities, I shall suggest, have not paid yet, and utility investors cannot safely disregard the notice.

It may be unusual for an experimental physicist to discuss utility investments with eminent financial experts; but it is not illogical. The economic options for profitably selling energy, especially electrical energy, are constrained by the thermodynamic laws governing its conversion and distribution. This paper will explore the logical consequences of this constraint for the value and security of utility investments. This physical perspective could hardly be more timely, for the utility business is increasingly pervaded by a sense of change and uncertainty. This uncertainty is not a temporary aberration, a political fluke, or only a mirror of broader macroeconomic shifts,[1] but an early symptom of the market's addressing a much deeper question: namely, does further central electrification in North America make economic sense? The answer, I shall argue here, is predictable and negative. To the questions: Should we expect a much bigger electric grid? In the long run, will most of the generating capacity be thermal plants? And will they be owned by the utilities?—I shall answer: Not if we do what is

* Adapted from invited remarks to the E.F. Hutton Fixed Income Research Conference on Public and Investor Owned Electric Utilities, held at the Hotel Pierre, New York, 8 March 1979. The arguments in this paper have been updated in A.B. Lovins and L.H. Lovins, *Energy War: Breaking the Nuclear Link* (San Francisco: Friends of the Earth, 1980, and New York: Harper and Row, 1981), and in A.B. Lovins, "How to Keep Electric Utilities Solvent," *Energy Journal* (1981, forthcoming).

economically rational, that is, cost minimizing, risk minimizing, and profit maximizing.

MARGINAL ELECTRICITY IS UNCOMPETITIVE

The argument for this proposition is set out more fully and in an international context in an annotated paper.[2] To summarize it briefly: electricity is a very special, high-quality form of energy; higher-quality even than the best fuels, for its effective temperature is infinite, and it is capable of doing difficult kinds of work such as driving motors, chemical reactions, and electronics. Making electricity conventionally requires large and complex machines, precisely made from costly materials. Accordingly,

Electric utilities are without doubt the most capital-intensive of all economic enterprises. The power industry today requires approximately $5 of investment per dollar of annual revenue. This is in sharp contrast to most manufacturing industries, which require $1 or less. . . . [3]

Another measure of this extraordinary capital intensity is that for a given rate of delivering energy (measured by heat content) to final users, central-electric systems are about a hundred times as capital intensive as traditional direct-fuel systems, and about ten times as capital intensive as Arctic, offshore, or synthetic oil and gas systems.[4-6] For this reason, President Ford's 1975 State of the Union Message energy programme would have required a 1976–1985 investment of over $1 trillion (1976 dollars), three fourths of it for electrification.[7] The requirement equals about three fourths of cumulative net private domestic investment over the same period.[8] It is doubtful that the market could have allocated capital in that way, so starving other sectors.

The high capital cost of electricity must be added to a high fuel cost, for a classical power station converts roughly three units of intermediate-quality fuel energy to one unit of high-quality electrical energy plus two units of low-quality waste heat. The result is that delivered electricity is an extremely expensive form of energy. The Electric Power Research Institute has estimated a short-run marginal cost of 3.7–4.4¢/kWe-h at the busbar,[9] and I have estimated[4,5] a conservative busbar cost of 3.7¢ and delivered price of 6.3¢/kWe-h, all in 1976 dollars. The busbar cost and delivered price correspond on a heat basis to buying oil at $60–$71 and at $101 per barrel respectively. The latter figure is about four times the early-1980 world contract price of crude oil. Even allowing for the end-use efficiency of an electric heat pump and an oil furnace, electricity cannot be an economically attractive marginal source of heat.[10] It can also be readily shown that even if unforeseen breakthroughs in battery technology make electric cars competitive in performance with fuelled cars, they will not be competitive economically, since the best present art in fuelled cars is five times more efficient than the present U.S. fleet average and can improve further.

In short, electricity is so expensive, especially at the margin, that it is worth using only for the premium applications that uniquely require it and that can use its high quality to advantage. But these electricity-specific end-uses—electronics, lights, motors, electrochemistry, electrometallurgy, arc-welding, and the like—are only 8 per cent of all delivered U.S. energy needs (7 per cent of Canadian). Further, they are already saturated, since we now supply 13 per cent of our end-use energy (over 15 per cent in Canada) in the form of electricity. (The difference is electricity used for low-temperature heating and cooling.) If, therefore, we make still more electricity, we shall have no choice but to use it for low-grade purposes where it is grossly uneconomic. This would be rather like using a forest fire to fry an egg, or cutting butter with a chainsaw. It would be inelegant, messy, dangerous, and extremely expensive. For the ~92 per cent of our delivered energy needs that are *not* electricity specific—the 58 per cent (69 per cent in Canada) that we need as heat (mainly at low temperatures) and the 34 per cent (24 per cent in Canada) that we need as portable liquid fuels for vehicles[2,4]—electricity is far too costly and too slow[2] to be a rational form of supply. Thus, arguing about what kind of power station to build is irrelevant to our energy supply problem—like debating the best buy in champagne when all we need, or can afford, is a drink of water. It is much cheaper to meet our heat and liquids needs directly, without going through electricity.

The high price of delivered electricity also places a premium on improving its productivity as a factor input to the economy.[11] Surprisingly, presently cost-effective technical improvements in end-use efficiency, using present technologies and not changing life-styles, can be as large for electricity as for heat and liquids—that is, *severalfold*. For example, such "technical fixes" can improve the technical efficiency of household electrical appliances and of industrial electric motors by average factors of four and two respectively.[12,13] Not only can we more than double the technical efficiency of the electricity-specific applications (which form just over half of present U.S. electrical demand); we can also squeeze out the 10 to 15 per cent of that demand that is pure waste (lighting offices at headache level), and can replace the ~32 per cent of demand—often far more in other countries—that provides low-temperature heating and cooling. (That can be done far more cheaply with weather-stripping, insulation, heat exchangers, window overhangs, venetian blinds, trees, passive and active solar heat, etc.) Thus, if we used electricity *at a level of efficiency that is now economically rational*, we could live just the same as now but use *less than a third* as much electricity as now.

This finding follows from elementary economic analysis[14] and from disaggregated physical analysis of how electricity is used. It stands in striking contrast to most utility projections, which rest on highly aggregated econometric extrapolations notably lacking in either economic[2,15] or engineering sophistication. Some prominent utility representatives still explicitly reject the possibility of *any* significant efficiency improvement, for they equate static electric supply with static economic output.[3,16] Others implicitly concur by stating that failure to maintain 4 to 5 per cent per year electrical growth for the next decade will mean blackouts.[17] As demand forecasts continue to plummet[18] and forecasting errors, to mount—the U.S. utilities' forecasts of peak-load growth between 1975 and 1979, made only one year ahead, averaged two and a half times the actual growth[19]—only a few analysts have the courage to admit the bankruptcy of traditional forecasting methods.[20] But a salutary warning of the dangers of economic projections that lack physical concreteness has just appeared. An impeccable physical analysis of opportunities for purely technical efficiency improvements in over four hundred end-use sectors of the British economy[21] has shown that a trebled real gross domestic product (GDP) embodying traditional industrial growth can be achieved with a slight *reduction* in national primary energy use during the next fifty years, rather than requiring a large increase as officially forecast. Further, if the efficiency improvements selected are to be cost effective, not against cheap North Sea gas but against new power stations and synfuel plants, they can readily result in the same trebling of real GDP while primary energy use *decreases by half* and electricity use per capita drops to less than one fourth of current North American levels.[22] Without detailed physical analysis, such striking opportunities remain invisible.

Strong confirmation of the inability of new power stations to compete at the margin comes from a recent and authoritative study[23] of what U.S. energy demand would have been in 1978 if the cheapest system had been bought at the margin between 1968 and 1978 in the knowledge of actual 1978 prices. The result: oil purchases would have fallen by 28 per cent (cutting U.S. oil imports in half), coal imports down by about 34 per cent, and electricity purchases reduced by about 43 per cent—implying that over a third of today's power stations, including virtually the entire nuclear programme, would never have been built. The total cost of such a "least-cost strategy" would have been about 17 per cent less than U.S. consumers actually paid for the same energy services they received in 1978.

WHAT IS THE ENERGY PROBLEM?

The fallacy of emphasizing homogeneous primary energy demand rather than heterogeneous end-use energy needs reflects the view that the energy problem is simply finding more energy to meet extrapolated, aggregated demands—treating one unit of energy the same as any other, regardless of their relative price, quality, or application, and assuming that the more energy we use, the better off we are. But the principles of good engineering—economy of means, the right tool for the job—would have us start by asking how we can meet our *heterogeneous* end-use needs with a minimum of energy supplied *in the most efficient way for each task*. This simple restatement of the problem, reasoning from ends to means rather than the other way around, has profound implications. It means that if we decide not to build a particular power station, we should not necessarily seek another source of big blocks of electricity: rather, we should seek the most effective way to do the end-use tasks that we would have done with the oil and gas if we had had them in the first place. (Their scarcity was our motive for building the station.) Thus, nuclear plants do not have to compete with coal plants but rather with batts of glass fibre. This they cannot possibly do.

Let me amplify this point. What is the competitor? We traditionally compare different power stations and synfuel plants with each other in cost of energy sent out. (This is itself fallacious because diseconomies of large scale[4] may cause minimum busbar cost to entail maximum delivered cost, and because what we really care about is the price of the delivered function, not of the energy that performs it, so we must count the cost and efficiency of the whole system including the end-use device.) But an even more striking asymmetry is that we traditionally require efficiency improvements and renewable sources to compete not with the marginal power plants and synfuel plants, but instead with the historically cheap and heavily subsidized oil and gas—which we are running out of and which all these investments are meant to replace. Thus the U.S. Department of Energy routinely rejects as "uneconomic"—costing more than $15 per barrel of oil—the costlier kinds of active solar heat and biomass liquid fuels at $20−$25 per barrel, but simultaneously proposes, out of our pockets, lavish subsidies[24] for turkeys like syngas at $30−$40 per barrel or electricity at $100 per barrel. That is just nuts (or, more formally, that leads to a misallocation). We ought to compare all our marginal investments *with each other*, not some with each other and some with the cheap oil and gas. If we do this, we find[4] (see Tables 1 and 2), as did the 1979 Harvard Business School Study (R. Stobaugh and D. Yergin, eds., *Energy Future: Managing and Mismanaging the Transition* (New York: Random House, 1979)) that the cheapest marginal energy sources in both capital and total cost are efficiency improvements, then transitional and "soft" (appropriate renewable) technologies, then synthetic fuels, and last—costliest by far—the central-electric systems. This comparison is based on empirical data reflecting the best present art, contains many conservatisms[4] in the sense least favourable to my case, ignores most of the important diseconomies of large scale,[4] and omits all externalities. Yet it shows a robust advantage in capital and total cost for soft technologies over central-electric systems in each unsaturated application at the margin.[25] (It also shows that if desired, electrical sources cheaper than central stations are now available.) I have argued elsewhere[4] that the best soft technologies now in or entering commercial service are more than ample if we use each to do what it does best; and that like efficiency improvements, they are not only cheaper but *faster* than central-electric systems, returning more energy (and money and jobs) sooner per dollar invested. Analysts sceptical of that conclusion must nonetheless agree that the present overcapacity and construction, and the quick and attractive 200+ GWe of industrial cogeneration potential,[26] offer an ample back-up to guarantee electric supplies during a gradual transition.

Table 1
APPROXIMATE CAPITAL INVESTMENT TO DELIVER NEW
ENERGY TO U.S. END-USERS

Energy System[a]	Op. Date	1976 $/(bbl/day)[b]	Form Supplied
Hard Technologies			
(Traditional direct fuels	1950s-60s	2-3,000	Fuel)
Arctic & offshore oil & gas	1980s	10-25,000	Fuel
Synthetics (coal/shale)	1980s	40,000	Fuel
Central coal-el.	1980s	170,000	El.
Nuclear-el. (LWR)	1980s	235,000	El.
"Technical Fixes" to Improve End-Use Efficiency			
New commercial bldgs.	1978	-3,000	Heat/(El.?)
Common industrial & architectural			
leak-plugging, better home appliances	1978	0-5,000	Heat & El.
Most heat-recovery systems	1978	5-15,000	Heat
Bottoming cycles; better motors	1978	20,000	El.
Very thorough bldg. retrofits	1978	30,000	Heat
Transitional Fossil-Fuel Technologies			
Coal-fired fluidized-bed gas			
turbine with district htg. &			
heat pumps (COP=2)	1982	30,000	Heat
Most industrial cogeneration	1979	60,000	El. & Heat
Soft Technologies			
Passive solar htg. (≤100%)	1978	<0-20,000	Heat
Retrofitted 100% solar neighbourhood-			
scale space & water heat	1985	20-40,000	Heat
Same, single house	1985	50-70,000	Heat
300°C solar process heat	1980	120,000	Heat
Bioconversion of farm & forestry			
wastes to alcohols/pyrolysis oil	1980	10-25,000	Fuel
Microhydroelectric plants	1980	30-140,000	El.
Solar pond/Rankine engine	1979	120,000	El.
Wind-el. (Schachle/Riisager)	1979	70-185,000	El.
Photovoltaics (Patscentre CdS)	1980	110,000	El.

a. Empirical cost & performance data except synfuels. *Source*: A.B. Lovins, "Soft Energy Technologies," *Annual Review of Energy* 3 (1978): 477-517.
b. Enthalpic (heat-supplied) basis neglecting potential end-use efficiency.
For further discussion of nuclear capital costs, see "Lovins on Energy." *Science* 201 (22 September 1978): 1077-78.

This argument, based on the most orthodox economic criteria, implies the following:

- That additional power stations are (in Marvin Goldberger's memorable phrase) spherically senseless—that is, they make no sense no matter

Table 2
APPROXIMATE DELIVERED ENERGY PRICES TO U.S. END-USERS, CALCULATED FROM EMPIRICAL COST & PERFORMANCE DATA (EXCEPT SYNFUELS) USING CONSERVATIVE ASSUMPTIONS THAT FAVOUR HARD OVER SOFT TECHNOLOGIES

Energy Form	Energy System	1976 $/GJ	1976 $/BBL[a]
Subsidized	Natural gas	1.9	11
Historic	Coal	2.1	12
Fuels (U.S.	#2 fuel oil	2.8	16
Regional Av.,	Propane	4.4	25
End 1976)	Taxed regular gasoline	5.3	31
	Electricity (residential)	8.7	51
Low-temperature	Improved end-use efficiency,	-0.2-1.4	-1-8
Heat (35% of	passive solar heat		
all U.S. end-	100% solar (neighbourhood)	1.2-2.5	7-14
use needs)	1976 natural gas/furnace	2.8	16
	1976 #2 fuel oil/furnace	4.1	24
	100% solar (single house)	2.9-4.1	16-24
	Synthetic gas/furnace	5.0-6.4	29-37[b]
	LWR/long-storage heat-pump	7.4	43
	LWR/resistive heater	17.5	101
Process heat	1976 direct coal	2.8	16
(23% of all	Synthetic gas/furnace	6.2	36[b]
U.S. end-use	300°C Winston solar/		
needs)	buffer storage	7.3	43
Portable liquid	Char-oil from Tatom mobile		
fuels (34% of	pyrolyzer free, 50%		
all U.S. end-	wet feedstock	0.8-1.1	5-6
use needs)	Alcohol bioconversion	1.4-4.8	8-28
Electricity	Microhydroelectric	1.7-8.0	10-46
(8% of all	Solar pond/Rankine engine[c]	6.9	40
U.S. end-use	Wind (Schachle/Riisager)		
needs)	with 10-h storage ($6.3/GJ)	10.5-17.2	61-100
	Photovoltaics (Patscentre		
	CdS) with 10-hour storage[c]	7.5	43
	LWR	17.5	101

a. $/barrel crude-oil heat equivalent (5.8 GJ) @ 100% conversion efficiency.
b. USDOE estimates are ~$40/BBL or $7/GJ.
c. Solar pond/Rankine and photovoltaics are not assumed in Lovins's soft-path analysis.
Source: A.B. Lovins, "Soft Energy Technologies," *Annual Review of Energy* 3 (1978): 477-517.

which way you look at them. They are a future technology whose time has passed, for they supply an extremely costly form of energy of which we already have several times as much as we can use to economic advantage. (The same is true abroad.[2])

• That even ignoring all constraints of logistics, politics, and capital allocation, the *maximum* potential economic role of any baseload electrical source is the baseload portion of electricity-specific end uses. That is only about 4 per cent of total end-use needs (about the same as U.S. firewood today) or about 10 per cent of total primary demand.

• That if the United States uses electricity in an economically rational way, demand will decrease so far that present hydro, readily available microhydro,[4] and a modest amount of wind capacity would suffice[27]—so the United States could be more prosperous than now, with no changes in life-styles, using no thermal power stations at all. Further, with a hydro-dominated grid, integration of intermittent renewable sources (wind, photovoltaics) would need no storage and hence becomes cheaper.[28]

• That a country that has just built a new nuclear power station will probably *save money* by writing it off and never operating it. Why? Because its output can only be used for low-temperature heating and cooling (other markets being saturated); but all that it is worth paying for these services is what it costs to provide them in the cheapest way: weather-stripping, insulation, heat exchangers, window shades and coatings and overhangs, trees, greenhouses, and the like. But these measures, intelligently done, cost less than the *running costs alone* for a new reactor (fuel-cycle, O&M for station and associated marginal grid, waste management, and decommissioning). One therefore saves money by not running the plant. Indeed, under U.S. tax conditions, one saves the already-sunk capital costs too! This remarkable result obtains[29] because the stream of unpaid federal tax subsidies plus utility profits, discounted to the day before commissioning, equals two thirds of the sunk capital costs, and can be offset as a saving against those costs if the plant is written off as a social rather than a private cost (e.g., through taxes). The remaining third of the capital costs can be covered simply by buying efficiency improvements and passive solar measures at a price about 0.3¢/kWh (about $5/bbl) cheaper than the plant's marginal running costs—implying an easily met price ceiling for the electricity-displacing measures of around $20–$25/bbl. Virtually all the thermal measures considered here are substantially cheaper than that.[30]

SOFT ENERGY PATHS

A further implication, described in detail elsewhere,[4,6] is that in place of a ''hard'' energy path based on ''Strength Through Exhaustion''— centralized conversion of depletable fuels to premium forms (especially electricity)—one can construct a ''soft'' energy path. It would evolve differently, relying on vigorous improvements in end-use efficiency and on the brief and sparing use of fossil fuels during a transition, over perhaps fifty years, to essentially complete reliance on diverse renewable sources. These would be relatively understandable to the user (though they can still be technically very sophisticated) and would supply energy of the appropriate scale and quality for each task. (See illustrative schematic graphs in Figure 1.) I have argued elsewhere[4] that a soft path is cheaper, faster, surer, safer, better for jobs and the economy, environmentally more benign, and politically much more attractive than a hard path. A soft path is not cheap or easy; only cheaper and easier than not doing it. No energy future is free of social problems, but we must choose which kinds of problems we prefer.

Although this argument has been framed in terms of lowest private internal cost, there is a subsidiary and independent line of argument[4,6] based on political economy. Historically, we have seldom, if ever, based major public policy decisions about energy on economics; rather, we have based them on political expedience and then juggled the subsidies to make the economics come out to justify what we just did. The real springs of action in energy policy are to be found (if one looks at the political forces signalled in the market-place) among such structural problems as centrism, autarchy, vulnerability, and technocracy—precisely the political problems that really define a hard path. A soft path has a more tractable set of political problems, mainly those of pluralism—of addressing the energy problem with millions of dispersed choices in the market rather than with a few centrally managed projects. Indeed, a soft path has such great economic and political advantages[4] that if they are allowed to show themselves, it will largely implement itself—as it is already starting to do very rapidly—through existing market and political process. To ensure timely realization of this economically efficient outcome, we should

• Clear away institutional barriers (market imperfections) that keep people from doing what is economically rational. Examples include obsolete building codes and mortgage regulations, split incentives between builders and buyers or landlords and tenants, restrictive utility practices, architectural fee structures that encourage inefficient design, poor information, and inequitable access to capital (see below).[31]

• Desubsidize the energy sector: 1978 U.S. subsidies to conventional fuels and power were about $100 billion (two thirds through rolled-in pricing and the rest through direct tax subsidies).[2]

• Move gradually and fairly[31] toward long-run marginal-cost pricing of depletable fuels or, pending rational pricing, install decision rules[31] that ensure equivalent allocation of capital, as I shall propose presently.

If we are not clever enough to solve these institutional problems—the first mainly at the state and local level, the second and third mainly at the federal level—I doubt that we shall be clever enough to solve the far more formidable institutional problems[32] of continuing with present policy.

Figure 1
ALTERNATIVE ENERGY FUTURES

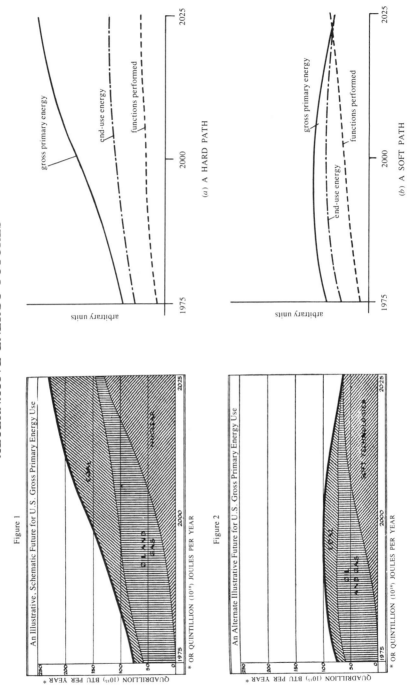

Figure 1

(a) A HARD PATH

Figure 1
An Illustrative, Schematic Future for U.S. Gross Primary Energy Use

* OR QUINTILLION (10^{18}) JOULES PER YEAR

(b) A SOFT PATH

Figure 2
An Alternate Illustrative Future for U.S. Gross Primary Energy Use

* OR QUINTILLION (10^{18}) JOULES PER YEAR

It is also important to note a difference of technical structure between hard and soft energy paths. Most of the growth in a hard path goes to conversion and distribution losses, especially the Carnot losses of electrification. Energy growth therefore produces little marginal welfare. But in a soft path, by supplying energy at a scale and quality matched to the task, the costs and losses of energy distribution and conversion, respectively, are largely squeezed out, so that primary and end-use energy converge. Meanwhile, because energy is properly priced and subsidies to supply are phased out, price elasticity of demand ensures that we gradually build up a conversion system that requires several times less energy per unit of output than now. Thus, the bottom dashed line in Figure 1(*a*) and 1(*b*), representing delivered goods and services, ends up higher in the soft path than in the hard path, even though the former uses much less energy. We would be doing more with less.

Let us now consider the financial risks of not appreciating and acting upon the foregoing arguments. Risk-averse investors should explore what may happen if those arguments are right but the utilities ignore them—and should ask how that risk can be hedged.

WHY UTILITIES GO BROKE

The cash flow of electric utilities is inherently unstable.[33] Electric utilities combine extraordinary capital intensity with long plant construction times (lead times). In particular, typical lead times are much longer than the time constant of short-run price elasticity of demand. Installed capacity is therefore likely to overshoot its economically sustainable level.[34] To put it in mundane terms, during the long construction period, a utility must generally raise its electricity price in order to finance the construction,[35] keep up its bond ratings and coverage ratios, maintain dividends, and keep common stock values near book. But consumers respond to the higher price (and even to expectations of it) long before the plant is finished, so when it *is* commissioned, it sells too little electricity for its revenues to cover its fixed charges. (This is inevitable because utilities that assume a non-zero price elasticity compute it on historic costs, not on marginal costs.) The shortfall of revenue makes it necessary to raise prices further, so that the rate of growth—and conceivably (as in Britain) the level—of demand is dampened further, requiring a still higher price to amortize existing plants, and so on into the "spiral of impossibility." For cash flow to collapse, it is not necessary that the actual level of demand should decrease;[36] it is sufficient that demand should persistently fall short of expectations and grow more slowly than capacity.[37]

The problem outlined here does not depend on whether the utility is regulated, perfectly regulated, or unregulated, nor on whether it is publicly or privately owned, nor on its methods of management. The cash-flow instability is *inherent* in the nature of building power stations. Further, it is a dynamic problem of cash flow, not an equilibrium problem of profit and loss, but the latter type of problem might also exist. For example, if long-run price elasticity of demand has an absolute value greater than one—as many analysts now suspect—then, absent compensatory increases in population or income, increasing the price *reduces* the utility's revenues, because the utility loses more on the number of kilowatt hours sold than it makes up by charging more for each kilowatt hour sold. If long-run revenue elasticity is indeed negative, as it may well be, then the utility business is fundamentally unprofitable at equilibrium, independent of its cash-flow problems, because any investment simultaneously requires higher revenues (to pay higher marginal costs) and produces lower revenues. As a conservatism, we shall ignore this unpleasant possibility and discuss only the dynamic problem; but the solution we shall propose for that problem also hedges the risk of equilibrium unprofitability.

If one considers utility cash flow using the metaphors of control theory, the extent of the overshoot in installed capacity, and hence the depth of the cash-flow crisis, should depend primarily on the ratio of time constants (lead time divided by response time to higher price). A higher growth rate should also increase the instability. How soon the cash flow starts to deteriorate should depend on the margin by which price elasticity is underestimated and on the safety margin built into the initial cash flow: utilities with low coverage ratio, high debt-to-equity ratio, low liquidity, or high reliance on non-cash income[38] have less room for manoeuvre. Conversely, utility managers alert and flexible enough to respond to early warning signals by drastically cutting back on construction (not merely deferring it) may be able to stay out of trouble, or at least to defer it. But the reflex response of most managers is to suppress those signals by seeking higher subsidies—faster depreciation, higher investment tax credit, more rolled-in pricing, and the like. This is exactly the wrong thing to do: it makes the utility crash harder (and somewhat later),[33] and it thereby greatly increases the risk to investors. Subsidy inflates demand beyond the level whose revenues can ultimately amortize the supply investment. For a regulated utility, which is required to provide adequate supplies and whose regulators are required to provide adequate return for it to do so, investment incentives are also superfluous and can only encourage overbuilding and over-reliance on debt. Present subsidies are very substantial: for example, they enabled U.S. private electric utilities during 1974—75 to reduce federal taxes from 12.7 to 1.8 per cent of revenues and to reduce their absolute amount by two thirds.[39] Accelerated depreciation, investment tax credits, and interest deductions allow most utilities to pay zero or negative state and federal income tax on revenues from marginal investment—a tax subsidy amounting to some 2¢/kWe-h (1977$) for marginal nuclear residential electricity. (Rolled-in pricing provides a further 1.5¢, lowering a social internal cost of 8¢ to an apparent private cost of

4.5¢.)[40] There is a cogent array of arguments, and widespread support, for eliminating these subsidies and the utility federal corporate income tax that gives rise to them.[41]

The instability of utility cash flow is not just the over-expansion risk normal in other industries in which the market corrects excessive capitalization. It is unique to monopolies with a regulated rate of return and an obligation to meet all demand, for this position reduces their incentive to be efficient and their degrees of freedom in responding to fiscal constraints. The acute phase of the instability may be triggered by real cost escalation, increased delays, or exogenous reductions in demand growth (increases in the absolute value of implicit price elasticity). Many second-order effects conspire to enhance the instability: for example, perceived capital shortage (reflected in higher cost of money and reliance on short-term notes), higher perceived risk (reflected in lower share prices and bond ratings, hence higher required rate of return to be able to sell debt or equity, hence higher consumer prices), higher federal budgetary deficits (hence inflation) arising from tax subsidies, loss of the present value of tax deferrals, political pressure on regulators to be less sympathetic to rate hikes,[42] and greater difficulty in attracting first-class managers to a sick company. Moreover, most utilities have a variety of motives for continuing to expand when it is clearly imprudent and contrary to their interest to do so[14,43]—force of habit, prestige, desire of executives for advancement within a bigger pyramidal hierarchy, the Averch-Johnson effect,[44] a perceived divine mission to carpet the earth with power plants, or simply fear that if growth slows, artificial income provided by the accelerated depreciation allowance will dry up and deferred tax for past years will fall due.[45]

PLANNING FOR UNCERTAINTY

Further, utilities are naturally afraid of underestimating demand and being caught short, vilified, and perhaps nationalized. Several recent studies[46] have conveniently reinforced the resulting belief that it is safer and cheaper to overbuild than to underbuild, especially with regard to baseload capacity. But it now appears that this conclusion is an artifact of the models' lack of sophistication in certain key respects.[47] The conventional wisdom that overbuilding entails only a small extra carrying charge whereas underbuilding risks catastrophic outage costs also ignores opportunity costs on both sides. Overbuilding means one has forgone the opportunity to spend the money in other ways, including efficiency improvements and alternative supplies, and that lost opportunity has a cost. Conversely, underbuilding means one still has the money and can spend it in ways that hedge the risk.

Most utilities today are continuing to overbuild in case demand growth picks up again. They explain that even if they will not need the power themselves, their neighbouring utilities will. The trouble is that the neighbours follow the same logic. In my travels around North America in 1978, I had a strong impression that many utilities are all planning to sell their surplus power to each other in a big circle. They evidently do not realize the high financial risks of overbuilding. In recent months, an impressive literature of computer simulations of utility finance has begun to quantify these risks. Preliminary results of these studies[48-50] suggest that at least for utilities that do not include construction work in progress (CWIP) in their rate base, the inevitable uncertainty in demand is best dealt with by building *less* than the forecast requirements for baseload plants. By clever choice not only of whether to build but of *what* to build, utilities can hedge their risk. The key variable is plant lead time: big plants take longer (twice as long for a 3-GWe as for a 0.5-GWe coal plant, for example), and long lead times increase both the inertia (inflexibility) of construction programmes and the uncertainties of forecasting.[50] The calculated financial risk to utilities, shareholders, and ratepayers is *greater* if baseload plants are overbuilt than if they are underbuilt,[33,48-50] because the extra depreciation and return on excess baseload capacity will cost more than extra operation of short-lead-time peaking and intermediate-load-factor plants (coal fired or even gas turbines). This result is quite insensitive to the extra fuel costs of such operation or even of importing electricity.[49] The general findings seem robust and probably conservative. Thus, it appears that doctrinaire notions of economies of scale need to be urgently re-examined,[51] and that if demand is uncertain, building as many baseload plants as forecasts indicate is an especially high-cost and high-risk strategy.[48-50] And though the models do not fully analyse this point, such overbuilding is likely to exacerbate the instability of cash flow.

FUNNY MONEY

Many utilities today appear heedless of these risks to themselves or to others to whom the risks are to be transferred. The ideological attachment to building large numbers of baseload plants is so powerful that some utilities are making strenuous efforts at ''creative'' financing of plants that the market, for excellent reasons, is unwilling to finance by normal means. These efforts are taking over more bizarre forms. To pick a few U.S. examples:

• Controversy continues over the effect of 1980 legislation that is meant to extend, via the Bonneville Power Administration (BPA), a federal guarantee to buy the output of any new plant proposed by a private utility in the large BPA service area if BPA decides (by a well-insulated process) that it is needed. The utility would pay the average BPA system price, rolling in its high marginal costs with historically cheap hydropower so that other BPA customers subsidized the new plant. Payment would be guaranteed even if the plant worked badly or not at all. By socializing the risk and making everyone pay for projects even if they are not needed or do not work, this scheme would eliminate all penalty to the utilities for overbuilding.

• Another federal bailout is now taking place via the 1973 amendments to the *1936 Rural Electrification Act* (REA). Distribution loans at 5 per cent per year interest (2 per cent per year in a few cases), federally guaranteed, were augmented in 1973 by 100 per cent, thirty-five-year federally guaranteed loans for power plants and transmission facilities, at interest rates currently running about 7.5 per cent per year.[52] The REA power supply borrowers have thus become such attractive partners for private utilities in large-scale coal and nuclear generating projects, often for remote urban supply, that *three fourths* of the REA loan guarantee commitments in FY 1977 were used for this purpose.[53] The REA loan programme rose from $0.62 billion in FY 1973 to $4.8 billion in FY 1977, and a 1976 Federal Power Commission report projected a cumulative total of about $20−$40 billion over the next decade.[54]

• The President of Public Service Co. of New Hampshire proposed[55] (unsuccessfully) an unprecedented arrangement whereby, in lieu of CWIP, the State of New Hampshire would guarantee the principal and interest on $400 million worth of Seabrook bonds.

• In a move reminiscent in intent of the Empire State Power Resources Inc. proposal to evade NYPSC jurisdiction and transfer risk,[56] the holding company, Middle South Utilities, is building a 2500 MWe nuclear plant in Mississippi, which is not on anyone's grid, effectively not under any PUC's jurisdiction, and not subject to any state need-for-power test. Middle South apparently trusts that at least one of its constituent utilities can be induced to buy any power produced. The investment is purely speculative in the sense that it bears no guaranteed rate of return, but nevertheless it was financed—initially—through bank loans.

To those of us who believe in the salutary discipline of the market-place, the effective regulation of natural monopolies, and the efficient allocation of risk, these examples of expedience are disturbing.[57] They suggest that some utilities' short-term zeal to continue building big baseload plants outweighs their prudence or their concern with the long-term preservation of an orderly, equitable, and low-risk financial structure for their industry. Such desperately inventive attempts to finance the unfinanceable fill me with foreboding. I fear that such misallocations not only represent important opportunities forgone, but set the stage for credit deterioration and for defaults on a scale beyond the capacity even of a flourishing private capital market or of the federal Treasury itself to bail them out.

But I am not a pessimist. I believe that if investors exercise their responsibilities constructively, the economic collapse of electric utilities can be avoided. Rather than making obvious points about utility rate reform, I should like to make a modest proposal that I think could restore the utilities to financial health, help solve our energy and equity problems, reduce capital shortages, increase profits and employment, increase the resilience of our energy systems[58] (and hence our national security), help clean up the environment, and reduce nuclear proliferation,[4] all at the very same time. I propose to do this with two rather small changes in the way utilities and regulatory commissions work within existing incentives.[31]

TRANSFERRING UTILITY CAPITAL

First, though several states have already mandated utility loans for energy conservation, I would refine and extend this system so that interest rates were fair and pay-back times not unduly restricted. Utilities (competing with other institutions—banks, oil companies, insurance companies, federal loan agencies, etc) should loan money at their internal interest rate[59] for fuel-saving investments—efficiency improvements and renewable sources—to householders and others in order of their current access to relatively cheap capital. The borrowers would spend the money at their own discretion and repay the principal and interest through their utility bills at or slightly below their own rate of return from the investment. They would thus capture the advantages of the fuel saving, but their bill would be no higher meanwhile than if they had done nothing. The utility would preferably treat the loan as below the line, not a rate-base item, and would receive a cash-flow benefit comparable to its return if it had used the money to build a new plant instead.

Utilities have three strong incentives for making such capital transfers:[31]

• The alternative investment requires several times less capital than a new plant would (see Tables 1 and 2).

• Since the borrower's investment typically takes days or months to build and a few years to pay back, compared to ten years to build and thirty to pay back for a power station, the utilities turn over their money much faster,[60] improve their cash flow, and do more work per dollar of working capital, while providing a profitable home for any excess revenues from rate reform: indeed, the revolving loan fund revolves so quickly that it can often be financed entirely from internally generated capital (totalling over $10 billion for U.S. investor-owned utilities in 1979), thus avoiding the high marginal cost of new capital—a major benefit to cash flow.

• By getting into a short-lead-time, fast-pay-back business at the margin, utilities remove the instability inherent in their cash flow and hence avoid overbuilding and eventual bankruptcy.

Such incentives led gas and electric utilities to invent conservation loans, and some utilities today find it advantageous to go even further than the system proposed here.[61] The economic point of this system of capital transfers is that utilities and consumers, supply and efficiency, hard and soft technologies, rich and poor people, would all have equitable access to capital. The political point is that utilities would be co-opted into the transitional process—as an opportunity, not a threat—and over the next fifty-odd years, as they gradually evolved into a distribution service something like the telephone company, they would have something to do at the margin that they do well, namely merchant banking, using their existing billing infrastructure at low transaction cost.

Second, Gus Speth[62] has proposed that the need-for-power test should be reoriented from projected electrical demand to end-use needs—the range of tasks we are seeking to do with the incremental energy. To obtain a certificate to build a new plant, the utility would have to prove to its regulatory commission that its proposed plant was the most economically efficient way to meet end-use needs. Specifically, it would have to show that within its service area, it had exhausted the efficiency, load-levelling, and renewable energy potential that (*a*) would compete with the proposed plant in end-use functions, (*b*) would compete with it economically in those functions, and (*c*) could have been financed by loaning out, on the equitable terms just described, the money that the utility proposes to spend building the plant. (To keep this comparison honest, if the utility passes this investment-balancing test and builds the plant, the real plant cost it puts in the rate base should not exceed that which it assumed when making this comparison.) This relatively minor change in the focus of the need-for-power test—similar to present practice in California and perhaps elsewhere—has a profound and decentralized economic effect: because it compares efficiency and solar investments not with temporarily cheap fuels but with the power plants representing their long-run replacement costs, it ensures that in the utility sector, and in the efficiency and renewable-energy sectors financed by capital transfers from the utility sector, capital is allocated *as if energy were already priced at long-run marginal cost—but without having to achieve those unpalatably high prices first.*

These two proposals are undoubtedly imperfect and I hope they will be improved. But my discussions around North America with utility managers and regulators suggest that they are generally on the right track and can probably be implemented at the state or provincial level even without federal facilitation. They would go a very long way towards restoring the utilities to the financial integrity that comes from economically rational behaviour, and would enable utility investors to participate in low-risk investments, many of which have the highest rates of return available anywhere.

Whether this actually happens depends ultimately on the quality of utility management. That, in turn, depends on how strongly investors want to maintain the value and security of their utility investments. the absurdity of spending three fourths of our marginal energy investments (and of our energy R&D money) on 8 per cent of the energy problem, and the strong potential role of the utilities as bankers for the energy transition that is already irresistibly underway—if I were to consider investing in a utility whose management had not demonstrated a thorough understanding of these arguments, I would certainly demand a quite immoderate risk premium.

NOTES

[1] See, for example, U.S. Congress, Joint Economic Committee, staff study, "U.S. Long-Term Economic Growth Prospects: Entering a New Era," 25 January 1978 (Washington, D.C.: Government Printing Office), and J.W. Forrester, "Changing Economic Patterns," D-2891-1, MIT, 8 May 1978 (paper to MIT Club of Chicago conference, "Management's Challenge in the 1980's," 4 May 1978).

[2] A.B. Lovins, "Is Nuclear Power Necessary?" (London: Friends of the Earth Ltd., 1979). Its Appendix deals particularly with trends in reactor capital cost since nuclear power left the turnkey era and started to drop the 'n'. See also A.B. Lovins, "Economically Efficient Energy Futures," in *Interactions of Energy and Climate*, edited by W. Bach *et al*. (Dordrecht: Reidel, 1980), pp. 1-35; expanded in A.B. Lovins, L.H. Lovins, F. Krause, and W. Bach, *Energy Strategy for Low Climatic Risks* (forthcoming).

[3] T.J. Nagel, "Operating a Major Electric Utility Today," *Science* 201 (15 September 1978): 985-93.

[4] A.B. Lovins, *Soft Energy Paths: Toward a Durable Peace* (New York: Harper and Row, 1979 (originally published 1977), and, more technically, "Soft Energy Technologies," *Annual Review of Energy* 3 (1978): 477-517.

[5] A.B. Lovins, "Lovins on Energy Costs," *Science* 201 (22 September 1978): 1077-78, and "Energy: Bechtel Cost Data," *Science* 204 (13 April 1979): 124-29, responding to J.M. Gallagher, "Lovins' Data Source," *Science* 202 (22 December 1978): 1242-44. See also Appendix 2, and C. Komanoff, *Power Plant Cost Escalation* (New York: Komanoff Energy Associates, 1981).

[6] Approximately three dozen critiques of a 1976 article (A.B. Lovins, "Energy Strategy: The Road Not Taken?" *Foreign Affairs* 55 (October 1976): 69-96) that was the precursor of reference 4 have been published, all with full responses. A readable version of all the substantive exchanges, edited into a point-counterpoint format by Hugh Nash, is *The Energy Controversy: Soft Path Questions and Answers* (San Francisco: Friends of the Earth, 1979).

[7] The August 1975 *Energy Supply Planning Model* (Bechtel Corp., PB-245 382 and -3, NTIS) calculated a direct construction cost (1974 dollars, 1974 ordering) of $559 billion, including work in progress but not yet finished in 1985. Interest, design, administration, and land bring the Bechtel total to $743 billion. Correcting to a 1976 ordering date and 1976 dollars is estimated by M. Carasso (then director of the Bechtel project) to yield over $1 trillion, neglecting all real escalation after 1976.

[8] Assuming that NPDI remains 7 per cent of GNP, which experiences real growth of 3.5 per cent per year.

[9] C.L. Rudasill, "Coal and Nuclear Generating Costs," EPRI PS-455-SR (April 1977).

[10] Details of the calculation are in reference 4; but note that one can also build a very efficient oil- or gas-fired heat pump. The argument is as valid for renewables as for oil.

[11] Strictly speaking, energy is treated in most economic formalism as an intermediate good rather than as a primary factor input such as labour and capital.

[12] J. Nørgård, *Husholdninger og Energi* (Copenhagen: Polyteknisk Forlaget, 1979). This engineering analysis shows that straightforward design improvements, paying back in six years 5.5¢/kWe-h, can improve average Danish appliance efficiency three and a half times. U.S. designs are less efficient.

[13] Industrial electric motors, accounting for over half of all electricity-specific demand, are normally thought to be ~90+ per cent efficient (their nameplate rating). *As actually used* throughout British industry, however, they are typically ~25−35 per cent efficient (W. Murgatroyd and B.C. Wilkins, *Energy* 1 (1976): 337-45), because they tend to be oversized and coupled through inefficient drive trains. Just good housekeeping roughly doubles the motor efficiency, and even larger improvements are obtainable, for example, through hydraulic drive. The Fichtner-Studie came to the same conclusion in the Federal Republic of Germany. The position in U.S. industry is generally similar.

[14] It can be shown with extremely conservative data that utility investment in efficiency improvements rather than in new plants is financially advantageous to the utility, its ratepayers, and its shareholders. See W.R.Z. Willey (Environmental Defense Fund, Berkeley, CA), Testimony to Ca. PUC, Exhibit 151, PG&E Applications 57284-5, April 1978 (argument accepted in 6 September 1978 ruling), and Testimony to Arkansas PSC on behalf of Attorney General, Docket U-2903, 15 May 1978 (economic argument accepted at p. 8 of 31 August 1978 order, though implementation rate considered too slow on a judgemental basis apparently not supported on the record; generic study of conservation strategies recommended).

[15] For critiques of some of the more common economic shortcomings, see, for example, R. Halvorsen (Dept. of Economics, University of Washington): testimony of 14 August 1978 to Ma. DPU (docket 19494), of 13 December 1978 to Wa. UTC (U-78-21), and of 20 December 1978 to NM PSC (1452). See also citations at pp. 2-3 of reference 2.

[16] G.W. Porter, "Energy: The Ostrich and the Eagle," *Rural Electrification* 37 (December 1978): 2-4.

[17] T.A. Vanderslice (GE), *Christian Science Monitor* (25 October 1978), p. 23.

[18] See, for example, A.N. Gordon Jr., Chairman, Northeast Power Coordinating Council, "The Waiting Game: A Report on Oil Dependence and Bulk Power Supply in New England and New York," 1978. Gordon shows at pp. 5-6 that forecast peak and total 1983 loads, made annually during 1974−78, declined by 8.1 and 8.4 per cent/year respectively for New York, 9.6 and 9.3 per cent/year for New England. Perversely, he concludes that the 3.6 per cent/year of plant slippage during the past 3 years may be disastrous for supply: his data show that during those 3 years, the peak load forecast for 1983 fell by 18.5 GW, so not only is it lucky that the 13 GW of supply increment for 1987 slipped; it should have been cancelled outright. For another example, see Ontario Hydro Load Forecast Report No. 780213 (1978). On the dominant East System (p. 14), for example, the forecast peak load for 1985 fell by an average of 7.1 per cent/year during 1974−78. The 1978−85 average growth rate was forecast as 6.3 per cent/year in 1977, 5.3 per cent/year in 1978, and 4.7 per cent/year in early 1979. The Ontario Energy Ministry has just forecast 3.7 per cent/year, and the chairman of the Ontario Royal Commission on Electric Power Planning, who was until recently thinking of a long-term 4 per cent/year, is increasingly talking of 2 per cent/year. Likewise, Manitoba's load growth in 1978 was 1.4 per cent; prices increased 78 per cent since 1974. Perhaps somebody will notice the connection.

[19] *Energy Daily* 6 (209) (1978): 3-4 (EEI data), cited in Dr. Kahn's paper at the E.F. Hutton Fixed Income Research Conference on Public and Investor Owned Electric Utilities, New York, 8 March 1979.

[20] W.C. Hayes, "What Happened?" *Electrical World* 190 (1 October 1978): 3; D. Sebesta, "Summer-Peak Growth Slumps to 2.3%," *Electrical World* 190 (1 November 1978): 54-55.

[21] G. Leach *et al.*, *A Low Energy Strategy for the United Kingdom* (London: International Institute for Environment and Development and Science Reviews Ltd., 1979).

[22] D. Olivier (London: Earth Resources Research Ltd.) analysis published mid-1981; also includes a U.K. renewable energy supply plan. The nearly sixfold efficiency gain applies to electricity as well as to total primary energy. For a broader survey of such studies, see the last two citations in note 2.

[23] R. Sant, "The Least-Cost Energy Strategy," (Arlington, Va.: Mellon Institute, Energy Productivity Center, 1979).

[24] Such as rolled-in pricing, federal loan guarantees, and an "all-events tariff" (which FERC granted but the courts reversed) whereby ratepayers would be liable to pay off the debt, and in certain circumstances, an unspecified part of the shareholders' equity too, even if the plant did not work. Even on such terms, synfuel plants do not seem easy to finance: perhaps the market is trying to tell us something.

[25] An apparently contradictory result was reported in *Energy Daily* 6 (188) (28 September 1978): 1-2, but turns out to be spurious: see M. Carasso (Ca. Energy Commission), letter to A.L. Alm (DOE), 15 November 1978. It relies on a model grossly unsuited to the task, omits efficiency improvements, caricatures a soft energy system, omits passive solar techniques, omits transmission and distribution savings, and obtains over a third of its total calculated soft-system capital cost from solar space conditioning that could be much more cheaply done by architectural means.

[26] R.H. Williams, "Industrial Cogeneration," *Annual Review of Energy* 3 (1978): 313-56.

[27] This ignores the large Canadian hydroelectric surplus if the Canadians were also efficient.

[28] E. Kahn, LBL-8388 (December 1978) and *Annual Review of Energy* 4 (1979): 313-52, discusses the integration problem from a slightly different perspective.

[29] D. Chapman, "Nuclear Economics: Taxation, Fuel Cost and Decommissioning," report A.E. Res. 79-26 from Department of Agricultural Economics, Cornell University, to California Energy Commission, October 1979, and personal communication, February 1980, provided the data used here for the streams of subsidies and profits.

[30] The economic argument presented here for writing off a newly built nuclear plant applies *a fortiori* to a nuclear plant that is only partly built or that is partly amortized. It also applies to non-nuclear plants, right down to the installed central-station capacity needed to meet electricity-specific needs at an economically efficient level of end-use efficiency. In North America, that level is zero, and in many European countries, it is close to zero.

[31] A.B. Lovins, "How to Finance the Energy Transition," *Not Man Apart* (September-October 1978): 8-10.

[32] Some of these go to the very roots of our social structure. See, for example, Luther Gerlach, "Energy Wars and Social Change," (Minneapolis: University of Minnesota, Dept. of Anthropology, 1978). The United States currently has over sixty such "wars" in progress, and

proposals to suppress their causes could incur prohibitive political costs: A.B. Lovins, "Democracy and the Energy Mobilization Board," *Not Man Apart* (February 1980): 14-15.

[33] Among the earliest and best analyses of this effect, which has partial analogues in the transportation industry, is the classic 1976 paper by E. Kahn *et al.*, LBL-4474 (Lawrence Berkeley Lab.). Their simple and transparent model fits the historic data remarkably well. See also the cover story in *Business Week* (28 May 1979), and the proceedings of the 18-19 April 1980 utility conservation finance colloquium sponsored by the California Public Utilities Commission (San Francisco), *Energy Efficiency and the Utilities: New Directions.*

[34] This is a familiar theorem of control theory—high-flux long-lag systems overshoot and collapse—but electrical engineers, who know control theory, are seldom in charge of utility finance, so the richness of the metaphor tends to go unnoticed.

[35] Pathological but still educational examples include the 1973–78 rate hikes of 126 per cent by Portland General Electric and 119 per cent by Public Service Co. of New Hampshire, respectively proprietors of Trojan and of the Seabrook site. (Environmental Action Foundation/Critical Mass, "Nuclear Power and Utility Rate Increases," 1978.) For many utilities, planned construction accounts for two thirds of proposed rate hikes.

[36] In recent LASL modelling (references 49-50), demand does not decline even with a price elasticity of -1.0 and zero delay (long-run elasticities as high as -1.69 have been estimated—R. Halvorsen, Paper 73-13, Department of Economics, University of Washington); but the financial sectors do not include such important effects as the loss of cash flow from accelerated depreciation or the negative feedback from rate of return to cost of money, so the result is inconclusive. For a practical illustration, however, see *The Times* (1 March 1980), p. 1, where the Chairman of the U.K. Central Electricity Generating Board states: "We face a disturbing prospect: a vicious circle of rising electricity prices causing further reductions in demand, which in turn would push up prices still more."

[37] Kahn *et al.*[33] point out that in capital-intensive industries that are essential or economically important, investment tends to determine demand because government subsidy can be relied upon in case of difficulty; this "locks consumers into a pattern of investment that does not reflect preferences expressed in the market."

[38] For example, allowance for funds used during construction. Kahn's paper at the E.F. Hutton Fixed Income Research Conference on Public and Investor Owned Electric Utilities, New York, 8 March 1979, shows that AFUDC then represented about 40 per cent of the net income of an average private electric utility (~70 per cent for some!) and that the market devalues accordingly. (By 1981, these figures were ~50 and >200.) See also I.C. Bupp, "The French Nuclear Harvest: Abundant Energy or Bitter Fruit?" *Technology Review* 82 (November/December 1980): 30-39.

[39] M.O. Bidwell (NYPSC), "Electric Utility Dynamics," attached to R. Batinovich (Ca. PUC), "Options for Utility Tax Policy," 24 November 1978 (CPUC, San Francisco).

[40] D. Chapman (Cornell University), "Taxation and Solar Energy" (Solar Energy Office, Ca. Energy Commission, 15 April 1978); a simplified version appears in Chapman's "Taxation, Energy Use, and Employment," *Creating Jobs Through Energy Policy*, pp. 45-52, Joint Economic Committee, U.S. Congress, 15-16 March 1978 (Washington, D.C.: Government Printing Office). A more recent calculation (personal communication, 11 October 1978) indicates that half the profit from a new nuclear plant is its negative income tax, ample to cover its own and other revenues.

[41] Quirks in the Internal Revenue Code encourage "phantom taxes": in 1977, for example, U.S. electric utilities paid $0.6 billion in federal taxes, but charged their customers $3.1 billion for it and pocketed the difference. Not only did ratepayers have to provide the $2.5 billion balance as an interest-free loan; they had to pay shareholders a return on it too. Federal corporate income tax of regulated utilities is effectively an excise tax, since utilities are guaranteed their *after*-tax rate of return, and the tax subsidies are effectively paid by the public without regard to electricity consumption. Abolishing the tax—and the subsidies therewith—leaves dividends still taxable, and a kWh tax can be substituted for the corporate income tax if desired. (S2008 and HR8897 in the 95th Congress, introduced by the late Senator Metcalf and by Rep. Stark respectively, proposed just that—a 0.4 m$/kWh tax that would raise as much revenue as the utility income tax.) Such a shift would make the tax law neutral as between debt and equity, encouraging a lower debt-to-equity ratio. See Batinovich,[39] Ca. PUC, "Arguments Regarding Proposed Exemption of Utilities from Tax," briefing document, 1978; testimony of public utility commissioners from Ca., Md., Me., Tn., and Wi. before the U.S. House Ways and Means Committee, 8 March 1978. It may seem like a windfall to abolish utility income tax, but in fact utilities pay little in *toto* and negative tax at the margin[40]—an anomaly that led the assistant comptroller of AT&T in early 1978 to *oppose* a reduction in the rate of corporate income tax because it would reduce the company's net tax benefits! Students of the controversial but highly successful Batinovich school of utility regulation should note that he does not consider a kWh tax necessary, would abolish the rate base, and would regulate return on equity (with minor necessary adjustments): personal communication, 8 November 1978. C. Davis provides an excellent analysis of utility tax and subsidy issues in *Harvard Environmental Law Review* 2 (1980): 311-58.

[42] In the first quarter of 1978, for example, utility commissions granted on average only half the rate relief sought (versus three quarters in 1974), the lowest level in fifteen years.

[43] A. Ford (LASL), "Is Growth Really Necessary?" *Electric Light & Power* (July 1976): 13-15.

[44] H. Averch and L. Johnson, "Behavior of the Firm Under Regulatory Constraint," *American Economic Review* 52 (December 1962): 1052-69; A. Kahn, *The Economics of Regulation* 2 (New York: Wiley, 1971), pp. 49-59. The effect is that guaranteed return on invested capital encourages overcapitalization at the expense of other factor inputs.

[45] The reform proposed in reference 41 would eliminate this worry by eliminating the deferred tax liability. Batinovich[39] comments further (at pp. 3-4): "Utilities have claimed that normalization [an accounting convention that permits "phantom taxes"—and thus on August 1978 Dean Witter Reynolds' data, raised the average 1973−77 market/book value ratio in the hundred largest U.S. utilities by 0.066] is preferable to flow-through [which passes through tax savings to consumers], because deferred taxes are kept by the utility in a deferred tax account 'if repayment becomes necessary' because of reduced growth. Flow-through, it is argued, is risky because funds for repayment are in the hands of consumers. This logic breaks down under closer examination, however. The deferred tax account is, of course, an accounting fiction; the funds are not liquid, but are invested in supposedly essential plant and equipment. As a result, utilities are now claiming that if growth were too slow and repayment became necessary, such repayment would have to come from the ratepayers (i.e., customers would 'pay' the same taxes twice). Utilities argue that continued growth is therefore necessary to avoid such an eventuality, "which would be politically damaging to regulators as well as utilities." It is also important to note that under present tax law, consumers never recover the full amount of investment tax credit,[41] and that with inflation, the value of deferred taxes grows indefinitely rather than being returned to the Treasury. At current growth rates, consumers' utility bills would have a *lower* present value if construction were financed "up front" rather than through tax subsidies.[41]

[46] E.G. Cazalet *et al.*, "Costs and Benefits of Over/Under Capacity in Electric Power System Planning," EA-927, EPRI, (October 1978) (Decision Focus Inc); "Planning for Uncertainty," *EPRI Journal* (May 1978): 6-11; J.G. Stover *et al.*, "Incorporating Uncertainty in Energy Supply Models," EA-703, EPRI (February 1978) (The Futures Group).

[47] For example, EPRI models assume that all forms of generating capacity are expanded at the same rate, so that baseload shortages automatically incur outage costs rather than extending the capacity or load factor of peaking or intermediate-load-factor plants. The use of planning reserve margin as the key dependent variable obscures the choice between plants of different lead times (LASL recommends changing the mix by underbuilding baseload plants of long lead time, then making up any shortfall with short-lead-time plants at lower cost). Low capital costs are assumed, so that even very large overcapacity does not greatly increase fixed costs. Outage costs are treated as homogeneous even though it would make more sense to market interruptible power to users with low outage costs. Uncertainties are assumed to be symmetrical with respect to over- or under-prediction. Of these deficiencies, the first is perhaps the most important, since it means that the plant-mix questions at issue[48-50] simply cannot be examined: plants are treated as homogeneous.

[48] R. Boyd (Division of Environmental Studies, University of California, Davis) and R. Thompson (Ca. GSA, Sacramento), "The Effect of Demand Uncertainty on the Relative Economics of Electrical Generation Technologies With Differing Lead Times," 1978, submitted to *Energy Systems and Policy* and discussed by E. Kahn, "Project Lead Times and Demand Uncertainty: Implications to Financial Risk of Electric Utilities," E.F. Hutton Fixed Income Research Conference on Public and Investor Owned Electric Utilities, New York, 8 March 1979.

[49] A. Ford (LASL) & I.W. Yabroff (SRII), "Defending Against Uncertainty in the Electric Utility Industry," LA-UR-78-3228 (Los Alamos, NM: Los Alamos Scientific Lab., December 1978).

[50] A. Ford (LASL), "Expanding Generating Capacity for an Uncertain Future: The Advantage of Small Power Plants," LA-UR-79-10, (December 1978), published as A. Ford and I.W. Yabroff, "Defending Against Uncertainty in the Electric Utility Industry," *Energy Systems and Policy* 4 (1-2) (1980): 57-98.

[51] See reference 4, particularly pp. 483-89 of the *Annual Review of Energy* survey; A. Ford, "A New Look at Small Power Plants," LASL-78-101 (January 1979); A. Ford & T. Flaim, eds., *An Economic and Environmental Analysis of Large and Small Electric Power Stations in the Rocky Mountain West* (Los Alamos, NM.: LASL, Energy Systems & Economic Analysis Group, Report LA-8033-MS, October 1979); and references 49-50.

[52] Northern Plains Resource Council (Billings, MT), "Basin Electric Power Cooperative and the Rural Electric Co-Ops," 13 June 1978.

[53] Hearings, U.S. Senate Committee on Appropriations, *Agriculture, Rural Development and Related Agencies, I—Justification, FY1979*, p. 886.

[54] FPC, *Factors Affecting the Electric Power Supply, 1980−85*, Executive Summary & Recommendations, December 1976, p. 43.

[55] C. Kenney & J. Worsham, *Boston Globe* (9 December 1978), p. 1.

56 D. Schoch, *The Power Line* 3 (9) (April 1978): 6 (Washington, D.C.: Environmental Action Foundation—a standard publication for utility reformers).

57 Some other examples are downright surreal. For example, Pat Smiley P.E. (NCAT, Butte, MT) reports that the Second Bacon Siphon, a new irrigation project in Washington State, requires about 500 MWe for pumping and to replace hydropower lost by the water diversion. Washington's share of the proposed Colstrip 3 and 4 plants in eastern Montana is also about 500 MWe. Washington farmers thus gain Siphon water and about break even on electricity. Montanans will lose coal, land, and water, gain smog, and effectively export water from them who hath not to them who hath already. In return they will not even get cheap BPA preference power. Another bizarre example of bad planning is the decommissioning charge being collected by nuclear utilities in at least seventeen U.S. states, which then allow the receipts to be spent on normal operating and construction costs rather than escrowed for later decommissioning.

58 A.B. Lovins, "Resilience in Energy Strategy," *New York Times* Section 4 (24 July 1977), p. 17; E. Kahn, "Reliability Planning in Distributed Electric Energy Systems," LBL-7877, October 1978, also will appear as "Reliability Planning for Solar Electric Power Generation," *Technological Forecasting and Social Change* (forthcoming).

59 This arrangement automatically passes through whatever tax subsidies the utility gets. Flow-through utilities (about eighteen states) should use an equivalent low interest rate such as several state commissions have recently set to ensure equity in loans.

60 As a conservatism, the cash-flow advantage is not reflected in Table 1. See E. Kahn, LBL-7876 (September 1978) and K. Bossong (Citizens' Energy Project, Washington, D.C.), "Banking on Solar," 1978.

61 Notably Pacific Power & Light and several neighbouring utilities, which offer zero-interest conservation loans with long grace periods and will insulate electric water heaters free—because conservation, the cheapest marginal source, saves them so much money. They can resell the saved electricity without having to build a costly new plant to generate it. The "Oregon Plan" of rate basing part or all of the loan, however, is considerably less efficient than the capital-transfer scheme suggested here.

62 *Editor's Note:* Gus Speth is former chairman of the U.S. Council on Environmental Quality.

Chapter Seven

7230
9212

Energy Policy and Consumer Energy Consumption

by
D. N. Dewees

I. INTRODUCTION

Whether or not there has been an energy crisis, it is clear that the 1970s have been a time of energy problems, for Canada, the United States, and the world. The international price of oil has risen fourfold in real terms. Both the experience and prospect of energy shortages and the recognition of high costs of new energy resources have aroused interest in energy conservation. It is often said that it is cheaper to save a barrel of oil by conservation than to produce a new barrel. When discussing energy conservation, economists argue that higher energy prices will automatically reduce consumption and therefore encourage conservation. Price elasticities have been estimated at −.5 for all energy in Canadian manufacturing (Fuss, 1977), −.47 for all energy in the commercial sector (Canada, 1977, Table 16), −.32 for all energy in the residential sector (Canada, 1977, p. 31), −.05 for gasoline in the short run, and −.25 in the long run (Dewees *et al.*, 1975). These elasticities all demonstrate that higher prices will yield less consumption, although the small magnitude of some of the elasticities suggests that large price increases would be necessary to achieve significant reductions in consumption.

Governments in both Canada and the United States, however, have chosen to insulate consumers from rising world energy prices. In Canada, the domestic price of oil during October 1980 was less than half the import and export prices. The price of natural gas is fixed by regulation in both countries and, particularly in the United States, it has been held well below market levels in interstate commerce.

While insulating consumers from rising energy prices, governments of both countries have instituted other policies designed to conserve energy, including subsidies, tax credits, consumer information, and regulations. Legislation in the United States specifies maximum fuel consumption for the

115

average automobile sold by any manufacturer for each year from 1978 to 1985,[1] and in Canada a co-operative agreement with the auto companies was proposed to pursue a similar set of standards.[2] Building codes in Canada have been modified to require more insulation.

Some have argued that energy problems could be solved, and vast governmental expenditures for regulatory personnel saved, by eliminating both domestic energy price controls and other energy regulations (Friedman, 1979*a*, 1979*b*). If consumers were faced with higher energy prices, it is argued, they could manage their own energy conservation without further assistance from governments. Not only would our energy consumption be reduced, but so also would the budgets of federal and provincial agencies that now encourage conservation by non-price means.

This paper will explore several major issues with respect to energy conservation by individual consumers. Three questions will be addressed. (1) If oil and gas prices rose to world levels, would there by any reason for other energy conservation policies? (2) If so, what policies make sense for influencing individual behaviour, and how can one choose among these policies? (3) Are non-price policies a substitute for or a complement to higher energy prices?

In the next section of this paper, we will develop some criteria for evaluating energy conservation policies, and discuss a rationale for such policies based on information problems. Section III will consider briefly the range of policy alternatives available for promoting energy conservation, and consider how they might be evaluated. Section IV presents three case studies to illustrate and test the principles developed in sections II and III. Section V will draw some conclusions from the principles and case studies.

II. CONSERVATION CRITERIA AND INFORMATION PROBLEMS

How Much Energy Conservation Is Desirable?

The economist's prescription for energy conservation is simple. One should undertake energy conservation projects, whether insulating the attic, choosing a smaller car, or reducing lighting levels, when the benefits or savings from energy conservation are greater than the costs. Since the savings will accrue over a period of time, it is necessary to take the present value of costs and benefits, discounting future items by an appropriate discount rate. An individual's decision, based upon his own welfare, will utilize his personal discount rate and the market prices that he must pay for energy and for any other investment. This decision must also include an evaluation of any changes in the services he consumes that result from the energy conservation project. From society's point of view, however, the values should include all costs and savings to the entire society. Furthermore, these costs must be measured at their social opportunity costs, which may not equal current market prices. For example, an individual contemplating home insulation as a means of saving fuel oil will value the savings at the market price that must be paid for that oil. In Canada, the fuel oil price is based on a domestic crude oil price that is well below the import or export price of a barrel of crude oil. The opportunity cost to Canada of a barrel of domestic oil consumption is not the domestic price but the international price, that is, the opportunity cost equals the revenues forgone from not exporting that barrel, or the price actually paid to import it or a substitute, as long as a reduction in consumption by an individual means an increase in exports or a reduction in imports, rather than a change in consumption by some other domestic consumer. Thus, the social welfare calculation should utilize world oil prices rather than domestic prices.

Determining the future savings from an energy conservation project requires some assumption about the future price of energy. Currently, many experts expect real energy prices to rise for the remainder of the century, so that proper evaluation of energy conservation should assume such rising prices.[3] This is, in part, an answer to critics who complain that a purely economic cost-benefit analysis of energy conservation ignores the long-run exhaustibility of the resource, or the prospect of running out in the foreseeable future. If the world is in fact likely to run out of oil in the foreseeable future, then we should expect its price to rise dramatically. Building this price into the cost-benefit analysis of an energy conservation project will lead to the socially correct result. In this way, the prospect of running out is not ignored but is implicitly considered in the analysis.

Under What Conditions Are Non-Price Conservation Policies Justified?

Consider a consumer about to purchase a durable good that will consume energy throughout its lifetime. The total cost of that good is the initial purchase price plus the cost of the energy consumed over its useful life. Suppose that the consumer is faced with a variety of models all of which offer identical performance characteristics, but differ in their initial price and in their energy consumption rates, such that more expensive models use less energy because of better design. Under what circumstances will the consumer choose the model that yields the optimal amount of energy conservation from society's point of view? If the utilization of the good varies among consumers, under what circumstances will the individual's utilization maximize social welfare as well as his own welfare?

From the preceding section, it should be clear that the socially optimal model is the one that minimizes the present value of the purchase price plus the stream of operating costs over the life of the product if the performance of all models is identical. Meeting this criterion requires a number of assumptions. First, the market price of energy must represent its social opportunity cost both now and in the future. If energy is underpriced, consumers will choose models that have a low initial cost and high energy use, and they will utilize them excessively. Second, the consumer's discount rate must equal the social discount rate. There is some evidence that individuals may behave as if they have discount rates well above the social optimum when purchasing consumer durables.[4] If the consumer's discount rate is above the social discount rate, he will consider future energy costs less seriously than would society, and will therefore choose a model with a lower purchase price and higher energy consumption than would be socially optimal.

A third requirement is that the consumer be perfectly informed about the characteristics of alternative models and the energy consumption that each choice implies. If the consumer is not perfectly informed, he may not choose the model that minimizes the social present value of the lifetime costs. If, for example, the consumer was completely uninformed about energy consumption characteristics of appliances, he would simply choose the model with the lowest purchase price, which might have the highest energy consumption because of inefficient design. He would also use the good more intensively than if he were aware of the energy consumption associated with its use.

The fourth requirement for private decisions to coincide with socially efficient decisions is that the welfare of an individual consumer must be unaffected by the product choices of other consumers. This assumption is most likely to be violated for products associated with social status, or products causing pollution or safety hazards to others. My purchase of a larger and more powerful and hence less economical automobile may cause some of my neighbours who were previously satisfied with their vehicles to consider trading theirs for larger and more powerful models in order not to fall in relative social status. My use of a polluting automobile makes my neighbours worse off because of the pollution. When this phenomenon can occur, private utility-maximizing decisions are no longer necessarily consistent with social welfare maximization (Henderson and Quandt, 1958, pp. 212-13). If the social optimum is to be achieved in such cases, it will generally be necessary to discourage individual's consumption of the socially undesirable characteristic by raising its price or restricting its quantity.

In cases where energy consumption is associated with social status, pollution, safety hazards for others, or other externalities, the assumption that consumer decisions are independent is likely to be invalid, and private decision making will no longer lead to a social optimum.

It should be noted that the requirements for efficient consumer decision making regarding energy consumption are not unique. They are analogous to those for product durability, product safety, or a host of other product characteristics where there is no impact on other individuals (i.e., no pollution or congestion effects). With proper prices, well-informed consumers, and private discount rates equal to the social discount rate, individuals should be able to choose products with durability or any other characteristic that is optimal from a social point of view.

Using hypothetical costs and prices, Table 1 shows how the failure of some elements in the perfect market can lead to improper choices. The table shows the purchase price for six hypothetical models of a 17 cubic foot refrigerator, identical except for energy use. In this example, the energy consumption decreases as the purchase price increases. Suppose that the social discount rate is 6 per cent, and that the opportunity cost of electricity will be $0.04 per kilowatt hour (kWh) over the fifteen-year life of the refrigerator. Reducing electricity use by 10 kWh per month produces a present valve saving in electricity of $46.60. In this case, model 6, which uses 100 kWh per month, minimizes the present value of the total cost of the refrigerator. Suppose, however, that the consumer uses a 10 per cent private discount rate rather than the 6 per cent social rate. In this case, his cost will be minimized by model 5, assuming an electricity price of $0.04. Alternatively, suppose that the consumer erroneously believes that the cost of electricity will be $0.03 per kWh, perhaps because the consumer price does not reflect all social costs. In this case, he will again invest less in efficiency, and choose model 5. Finally, if the consumer makes both mistakes, and uses a 10 per cent discount rate and a $0.03 per kWh electricity price, he will choose model 4. Of course, if the consumer were uninformed about energy consumption, he might simply purchase the model with the lowest initial price, model 1. In this case, the social value of his lifetime costs would be increased by $118.00 over the lifetime cost of the socially optimal model.

Table 1
LIFETIME APPLIANCE COST AT VARYING EFFICIENCY

Hypothetical Model	Price	Electricity Use	Present Value of Price Plus Electricity			
			@ 3¢/kWh		@ 4¢/kWh	
	($)	(kWh/MO.)	6% ($)	10% ($)	6% ($)	10% ($)
1	800	150	1324	1209	1499	1345
2	810	140	1299	1192	1462	1318
3	825	130	1279	1180	1431	1297
4	845	120	1264	1172	1404	1281
5	875	110	1259	1175	1388	1274
6	915	100	1264	1188	1381	1278

III. SOURCES AND SOLUTIONS FOR INFORMATION PROBLEMS

How Information Problems Arise

Most Canadians who have purchased refrigerators do not know the monthly energy consumption of the model they purchased, or of alternative models. On the other hand, most motorists have some idea of the rate of fuel consumption of their automobile. It is useful to look at information problems more specifically in order to see what distinguishes these cases and other intermediate cases.

Akerlof (1970) demonstrates that in a situation where perfectly informed buyers and sellers would buy and sell, the absence of information by purchasers about the products' quality might lead to a complete failure of the market—nothing would be bought and sold. Akerlof employs automobiles as his example, with the quality of used cars being the unknown variable. He demonstrates that in a used car market, where there is a wide variation in the quality of used cars, perfectly informed consumers would pay more for a high-quality car than a low-quality car. If, however, it is impossible for consumers to determine the quality of cars before purchase, then they will be willing to pay no more than the value of an average-quality car. Sellers, however, will not sell a high-quality used car for the price of a merely average-quality used car, so only cars of average quality or less would be offered for sale. Buyers will revise downward their willingness to pay, and by a kind of Gresham's law of products, the reduced willingness to pay of buyers will drive out the better used cars. Akerlof's example shows that when quality is not known to buyers, low-quality goods will flood the market and drive out high-quality goods by selling at a lower price. This may operate to the detriment of both buyers and sellers.

There are several important dimensions to consumers' knowledge or ignorance about the quality of the products they purchase. The most extreme case would be goods whose quality is unknown to the consumer before purchase, and cannot be determined after purchase. A second type of commodity is "experience goods," for which the consumer cannot determine the quality before purchase, but can determine quality while using the good after purchase; for example, the sharpness and durability of razor blades. If the consumer buys an "experience good" on a regular basis, his experience with successive purchases can lead him to the price-quality combination that best satisfies him. If the good is purchased infrequently, then the buyer's post-purchase learning may be of little use to him, although he can pass on the information he learns to his acquaintances.

Another category of goods is "search goods," for which the consumer can determine characteristics directly before purchase. Numerous characteristics of many goods fall into this category. For example, one can easily determine the riding comfort and passenger and luggage-carrying capacity of an automobile before purchase by sitting in it, inspecting the trunk, and taking a test drive. Similarly, one can determine the size of a refrigerator by careful examination of show-room models.

One might expect consumers to have little difficulty in making choices that best satisfy their needs for those characteristics that can be directly observed prior to purchase. Even with respect to these types of characteristics, however, consumers do not always make ideal choices given their own objectives. Clothing may languish in the closet because experience demonstrates that it does not in fact enhance the purchaser's good looks as much as it appeared to in the store. People sometimes complain about seating accommodations in automobiles, although they could easily have identified seating limitations without leaving the show-room floor. Thus, when we refer to a well-informed consumer, we do not necessarily mean one who always makes perfect choices.

Many goods that are important from an energy consumption point of view have characteristics that fall between the extreme cases described above. The common case is one where energy consumption is not directly measurable, but is partly correlated with characteristics that are measurable. For example, a homeowner cannot measure directly the energy consumption of his refrigerator. Yet most homeowners would probably assume, correctly, that larger refrigerators use more electricity than smaller ones, other things being equal. They would probably assume, also correctly, that an "energy saver" switch on a refrigerator will reduce energy consumption relative to identical models without such a switch. The thickness of refrigerator insulation can be guessed at by looking at the thickness of the refrigerator walls, although consumers may make little sense out of competing claims regarding the efficiency of different insulating materials.

The information that can be directly gathered by consumers from observation of the goods before purchase, or experience after purchase, may be supplemented by information provided by the manufacturer or retailer. Such advertising information, however, may be suspect because complete enforcement of legislation prohibiting false or misleading advertising is impossible. Consumers are thus appropriately wary of advertising claims that they cannot independently verify. While advertising sometimes points the way to a better product, it often points the way only to a more clever advertising campaign. One cannot rely on advertising by itself to solve information problems.

Solutions to Information Problems

There are several ways to deal with the problem that arises when consumers cannot adequately inform themselves about a product before purchase. The government can provide information, or require or subsidize its provision. Governments may require labels on products that provide specific information, or require that advertising material contain specified information, such as automobile fuel economy figures. Governments may provide information themselves, as when government agencies release test reports to the public describing the characteristics of alternative goods. Governments might subsidize the provision of information, perhaps paying for the publication of test results or subsidizing a testing programme itself. Finally, governments can regulate information by requiring a certain level of truthfulness in advertising.

A second approach is for non-governmental agencies to provide information. The Consumers' Association of Canada and Consumers Union in the United States both test a large number of consumer products and report these results to their members in regular publications. Such reports can provide a wealth of information to consumers who seek it. Still, such information is not perfect, as one can learn from occasional apologies or retractions by the testing services. In addition, the readership of the publications is a small fraction of the population. One should expect the provision of information by private sources to be inadequate from society's point of view because of the public good nature of information. The gains are inappropriable by the provider. People can read library copies of the magazine for free, or borrow it from a friend without paying.

A third approach is to abandon information provision entirely and set product quality standards. In the field of product safety, both Canada and the United States have a large number of standards imposed by federal government agencies. Private testing agencies such as the Canadian Standards Association and Underwriters Laboratories have procedures for setting product quality standards in co-operation with the manufacturers themselves. Such standards have traditionally dealt with safety, but they deal increasingly with energy conservation.

A fourth approach is to create price incentives favouring ''better'' products through taxes or subsidies. Excise taxes on auto air conditioning, and cars exceeding certain weights, are examples of policies that change relative prices to achieve public goals.

The choice among the above types of policies should depend, in principle, on which one achieves the greatest excess of savings over cost, once again evaluated on a present value basis. Applying this principle, however, is extremely difficult, since even after the fact it is not easy to measure the increased manufacturing cost from a particular programme, to say nothing of measuring the cost of administering and monitoring the programme. One might support quality standards where serious safety hazards are present on the grounds that society is not prepared to allow individuals to choose the risk of injury or death from certain types of product design. This principle quickly breaks down, however, when it becomes apparent that most human activities involve some risk of injury or death, however remote, and there is therefore no clear dividing line between potentially hazardous products and very safe products.

Another basis for choice would be to use quality standards when the process of disseminating intelligible information to consumers would be difficult or costly, when the technology for meeting the standards is well known, and when there is likely to be little debate about the harmful effects on the industry from imposing standards. Alternatively, one would provide information where this information would not be excessively costly to supply, could be readily understood by consumers, and where the technology for improving product quality is uncertain or likely to advance rapidly. The advantage of information programmes where technological progress is anticipated is that an effective information programme can create a demand for better quality and yield a constant pressure to improve. A quality standard, however, once it is fixed, provides no incentive for further improvement. In addition, where the technology is uncertain, it may be difficult to design a standard that forces manufacturers to improve from present technology. It is often easier to use market demand based on consumer information to induce better quality than to rely on direct regulation.

Even an effective information programme will not reach all consumers. There will always be some consumers whose limited analytical ability or lack of interest will leave them uninformed when making product choices. If information rather than quality standards is used, these consumers will be prey to manufacturers offering cheap products of low efficiency and high lifetime costs. This raises the question of the extent to which governments can or should protect such consumers from themselves by setting minimum standards or taxing low-quality products to make them unattractive. If one views information programmes as an effective mechanism for moving most of the industry to improve most of their products, then the purpose of a minimum quality standard is to protect ignorant consumers, not to force better technology, and the standard should be evaluated on that basis. Such a standard need not be overly strict and might be worked out in a co-operative programme with the industry and one or more testing agencies. In contrast, if one believes information programmes will be largely ineffective, so that quality standards are the only means of raising quality, then such standards will have to force better technology and will usually be resisted vigorously by the industry. The main purpose of the standards then is not to protect the ignorant few, but to achieve major quality improvements.

Finally, one should consider the possibility that, in some situations, the best solution to an information problem is to do nothing. It is easy to imagine cases where the cost of a programme to provide better consumer information or to establish and enforce minimum product quality standards would be greater than any conceivable benefits from the programme. It is not the case that a safer product is always a better product: the safest lawn-mower would have no blades, yet a government that required this design in lawn-mowers would undoubtedly be subjected to furious attack by indignant owners of unmowable lawns. Thus, one should soberly consider, after identifying market imperfections, whether proposed remedial programmes will in fact yield benefits larger than all of their costs.

IV. THREE CASE STUDIES

Refrigerators

Refrigerators are typical of major electrical home appliances, in that their lifetime energy cost is significant compared to the initial purchase price, there is substantial variation in available efficiencies, and consumers have in the past been ill-informed about those efficiency levels. Table 2 shows variations in energy consumption among Canadian and American refrigerator models for the year 1976. Even within the 15−17 cubic foot size range, the highest energy consumption of a Canadian model is 50 per cent more than the lowest, while the highest in the United States is more than three times the lowest. Hoskins and Hirst (1977) report on a variety of technological modifications that reduce refrigerator energy consumption by up to 50 per cent, most of them requiring no sacrifice in refrigerator performance. Most, if not all, of these modifications would be enormously attractive to an informed consumer—they would pay for themselves within a very few years in operating cost savings. Hirshhorn (1979) also concludes that reducing energy consumption of a refrigerator by up to 55 per cent would reduce the

Table 2
REFRIGERATOR ENERGY CONSUMPTION—1976 MODELS

Models	Electricity Consumption (kWh per month)	
	Minimum	Maximum
All models and sizes		
Canadian (limited data)	102	189
U.S.	50	245
15−17 cubic foot		
Canadian (limited data)	114	163
U.S.	50	175

Source: Canadian data are for those models for which information was available from the manufacturers, representing less than 50 per cent of all models available in Canada.
U.S. data are from Association of Home Appliance Manufacturers, *1976 Directory of Certified Refrigerators and Freezers,* 2nd ed. (Chicago, AHAM, June, 1976).

present value of the lifetime cost by saving more in electricity costs than the increased manufacturing cost (after discounting at a 10 per cent rate). These savings come almost exclusively from the use of more sophisticated and efficient components, rather than from changing visible product characteristics such as size. In Canada, evaluation of the benefits of more efficient refrigerators must recognize that during the heating season, the energy "wasted" by an inefficient model provides heat to the kitchen, so the benefits from greater efficiency are less than if extra heat was unwanted.

The above data suggest that, in the past, consumers in Canada and the United States have not chosen refrigerator models that would maximize social or private welfare. They have not invested in the level of energy conservation that would be best from the point of view of society as a whole, or even from their own private point of view. By looking at information flows in the refrigerator market-place, we can see that this result is not only plausible, it is almost inevitable.

In the past, Canadian consumers have had no information available on the absolute or relative efficiency of refrigerators offered for sale. Advertisements and brochures have offered no data that would assist a consumer in comparing the efficiency of alternative models. Even if a manufacturer had advertised the energy efficiency of a single model, the consumer would have no way of verifying that claim, and would have to rely on government or consumer policing of laws governing truth in advertising. The technical features that affect efficiency, such as the type of insulation, the design of the motor and compressor, and the efficiency of the condenser, are not subject to inspection and evaluation by the consumer. The energy use depends in part on the frequency of door openings and the load in the refrigerator. Once a refrigerator is installed, there is no way to determine its energy consumption. There is no separate electrical meter for that appliance, and a refrigerator uses a sufficiently small portion of total household electricity that one could not possibly determine whether a new model used more or less than an old model simply by comparing electrical bills. Thus, the energy consumption of refrigerators is a characteristic that cannot be determined by inspection beforehand, and cannot be verified in any way after purchase.

In this situation, there is little incentive for a manufacturer to produce a higher efficiency model. If it costs more, the manufacturer will be unable to recoup that cost from the consumer, because the consumer cannot himself verify that its efficiency is actually higher than that of inferior models. A comparison of the relative prices of efficient and inefficient Canadian refrigerators sold in 1976 showed no correlation between prices and energy consumption (Dewees, 1977*b*). It might pay to advertise efficiency, but it would not pay to produce it, except to remain within the broad confines of legislation governing truthful advertising. Rising electricity prices by themselves might lead consumers to purchase smaller refrigerators or to foresake visible energy-consuming features such as automatic ice makers, but could not lead to significant demand for invisible conservation features. This is a classic situation in which either consumer information or product quality standards might help to improve the efficiency of the product and achieve significant energy conservation.

Refrigerators were the first appliance to be subject to the requirement of the federal Energuide programme. This programme, conducted co-operatively by the federal Department of Consumer and Corporate Affairs and the Canadian Standards Association, provides for a common test to determine refrigerator energy consumption, and requires all refrigerators to bear a label stating clearly their energy consumption in kilowatt hours per month (CSA, 1978, pp. 1-3). These labels began appearing in refrigerators on sales-room floors about the beginning of 1979.

An evaluation of the Energuide programme would have to consider both the costs of the programme and the value of the energy savings from the programme. Hirshhorn (1979) evaluated these costs and benefits before the fact, and concluded that the programme would yield net benefits worth $100−$200 million in 1978 dollars.[5] This figure is only an estimate, and it would be useful to do follow-up studies to determine actual costs and benefits; but the potential savings here are so much larger than the costs that it seems most likely that this programme will turn out to be highly beneficial. It is still too early, however, to evaluate the actual effect of the programme.

Should further steps be taken, such as imposing minimum energy consumption standards for refrigerators? The answer to this question depends in large part on the response to the Energuide programme. If there were little or no response to Energuide, so that after a few years the design and market shares of refrigerators were unchanged, one would have to think seriously about looking for stronger programmes to achieve the vast benefits that seem to be available. If there appeared to be a reasonable response to the existing programme, however, there would seem to be little need for government-imposed minimum quality standards unless one wanted to protect consumers who failed to respond to the information programme. The imposition of minimum standards would create a climate of confrontation that might actually damage the operation of the information programme itself. Furthermore, tough minimum standards would likely affect firms rather differently, and would therefore be fought vigorously by those most seriously affected. More promising would be to encourage the industry to agree to a set of minimum standards that would be self-enforced. So long as this was not the major force for efficiency improvement, it might be acceptable as a backstop.

Setting minimum standards for refrigerators is not simple because energy consumption naturally varies with size and a number of other important features. There is no single energy use or efficiency level that diverse models could reasonably be expected to achieve. Setting minimum standards would require a complex table applying specifically to a wide range of sizes and features, and its complexity would necessitate a slow process of promulgation and adoption. In the United States, minimum energy consumption standards have been set for refrigerators and a number of other home appliances,[6] and have caused a serious battle with the manufacturers. Canada's labelling programme is ahead of that of the United States in part because it did not attempt the difficult and perhaps unproductive task of setting minimum standards.

If an information programme for refrigerators proved to be ineffective, the sensible next step would be to impose an energy conservation excise tax inversely related to energy use. This could be computed on the basis of the energy-use data generated for the information programme, and would increase the relative attractiveness of efficient models to both consumers and producers. Tax schedules could be devised that would generate the same revenue as the current 11 per cent federal excise tax, yet provide a powerful incentive to select more efficient models. Such a tax could work even if consumers had no understanding of the energy consumption of the models.

An information programme for refrigerators, or for other major home appliances, works best when energy is properly priced. If regulations or subsidies artificially depress the price of electricity to protect consumers from high prices, this would weaken the effectiveness of an information programme. If electricity is less expensive, Table 1 demonstrates that even a well-informed consumer will buy a less efficient refrigerator than if energy is more expensive. Therefore, subsidizing electricity consumers would be counter-productive to an appliance energy consumer programme that was based primarily upon improved consumer information.

Domestic Furnaces

The opportunities for reducing energy consumption by modifying the design of domestic furnaces represent a smaller percentage saving than for domestic refrigerators, but a much larger total energy saving. Field experiments have adjusted and modified oil and gas furnaces already in use, with savings in fuel use ranging from 10 to about 25 per cent (Dewees, 1977*a*). The average saving that could be achieved on the existing population of furnaces is undoubtedly smaller than this, since the field studies tend to focus on units where poor installation, design, or maintenance makes large savings possible. Some models of new oil furnaces are said to use 20 per cent less fuel than typical previous units, and gas furnace manufacturers are claiming savings of 15 to 20 per cent for high efficiency models today, with 30 per cent savings potentially achievable if a "condensing" furnace becomes feasible. Studies of a range of improvements to oil and gas furnace design suggest that a number of efficiency-improving modifications would be economically attractive to homeowners at any reasonable discount rate— even assuming constant real prices of fuel at 1979 levels, which is an unrealistically optimistic asssumption (Dewees, 1979, Table 5).

To see the impact of improved furnace efficiency, one can compare scenarios of future furnace sales assuming more or less rapid introduction of higher efficiency models. Assume, arbitrarily, for the base scenario, that in the absence of government programmes, several higher efficiency models capture 2 or 3 per cent of the new furnace market in 1980 and add 2 or 3 per cent to their shares every year until these shares reach a maximum of 30 or 35 per cent of the total market. In contrast, assume an accelerated scenario in which a government programme increases the market shares and annual increments to 2 to 5 per cent for improved models, and that improved models grow to entirely displace older models from sales after eight to twelve years. The improved models sold during the twelve years starting in 1980 would, over their useful lifetimes, save 416×10^{12} btu's. This is worth over $1.1 billion, at 1979 prices, compared to the base scenario. Even after allowing for the higher prices of the improved furnaces, the net savings would total over $1 billion, with a present value in 1979 of about $400 million. About 89 per cent of the fuel savings are in natural gas, representing 370×10^9 cubic feet of gas. Thus, furnace efficiency improvements could save large amounts of natural gas for Canadian homeowners.

The above analysis suggests that consumers are not well informed about the energy consumption and efficiency of furnaces they purchase relative to other models, since they have failed to embrace economically attractive, efficiency-improving designs. This might seem surprising, since most homeowners can inform themselves about their annual fuel use, can compare their heating bills with their neighbours, can request the furnace serviceman to conduct a simple efficiency test, and can compare this information with that of their neighbours or with published data.

Part of the problem of low-efficiency furnace purchases arose in the past when oil and gas prices were sufficiently low that homeowners had little interest in learning about efficiency because the dollar savings were small. Another explanation for the continuation of this problem is that the standard efficiency test currently in use (i.e., the steady-state heat loss analysis) is only a partial measure of furnace efficiency. This test does not record heat losses through the chimney when the furnace is off, or when the furnace is first turned on and may be operating inefficiently. There is, at present, no accepted procedure for conducting a seasonal efficiency test in the home that would determine the performance of the furnace throughout the entire heating season under actual operating conditions.

It is also difficult to determine furnace efficiency from fuel bills alone. Even if one uses oil heat, the fuel bill depends upon a combination of furnace efficiency, home insulation, and the severity of the winter. There is little value in comparing one's fuel bill with that of one's neighbours unless the houses and life-styles are essentially identical. Even if a homeowner replaces an old furnace with a new one, comparing fuel consumption before and after will not provide an accurate guide to the efficiency improvement (if any) until several years have passed, because differences in winter severity can significantly influence fuel consumption. With gas furnaces, if other appliances such as a hot water heater or stove are also gas fired, the heating portion of the gas bill is virtually impossible to isolate. Thus, most homeowners are largely ignorant about the absolute and relative efficiency of their current furnace, and are in no position to evaluate the relative efficiency of a replacement model.

In the past, furnace manufacturers have tended to advertise their products as being safe, clean, and efficient, without providing specific data on relative efficiencies. Some furnace brochures provide input and output heat rates that could be used to determine the steady-state efficiency, although most homeowners could not identify or use these data, and the steady-state efficiency is only part of the seasonal efficiency. Some manufacturers have recently advertised high-efficiency models and as-sociated savings, but these claims will be difficult for homeowners to assess because the information problems discussed above make it very difficult to evaluate such savings after the fact. These advertising claims frequently compare the new model with an old model already in use, and therefore provide the homeowner with little basis for choosing among alternative new models. Finally, a large proportion of all furnaces are purchased by building contractors who have no reason to care about the efficiency of the furnaces they install in homes they will sell. Since the ultimate buyer of the home will be unable to assess the relative efficiency of the furnace, he will be unwilling to pay a higher price for the home because of an allegedly efficient furnace installed in it. The builder, therefore, has every incentive to install the cheapest unit that will heat the home satisfactorily. Thus, an analysis of information availability confirms what the review of economic and technological data suggest—there has been little incentive for manufacturers to improve the efficiency of their units.

Current efficiency rates vary no more than they do, in part, because the testing agencies, including the Canadian Standards Association and the Canadian Gas Association, which must evaluate and improve furnaces for safety purposes, also set minimum efficiency standards. The minimum steady-state efficiency for oil furnaces has recently been raised to 80 per cent from 75 per cent. Gas furnaces are uniformly rated at 80 per cent efficiency, and while the actual test efficiencies are not available to the public, it is generally assumed that these efficiencies fall in the range of 75 to 80 per cent (Dewees, 1979).

Because a substantial portion of the savings that might be achieved, particularly with gas furnaces, comes from improving the seasonal efficiency rather than the steady-state efficiency, it seems unlikely that improvements can take place until a seasonal efficiency test is devised and agreed upon by the industry. Such a test has been proposed and is in the process of being adopted in the United States today.[7] It has been estimated that the incremental cost of the seasonal rather than the steady-state test might raise the cost of new furnaces by a few dollars.

Once a test has been adopted, one might adopt an information programme, a minimum-efficiency programme, or both. Both have been pursued in the United States, with fierce battles ensuing over the design of the test and the requirements of the standards. Once again, it would seem prudent to begin with an information programme, and assess whether in fact that programme can lead to satisfactory performance. An information programme for furnaces should include a large permanent label on the furnace itself stating the seasonal efficiency in characters clear enough so that the buyer of the furnace or the purchaser of the home could easily observe the rating. This would allow contractors to charge a higher price for homes with high-efficiency heating systems, since they could demonstrate, by pointing out the rating, that the furnace was in fact a high-efficiency model. It would also allow homeowners to confirm that the unit installed was the unit that the contractor had promised. It might also be desirable to require that brochures and catalogues state clearly the rated seasonal efficiency of all models. As in the case of refrigerators, fuel prices must not be artificially depressed or subsidized if an information programme is to have the proper effect.

Another type of information programme might assist homeowners in purchasing efficient furnaces and evaluating other home-heating conservation projects such as insulation. The government could require that heating oil and gas distributors make available to consumers a summary of their fuel consumption during the current year and the previous five years, measured in dollars (to attract attention) and in physical units such as Btu's. The same report should show the winter severity in degree days. The quotient of these two types of data would be a measure of the efficiency of the overall heating and insulation system. While winter severity may vary substantially from one year to the next, the ratio of fuel use to severity should be reasonably constant. This energy report would allow the homeowner to evaluate a new furnace after only one or two years, rather than waiting many years for the winter severity to average out. It would also allow him to determine the savings from attic insulation, improved storm windows, or any other changes he might make in the home. Since most oil and gas distributors have computerized their billings, it would not be expensive to produce the proposed report. Distribution of the report could be limited to only those customers who request it, to avoid confusing consumers with information that is unwanted or difficult to understand. The energy report would provide the consumer with another means for assessing new furnaces, in addition to the furnace efficiency rating, and would therefore provide a means for double checking the rating system itself.

Should minimum efficiency standards for furnaces be used as a major policy for raising efficiency? Here again, there would be difficulties in using a minimum standard to force the industry rapidly to modify its products. The number of oil and gas furnace manufacturers in Canada has been declining steadily over the last decade, and now there are a total of fifteen to twenty manufacturers. The firms are not now oriented toward research and development, or to a high rate of product innovation. Forcing rapid improvements in efficiency would undoubtedly provoke protests that some firms would have to abandon the business and that jobs would be lost. If any substantial investment in either design or equipment was required by the standards, that protest would be justified for some marginal firms. There would be serious problems in making a "technology-forcing" minimum standard system work. It seems much more attractive to let an information programme generate a demand for high-efficiency models that can be sold at a premium price, and allow the manufacturers to respond to this demand by moving after the higher profits than to wield the stick of minimum standards with fines against manufacturers of substandard products.

If the information programme is insufficient, a sales tax on furnaces could be tied to efficiency ratings, imposing no tax on highly efficient models and high taxes on inefficient ones. There is one argument for encouraging the testing agencies to be vigorous in setting minimum standards for seasonal efficiency—it would keep the worst performers within a reasonable distance of the industry average. It is unlikely that a formal system of minimum efficiency regulation could achieve more than this, however.

In the case of home furnaces, as with refrigerators, the Canadian market is reasonably independent of the United States. Only a small fraction of Canadian furnaces are currently imported. Thus, Canadian policy can be reasonably independent of U.S. policy, since American technology will not necessarily appear at once in the Canadian market-place. In fact, one might hope, in time, that a progressive Canadian furnace industry could create an export market to the United States, since Canadian firms could capitalize on the image of knowing what the home heating business is about given the inherently harsh nature of our climate.

Automobiles

If one compares the rated fuel economy of all automobiles available on the Canadian market, one finds a range from approximately 25 kilometres per gallon to over 75 kilometres per gallon (Canada, 1979, pp. 31-41). The Diesel Rabbit, however, is hardly comparable to a Lincoln Continental, so a more sensible comparison is among vehicles of the same size. Here, one can easily find variations of 25 per cent in the fuel consumption of vehicles that have approximately the same passenger and luggage-carrying capacity. Furthermore, as Detroit proceeds with its "downsizing" programme, new models are appearing that save 25 per cent of the fuel consumption of their predecessors. For a typical new car buyer travelling 24,000 kilometres per year and paying a dollar per gallon for gasoline, an improvement from 24 to 32 kilometres per imperial gallon would lower his annual fuel cost from $1,000 to $750 per year. The present value of this fuel saving over a 150,000 kilometre vehicle life at a 6 per cent discount rate is over $1,000—not an insignificant sum compared to the purchase price of the vehicle itself.

When motorists purchase and use an automobile, are they well informed about its energy consumption and cost? Once an automobile is purchased, it is possible to keep careful records of fuel consumption and mileage, and to determine quite accurately the cost of using the vehicle. While not all motorists are capable of or interested in making this comparison, those who are interested can be well informed after purchase about the fuel consumption of their vehicle. However, it is difficult to compare this type of information with other vehicles one might have purchased or might purchase in the future. One can compare fuel consumption with that of one's friends and neighbours; but only if their driving styles are similar do these comparisons carry much value. One cannot easily compare recorded fuel consumption with the published ratings for new cars, since such ratings are known to be better than most motorists will achieve under any conditions. Even an individual who purchased a new car every year would require many years to accumulate a substantial body of reliable information about the relative fuel economy of alternative makes and models. At a time such as today, when vehicle characteristics are changing rapidly, fuel consumption for a given model may be quite different from one year to the next. Consequently, an individual's ability to keep accurate records is almost worthless for helping him choose a vehicle.

The potential purchaser has access to several other sources of information. Most people understand that fuel consumption increases with vehicle weight and engine size. They also understand that air conditioning and automatic transmissions may increase fuel consumption. Thus, many individuals would be able to make a crude ranking of the fuel consumption of a number of makes and models of vehicles after observing a few obvious characteristics. This would leave a large margin of error attributable to subtleties of design such as aerodynamics, efficiency of carburation, and sophistication of the transmission.

The interested consumer, however, has still more information available. Consumers Union in the United States tests a large number of vehicles every year, and records their fuel consumption on a standard trip. All these data are presented in an annual buying guide, which allows accurate fuel consumption comparisons. While Canadian data differ from the American data somewhat, the American data are a good guide to the relative fuel economy of most makes and models. In the past, competitions such as the Mobilgas Economy Run provided a standardized opportunity for vehicles to compete, and the results of those runs were widely used in advertising and articles in automobile-related magazines. Finally, half a dozen popular automobile or science magazines test drive numerous cars every year and report on their fuel consumption as well as their acceleration and other characteristics. Thus, the serious consumer who is interested in fuel consumption information can find a variety of public sources.

During the last decade, the U.S. Environmental Protection Agency (EPA) began producing fuel consumption ratings as a by-product of its pollution emissions testing procedure. As interest in energy consumption has heightened, manufacturers have been required to publish their EPA fuel economy ratings in advertisements in the United States, and corresponding numbers developed for the federal Department of Transport in Canada are required to be made available to Canadian consumers. Thus, in both countries, there is a set of standardized numbers that permits detailed comparisons among all makes and models of vehicles. A booklet is prepared annually by the Canadian federal government listing all vehicles and their mileage ratings. While individuals will find it difficult to compare these numbers to their present vehicle, they do provide accurate comparative evaluations of alternative new models.

The existence of the fuel economy ratings raises the immediate question whether this consumer information programme has benefits that exceed the costs. The answer to this question is less obvious than for refrigerators and freezers because of the other information sources that are available to automobile buyers and the consumer's ability to gather some information after purchase. It has been argued that mileage ratings are useless because people discover that their own mileage never approaches the rated figures, and then disregard the numbers, failing to use them even on a relative scale. On the other hand, the ratings do provide a single simple system for making such comparisons. To determine whether the information programme by itself is justified would require a careful study of the costs to the federal government and to the manufacturers of producing the ratings, and of the incremental effects of those ratings on consumer purchases. It seems likely that the costs are small on a per vehicle basis, and the fuel savings substantial, but one could not be certain without empirical study.

Public policy, however, has moved beyond information to standards in the United States, where legislation requires that each automobile company determine the sales-weighted, average fuel economy (corporate average fuel economy, or CAFE) of their vehicles, and that this fuel economy not fall below a minimum standard, which is increasing over time. The standard was 20 miles per U.S. gallon for the 1980 model year, and will rise to 27.5 miles per U.S. gallon by 1985. There are penalties for failing to meet the standard, imposed at the rate of $5 for every vehicle sold per 0.1 miles per gallon by which a manufacturer's fleet average exceeds the standard.[8] In Canada, a co-operative programme between the federal government and the auto manufacturers is based on U.S. standards, with the actual numbers adjusted for the larger size of the imperial gallon.

Given the substantial information available to consumers, is there any argument in favour of CAFE or a Canadian equivalent? One justification would be that CAFE is designed specifically to override consumer preferences, and to force economical cars upon consumers whether they want them or not. Economists tend to scoff at such paternalistic legislation on the grounds that the government rarely knows better than the public what is good for the public. In the case of automobiles, however, it is plausible that people may tend to buy larger and less economical cars for status reasons, so there is a theoretical argument for public policy that raises the price or limits the availability of less economical cars. This goal could be reached just as well by raising the price of gasoline, which would also tend to reduce the congestion and air pollution effect of motoring. While it would be impossible to prove whether CAFE moves us to the social optimum, it is also impossible, given interdependent utilities, to prove on purely theoretical grounds that these standards are an economically unjustified intervention in consumer choice. Similarly, if consumer discount rates are well above the social rate, one can justify a policy to raise the fuel efficiency of vehicles purchased.

A second justification for minimum fuel economy standards would be that the domestic price of oil is held artificially low to protect consumers, and that CAFE is designed to push consumers to purchase the same vehicle they would purchase if the price of oil, and hence gasoline, were allowed to rise to market levels. It would be astonishing, of course, if either the American or Canadian federal government could determine accurately what mix of cars consumers would purchase if domestic oil were sold at world prices, although the fuel economy standards undoubtedly start off in the right direction. Worse, however, CAFE would fail to reproduce some of the important effects produced by higher gasoline prices. These higher prices will lead consumers to purchase more economical vehicles and also to use them less. There is nothing in CAFE to encourage reduced driving. In fact, if it leads manufacturers to produce more efficient vehicles, people may actually drive more, since their cost per mile will decline.[9] Furthermore, there are close substitutes for automobiles, such as vans and light trucks, which are subject to less strict fuel economy standards. Sales of these vehicles have been booming for the last few years in both Canada and the United States, although they have dropped off in 1979 in the United States as a result of rising gas prices and declining gas availability. Corporate average fuel economy does nothing to encourage economy in the use of these vehicles and may, in fact, drive consumers to use them more if they are dissatisfied with the smaller size of automobiles. Since vans and light duty trucks comprise a quarter of the North American motor vehicle market, their availability poses a serious threat to the overall effectiveness of CAFE. It is possible that CAFE standards could be met in every year, approximately cutting in half the fuel consumption per vehicle mile of a typical automobile, yet total gasoline consumption might fall very little because of increased motoring and substitution of these large gas-hungry vehicles for increasingly thrifty automobiles.

Finally, there is an argument that the purpose of CAFE is to guarantee manufacturers a market for more economical vehicles so that they will be prepared to risk investing billions of dollars in new designs and production technology. There may be some validity to this argument given the high costs that manufacturers are experiencing in changing their designs, and the conservatism of the North American automobile industry. Even if the argument is valid, however, it applies to the United States and not to Canada. The menu of vehicles available in Canada is determined primarily in the United States and is based primarily upon demand in the United States. Whether or not Canada imposes similar requirements is unlikely to have any significant effect on the design decisions of U.S. manufacturers. Any justification in Canada for CAFE cannot rest on forcing the development of more economical technology.

It is difficult to determine how much of the current change in automobile design and consumer purchasing in the United States is attributable to CAFE and how much to changes in gasoline price and availability. General Motors began its planning for increased fuel economy in 1974 before the CAFE legislation was passed (*Motor Trend*, 1979, p. 43). The manufacturers had expected in 1979 and 1980 just to meet their CAFE requirements, but the gasoline shortage and price increases in the United States have shifted consumers from large to small cars, so that all manufacturers will probably be comfortably above the standard. In Canada, sales patterns through 1979 did not shift as in the United States because we suffered no gasoline shortages, had no significant gasoline price increases in 1979, and no effective CAFE. With these facts, we cannot separate the influence of CAFE from the influence of fuel price and availability. If a smaller company, such as Chrysler, had difficulty meeting the fuel economy standards and argued that these standards could cost jobs, one wonders if the standards might not be relaxed. We are faced with the spectre of tough legislation being enacted, which may then be postponed or weakened if in fact it is difficult to meet. If this is the likely fate of such legislation, one must ask whether it is worth the bother even in the United States. In Canada, there is no justification for such a programme, except as a statement of objectives by a government or as an attempt to take credit for what it is fervently hoped will happen anyway.

If income distribution concerns are a political barrier to raising domestic energy prices to world levels at once, it is still possible to achieve the energy conservation benefits of such a price increase. The federal excise tax on motor fuels, including gasoline and diesel fuel, could be raised to a level that gave the same effect as raising domestic oil prices to world levels. In October 1980, the domestic oil price was about $20.00 per barrel Canadian below the price of imported oil, and if such a price increase were passed directly to consumers, it would raise gasoline prices by about $0.60 per gallon. If the gap between domestic and world prices decreases in the future, this "conservation tax" could be decreased.

It should be noted that, in addition to energy conservation, there are independent justifications for added taxes on highway fuel. In the past, Canadian motorists have not paid all highway costs, nor have they paid for social costs such as pollution, noise, and congestion (Dewees *et al.*, 1979). In the 1970s, gasoline excise tax rates did not rise as rapidly as inflation, so the deficit of highway revenue is even greater than when previous empirical studies were performed. Finally, the current decrease in the fuel consumption rate of new cars will necessarily reduce road taxes per vehicle mile travelled. Since many highway construction and maintenance costs are not sensitive to automobile weight and fuel economy, both federal and provincial taxes must rise in cents per gallon to maintain constant taxes per vehicle mile travelled.

Suppose that good consumer information and higher fuel prices (based on world oil prices) are regarded as providing an insufficient inducement to automotive fuel conservation because of high private discount rates and interdependent utilities. A complete solution would be to tax motor fuel even more heavily, thus raising the lifetime operating cost of uneconomical vehicles still further. Any degree of discouragement for uneconomical vehicles could be achieved by some tax rate. If higher fuel prices were thought to be too discouraging to the use of all vehicles, one could substitute an excise tax on new vehicle sales that was directly proportional to fuel consumption. This tax should not be based on weight, which is only one determinant of fuel consumption, but on fuel consumption itself. It should not increase in big increments at discrete fuel consumption levels, but should be a continuous function of fuel consumption. To simulate the effects of a $0.10 per gallon gasoline tax on a 34 kilometre per gallon vehicle (given a 6 per cent discount rate) would require a new car tax of $325—not a trivial amount. Nevertheless, some combination of fuel tax and new car economy tax could achieve any desired encouragement to purchase economical vehicles and to drive economically. New car fuel economy standards in Canada appear to achieve nothing that cannot be achieved as well by fuel price increases, and perhaps fuel and/or vehicle taxes.

V. CONCLUSIONS

This paper has argued that where consumers may be ignorant of the energy-consuming characteristics of products they purchase, economists cannot rule out, on theoretical grounds, the possibility that public policies to encourage energy conservation may be welfare improving. It is theoretically possible that even when energy prices are set at their socially correct levels, consumer decision making may be improved by information programmes, financial incentives, and/or regulations. Certainly the converse cannot be demonstrated theoretically, but would have to be argued on the facts of each case. Demonstrations that higher prices yield lower energy consumption do not prove that other policies are unnecessary, only that there is some role to be played by prices themselves.

An examination of three cases suggests that in two of these cases, domestic refrigerators and domestic furnaces, the benefits of a reasonable information programme seem quite likely to exceed the costs of such a programme, perhaps by a wide margin. The exact magnitude of both costs and benefits is difficult to determine beforehand because we have little empirical basis for predicting consumer responses to information programmes. However, when the possible benefits appear to be worth many times the costs, it would seem a prudent gamble to try such a programme. Five years from now, sufficient empirical data should be generated by these programmes to allow an assessment of their costs and benefits. Since gas furnaces dominate the new furnace market and since the potential efficiency gains from gas furnaces are greater than for oil, any furnace-related programme should be oriented primarily toward gas furnaces.

In the case of automobiles, the argument for a government information programme is less compelling, in part because private markets already provide a considerable amount of information to interested consumers. Even when consumers cannot gather information directly from their own experience, it is possible for the market to supply it if there is sufficient interest. While the low cost of the present government-directed automobile fuel economy information makes it likely that benefits exceed costs, one would have to conduct a careful cost-benefit study to determine the merits of this programme.

Even if a successful information programme can achieve significant reductions in energy consumption, some consumers will undoubtedly be unaffected. One could consider, for each product type, whether a minimum standard that was not technology forcing would provide sufficient protection to such consumers to warrant the cost of implementing the standard. Such a standard would be unworkable in the case of automobiles, because of the great diversity of sizes and types, but might be feasible for home furnaces. In fact, the furnace industry is currently subject to minimum, steady-state efficiency standards imposed by the testing agencies. This kind of co-operative arrangement between manufacturers and testing agencies seems quite appropriate in the cases where a minimum standard is felt desirable.

In some cases, an information programme alone may have effects that fall far short of achieving an optimal trade-off between capital cost and energy consumption over the lifetime of the product. Such cases could be handled satisfactorily with an energy-related excise tax on the product. The tax would be imposed at the time of sale, and would be inversely related to energy consumption, reducing or eliminating the purchase price advantage of inefficient models. An energy-related excise tax would provide signals to consumers inducing them to buy more efficient models, and would provide incentives for manufacturers to develop more efficient products.

Setting rigid, minimum efficiency standards in Canada that were designed to force manufacturers to the outer limits of current technology and beyond would be a serious mistake. The damaging and counter-productive effects of such programmes in the United States are well known in the case of automobile pollution control, automobile fuel economy and, to a lesser extent, appliance efficiency. Under the American system, manufacturers have every incentive to conceal technological improvements from the regulatory authorities. Past history suggests, and in some cases the legislation requires, that when technological improvements are discovered, manufacturers must adopt them. In the long run, technological progress can offer enormous possibilities for improved energy efficiency, and policies that might in any way hamper this progress should be firmly rejected. In addition, it is very costly for regulatory bodies to determine what is technologically feasible, both now and in the future. Finally, manufacturers may be willing to comply voluntarily with an information programme, or even a modest excise tax based on energy consumption, but can be counted on to fight vigorously against rigid minimum standards. One cannot overestimate the delays that may be imposed, and costs that may be incurred, in fighting a recalcitrant industry as compared to the case where the majority of the industry is reasonably compliant.

In the automobile case, there is no justification for standards in Canada, since the technology will be determined more or less independently of the Canadian market. Furthermore, there is substantial opportunity for consumer evasion of the standards by the purchase of vehicles such as light duty trucks and vans. All of the benefits from a corporate average fuel economy programme in Canada could be achieved better by raising the price of gasoline and taxing fuel-inefficient vehicles.

To be effective, all of the information programmes rely on energy being priced at its opportunity cost. In the furnace and automotive cases, even a regulatory policy or energy excise tax requires proper energy pricing to be effective. Consumers can choose how high to set their thermostats and whether to improve the insulation and storm doors and windows in their homes. These choices will be made sensibly only if oil and gas prices reflect social opportunity costs. Consumers have choices about the intensity with which they use their automobiles and how many automobiles to own. Simply requiring efficient design will ignore a very important means of fuel saving: driving less or driving more economically. Every day that domestic energy prices are below world levels is one more day in which Canadians waste part of their finite stock of energy.

In the case of refrigerators, a direct regulatory programme or energy excise tax would not need higher energy prices as a companion in order to ensure the desired effect, since consumers cannot significantly vary the intensity of use of refrigerators. However, if low energy prices meant little consumer enthusiasm for more efficient goods, the difficulty of enforcing standards would be increased. Thus, even here, opportunity cost prices for energy would facilitate the effectiveness of a direct regulatory scheme.

We conclude that governments may play an important role in energy conservation. That role, however, is to improve the operation of markets, not to replace them. We cannot hold the price of energy below social cost and achieve the same conservation results as if those prices reflected social cost, no matter what policies are adopted. A serious energy conservation programme requires energy prices at social costs, information programmes for a few major energy-consuming products, perhaps an energy excise tax for some products and, in isolated cases, co-operative minimum efficiency standards that are not technology forcing. This is not massive regulation but mini-regulation. In the case of small appliances that use limited amounts of energy, the best solution is no formal intervention. By sensibly applying the principles suggested above on a case-by-case basis, we should be able to design energy conservation programmes in Canada that achieve considerable reductions in energy consumption and, at the same time, reduce the lifetime cost of major, energy-using, consumer products without creating huge new government bureaucracies.

NOTES

* The research reported on here was supported in part by a grant from the federal Department of Energy, Mines and Resources jointly awarded to the Institute for Policy Analysis and the Institute for Environmental Studies of the University of Toronto. The opinions expressed here are those of the author and not those of the Department of Energy, Mines and Resources or the University of Toronto. I would like to thank Mike Berkowitz, Frank Mathewson, Mike Trebilcock, and Len Waverman for helpful comments on earlier drafts of this paper.

[1] *Motor Vehicle Information and Cost Savings Act,* Title V. "Improving Automotive Efficiency," 15 U.S.C. 2002.

[2] Legislation establishing formal standards in Canada was not passed, so that if fuel economy in Canada falls below the contemplated standards, no sanctions will be imposed.

[3] See various papers presented at the International Association of Energy Economists Conference, Washington D.C., June 1979.

[4] Hausman (1979). This study ignores the problem of imperfect information, and may therefore have attributed to high discount rates choices that were in part governed by poor information about energy uses.

[5] A survey of consumers in early 1979 showed little shopper understanding of the labels and little interest in energy consumption. Still, manufacturers seem to be making some design changes to improve efficiency. Discussions at the Department of Consumer and Corporate Affairs, Ottawa, 5 October 1979.

[6] *Energy Policy and Conservation Act*, Title III. "Improving Energy Efficiency," 42 U.S.C. sections 6291-6309.

[7] A proposed seasonal efficiency test is described in U.S. Department of Energy, "Test Procedures for Furnaces and Vented Home Heating Equipment," *U.S. Federal Register* 43, No. 81 (10 May 1978), pp. 20155-81.

[8] *Motor Vehicle Information and Cost Saving Act*, Title V. "Improving Automotive Efficiency," §508(b)1, 15 U.S.C. 2008.

[9] A Rand study concluded that, in the short run, only an increase in the fuel prices could significantly reduce fuel consumption, and that, in the long run, the production of more economical vehicles caused by CAFE would lead to increased motoring in the absence of higher fuel prices (Wildhorn, 1976).

REFERENCES

Akerlof, G. (1970) "The Market for 'Lemons': Quality Uncertainty and the Market Mechanism." *Quarterly Journal of Economics* 84 (August):488-500.

Canada, Department of Energy, Mines and Resources. (1977) *Energy Demand Projections: A Total Energy Approach*. Report ER77-4. Ottawa: Minister of Supply and Services Canada.

Canada, Department of Transport. (1979) *Fuel Consumption Guide 1979*. Ottawa: Minister of Supply and Services Canada.

Canadian Standards Association (CSA). (1978) *The Consumer* 78-21 (Summer).

Dewees, D.N. (1977a) "The Economics of Home Furnace Efficiency." Report No. 8. Toronto: University of Toronto, Institute for Policy Analysis.

Dewees, D.N. (1977b) "Energy Conservation in Home Refrigerators." Report No. 9. Toronto: University of Toronto, Institute for Policy Analysis.

Dewees, D.N. (1979) "Energy Conservation in Home Furnaces." *Energy Policy* 7 (June): 149-62.

Dewees, D.N.; Hauer, E.; and Saccomano, F. (1979) ''Urban Road Use Subsidies: A Review of Theory and Measurement.'' Paper No. 4. Toronto: University of Toronto-York University Joint Program in Transportation.

Dewees, D.N.; Hyndman, R.M.; and Waverman, L. (1975) ''Gasoline Demand in Canada, 1956–1972.'' *Energy Policy* 3 (June): 116-23.

Friedman, M. (1979*a*) ''Blaming the Obstetrician.'' *Newsweek* (4 June), p. 70.

Friedman, M. (1979*b*) ''What Carter Should Do.'' *Newsweek* (18 June), p. 66.

Fuss, M.A. (1977) ''The Demand for Energy in Canadian Manufacturing.'' *Journal of Econometrics* 5 (January): 89-116.

Hausman, J.A. (1979) ''Individual Discount Rates and the Purchase and Utilization of Energy-Using Durables.'' *Bell Journal of Economics* 10 (Spring): 33-54.

Henderson, J.M. and Quandt, R.E. (1958) *Microeconomic Theory*. Toronto: McGraw-Hill.

Hirshhorn, R. (1979) ''A Case Study: Energy Consumption Labelling Requirements for Refrigerators.'' Studies on Government Regulatory Activity, No. 7. Ottawa: Treasury Board Secretariat and Department of Consumer and Corporate Affairs.

Hoskins, R.A. and Hirst, E. (1977) ''Energy and Cost Analysis of Residential Refrigerators.'' ORNL/CON-6. Oak Ridge, Tenn.: Oak Ridge National Laboratories.

Motor Trend. (1979) ''General Motors X-Cars.'' 31 (May).

Wildhorn, S., with Burright, B.K.; Enns, J.H.; and Kirkwood, T.F. (1976) *How to Save Gasoline: Public Policy Alternatives for the Automobile*. Cambridge, Mass.: Ballinger.

Chapter Eight

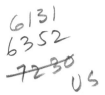

6131
6352
7230 US

Time-of-Day Pricing of Electricity Activities in Some Midwestern States

by
J. Robert Malko, Dennis J. Ray, and *Nancy L. Hassig**

INTRODUCTION

Because of the energy dilemma, general inflation, increases in system peak demand, capital shortages, and growing concerns for environmental and consumer interests, the design of electricity rates has become an *important* issue for regulatory commissions and utilities during the 1970s.[1] Since the conclusion of an important and forward-looking Madison Gas and Electric Company case in August 1974, regulatory commissions, utilities, and interveners have been examining the theoretical foundations for, the desirability of, and the feasibility of implementing time-of-day pricing of electricity.[2] In order to comply with the recent *National Energy Act of 1978*, each state regulatory commission must consider and make a determination concerning the electricity rate-making standard of time-of-day rates.[3]

The primary objectives of this paper are (1) to present a brief review of the *Public Utility Regulatory Policies Act*, which is part of the *National Energy Act of 1978*, emphasizing the provisions concerning electricity rate-making standards, including time-of-day rates; (2) to discuss actual time-of-day pricing of electricity activities in four midwestern states, Wisconsin, Michigan, Illinois, and Ohio; and (3) to propose a framework for considering and implementing cost-based, time-of-day rates. Time-of-day pricing is an indirect form of load management that prices electricity to reflect the differences in cost of providing service by time of day.

PUBLIC UTILITY REGULATORY POLICIES ACT OF 1978

This section of the paper is divided into two parts. First, there is a brief review of the background and basic features of the *National Energy Act (NEA) of 1978* and of one of its major components, the *Public Utility Regulatory Policies Act (PURPA) of 1978*. Second, a brief discussion is presented of the electricity rate design standards specified by PURPA for consideration by each state regulatory commission.

Background and Basic Features

After approximately a year and a half of deliberation, the *National Energy Act* (NEA) was passed by the U.S. Congress on 15 October 1978 and became law on 9 November 1978.[4] After congressional passage of NEA, President Carter stated: ''We have declared to ourselves and the world our intent to control our use of energy, and thereby to control our own destiny as a nation.''[5]

The NEA has the following five major components: (1) the *National Energy Conservation Policy Act of 1978*, (2) the *Powerplant and Industrial Fuel Use Act of 1978*, (3) the *Public Utility Regulatory Policies Act of 1978*, (4) the *Natural Gas Policy Act of 1978*, and (5) the *Energy Tax Act of 1978*.

The PURPA contains provisions concerning the following important matters: (1) rate-making standards for electric utility rate structures, (2) cogeneration provisions, (3) wholesale provisions, (4) aid to the states and consumer representation provisions, (5) gas utilities, (6) small hydroelectric facilities, (7) crude oil transportation systems, and (8) significant miscellaneous provisions such as the authorization of funding for the National Regulatory Research Institute, establishment of three additional university coal research laboratories, and clarification of natural gas transportation policies.

Electricity Rate-Making Standards

To address the problem of increasing electricity costs during the 1970s, the U.S. Congress decided that retail electricity rates should be designed or structured to encourage the following: (1) conservation of energy supplied, (2) efficient use of facilities and resources, and (3) equitable rates to consumers.[6]

The PURPA explicitly specifies that each state regulatory commission should *consider* and make a *determination* concerning the appropriateness, relative to the above three purposes and applicable state law, of implementing the following *six rate-making standards*:

(1) *Cost of Services*—Rates charged by any electric utility for providing electric service to each class of electric consumers shall be designed, to the maximum extent practicable, to reflect the costs of providing electric service to such class, as determined under section 115(a).

(2) *Declining Block Rates*—The energy component of a rate, or the amount attributable to the energy component in a rate, charged by any electric utility for providing electric service during any period to any class of electric consumers may not decrease as kilowatt-hour consumption by such class increases during such period except to the extent that such utility demonstrates that the costs to such utility of providing electric service to such class which costs are attributable to such energy component decrease as such consumption increases during such period.

(3) *Time-of-Day Rates*—The rates charged by any electric utility for providing electric service to each class of electric consumers shall be on a time-of-day basis which reflects the costs of providing electric service to such class of electric consumers at different times of the day unless such rates are not cost-effective with respect to such class, as determined under section 115(b).

(4) *Seasonal Rates*—The rates charged by an electric utility for providing electric service to each class of electric consumers shall be on a seasonal basis which reflects the costs of providing service to such class of consumers at different seasons of the year to the extent that such costs vary seasonally for such utility.

(5) *Interruptible Rates*—Each electric utility shall offer each industrial and commercial electric consumer an interruptible rate which reflects the cost of providing interruptible service to the class of which such consumer is a member.

(6) *Load Management Techniques*—Each electric utility shall offer to its electric consumers such load management techniques as the state regulatory authority (or the nonregulated electric utility) has determined will—

(A) be practicable and cost-effective, as dertermined under section 115(c),

(B) be reliable, and

(C) provide useful energy or capacity management advantages to the electric utility.[7]

The PURPA additionally specifies the following *special rules* for considering and making a determination concerning three rate-making standards:

(a) *Cost of Service*—In undertaking the consideration and making the determination under section 111 with respect to the standard concerning cost of service established by section 111 (d) (1), the costs of providing electric service to each class of electric consumers shall, to the maximum extent practicable, be determined on the basis of methods prescribed by the state regulatory authority (in the case of a state regulated electric utility) or by the electric utility (in the case of a nonregulated electric utility). Such methods shall to the maximum extent practicable—

(1) permit identification of differences in cost-incurrence, for each such class of electric consumers, attributable to daily and seasonal time of use of service and

(2) permit identification of differences in cost-incurrence attributable to differences in customer demand, and energy components of cost. In prescribing such methods, such State regulatory authority or nonregulated electric utility shall take into account the extent to which total costs to an electric utility are likely to change if—

(A) additional capacity is added to meet peak demand relative to base demand; and

(B) additional kilowatt-hours of electric energy are delivered to electric consumers.

(b) *Time-of-Day Rates*—In undertaking the consideration and making the determination required under section 111 (d) (3), a time-of-day rate charged by an electric utility for providing electric service to each class of electric consumers shall be determined to be cost-effective with respect to each such class if the long-run benefits of such rate to the electric utility and its electric customers in the class concerned are likely to exceed the metering costs and other costs associated with the use of such rates.

(c) *Load Management Techniques*—In undertaking the consideration and making the determination required under section 111 with respect to the standard for load management techniques established by section 111 (d) (6), a load management technique shall be determined, by the State regulatory authority or nonregulated electric utility, to be cost-effective if—

(1) such technique is likely to reduce maximum kilowatt demand on the electric utility, and

(2) the long-run cost-savings to the utility of such reduction are likely to exceed the long-run costs to the utility associated with implementation of such technique.[8]

By November 1980, each state regulatory commission must have begun the *consideration* or have set a hearing date for the consideration of the six rate-making standards with respect to each electric utility for which it has rate-making authority *and* for which PURPA applies.[9] *By November 1981*, each state regulatory commission must complete the consideration process and make a determination concerning the appropriateness of implementing the six rate-making standards.[10] In short, the decision concerning the appropriateness of implementation is the responsibility of each state regulatory commission based on state law, as supplemented by specific provisions of PURPA.

This paper concentrates on the rate-making standard of time-of-day pricing of electricity in order to attempt to provide some guidance and insight for regulatory commissions and electric utilities that are currently examining the desirability and feasibility of implementing time-of-day rates. The next section of this paper discusses some actual time-of-day pricing activities in four midwestern states, and the following section proposes a framework for consideration and implementation of cost-based time-of-day rates.

SOME ACTUAL TIME-OF-DAY PRICING ACTIVITIES

Before PURPA became law in November 1978, various regulatory commissions and utilities had been examining and implementing time-of-day pricing of electricity on a mandatory or voluntary basis for sets of customers during the 1970s. This section of the paper examines actual implementation activities of time-of-day rates for industrial, commerical, and residential customers during the 1970s in the following four midwestern regulatory jurisdictions: Illinois, Michigan, Ohio, and Wisconsin.

Illinois

Time-of-day pricing implementation in Illinois has consisted of placing large industrial and commercial customers on mandatory time-of-day prices and of experimenting with time-of-day pricing options in the residential class. Commonwealth Edison's large industrial and commercial customers, numbering approximately 700 with maximum demands exceeding 1500 kilowatts, have been on time-of-day prices since 26 November 1977.[11] In the fall of 1978, a residential pricing experiment was begun that affected approximately five hundred and sixty Commonwealth customers.[12] Other Illinois utilities are considering residential pricing experiments.

Commonwealth Edison's industrial and commercial rates were derived from three cost-of-service studies. One study used fully allocated costs, whereas the other two studies used different methods of measuring marginal costs (the peaker method and the cost-of-planned-additions approach). Since the fully allocated cost study indicated that these customers were already contributing a relatively higher proportion than other customer classes to cover costs, and that changing to a flatter rate structure could further increase this contribution, the Illinois commission approved final prices that would have minimal impact on their bills. Bill impact was estimated using load data from recording meters installed before time-of-day pricing began. Only six customers were expected to receive bill increases exceeding 4 per cent per year, assuming that customers' consumption patterns remained the same after the prices went into effect.

Under the rates effective on 14 December 1978, a basic energy charge is calculated from a seven-step, declining-block rate structure.[13] A time-of-day price adjustment is made by adding $0.408 per kilowatt hour of use during peak hours and giving a credit of $0.414 per kilowatt hour used during off-peak hours. A five-step demand charge with seasonal prices applies only to the maximum thirty-minute demand measured during peak hours from 9 A.M. to 10 P.M., Monday through Friday (except holidays).

The residential pricing experiment by Commonwealth Edison is being conducted to determine how many customers would volunteer for time-of-day prices *and* what the load characteristics and price responsiveness are for such volunteers. Six different rate schedules with a variety of prices and peak-period definitions were offered to randomly selected customers. Since the test extends to 1993, customers will be able to benefit from investments in equipment such as timers and thermal storage systems, which permit shifting of electricity use to the off-peak periods. The customers have been informed that the prices and peak-length definitions are subject to change at any time during the test.

Two other Illinois utilities, Central Illinois Light Company and Central Illinois Public Service Company, may be offering twelve (six each) rates to a total of 600 (300 each) residential customers on a mandatory basis. The purpose of these experiments, set up identically for each utility, will be to determine price elasticity so that cost-benefit analysis can be performed.

Michigan

Time-of-day pricing's introduction to Michigan consumers has focused on making optional prices available to industrial, commercial, and residential

Table 1
TIME-OF-DAY PRICING FOR MICHIGAN'S LARGEST UTILITIES

Utility	Rate Schedule	Affected Customers	Number	Estimated Percentage of Total Electric			Initial MPSC Order and Date
				Sales	Revenue	Peak Demand	
Detroit Edison Company	Primary Supply Rates D6 and D61	Medium to Large Industrial and Commercial	2,194	37.8	36.7	35.1	U-4807 March 30, 1976
	Bulk Power Rate D-7	Large Industrial (Demand Greater than 50,000 kW)	5	11.6	8.0	8.4	U-4807 March 30, 1976
	Interruptible Supply Rate D8	Voluntary for 10,000 kW or Larger Interruptible Loads	0	0	0	0	U-4807 March 30, 1976
	Domestic Test Service Rate D1.2	Voluntary for Selected Residences	66	Less than .1%	Less than .1%	Less than .1%	U-5161 October 25, 1976
	Experimental Domestic Space Heating D2.1	Optional for Eligible Residences	95	Less than .1%	Less than .1%	Less than .1%	U-5108 May 27, 1977
	Experimental General Service Rate D3.2	Optional for Eligible Customers	0	0	0	0	U-5108 May 27, 1977
	Primary Pumping Rate E4	Pumping Demand Greater than 50 kW	60	1.5	12	1.4	U-5502 September 28, 1978

Company	Rate	Description					Docket/Date
	Optional Domestic Service Rate D1.4	Full-time Farm and Space-heating	0	0	0	0	U-5502 September 28, 1978
Consumers Power Company	Primary Service Rates D and R-3	Medium to Large Industrial and Commercial	1,653	41.3	33.8	37	U-4840 April 13, 1976
	Primary High Load Factor Service Rate F	Medium to Large Industrial and Commercial	37	7.1	4.7	4.4	U-4840 April 13, 1976
	Primary Electric Furnace Service Rate J	Electric Metal-melting Loads	31	3.1	2.3	2.5	U-4840 April 13, 1976
	Primary Interruptible Service Rate I	Medium to Large Industrial and Commercial	0	0	0	0	U-5305 March 7, 1977
	Experimental Residential Rates X-1 and X-2	Voluntary for Selected Residences	208	Less than .1%	Less than .1%	Less than .1%	U-5306 April 4, 1977
	Optional Primary Public Pumping Service Rate PS3	Public Pumping	4	Less than .1%	Less than .1%	Unavailable	U-5331 July 31, 1978
	Optional Residential Service Rate A-3	Full-time Farms and Space-heating	0	0	0	0	U-5331 July 31, 1978
	Experimental General Service Rate X-4	Voluntary for Selected Commercial	0	0	0	0	U-5306 April 4, 1977

customers. The Michigan Public Service Commission describes its approach to time-of-day pricing implementation as follows:

> The Commission finds that a conservative, multi-faceted approach to the pervasive question of time-of-day pricing should be adopted. This approach will combine a variety of investigative and actual experiences to determine the most appropriate long-range treatment of this pricing concept.[14]

Investigative and actual experience has accumulated from mid-1976, when the large industrial and commercial customers of Michigan's largest utilities, Consumers Power and Detroit Edison, began paying time-differentiated demand and energy costs, to the present, when residential and small commercial customers are involved in experimental programmes and optional prices. This experience is being incorporated in a current generic study to determine the effectiveness of electric service with respect to load management including time-of-day pricing.[15]

Table 1 overviews the time-of-day pricing offerings for the major utilities and shows that a large portion of the kilowatt-hour sales and peak demand is from large industrial and commercial customers paying time-related prices. Prices include a 3 mill differential between on- and off-peak energy use and a generation capacity charge for on-peak billed demand. Transmission and distribution capacity costs are covered by a charge for the maximum fifteen-minute demand occurring at any time during the month. Peak-price hours differ between the two major utilities. Detroit Edison's hours are from 9 A.M. to 9 P.M. on non-holiday weekdays. Consumers Power's peak hours extend from 5 to 9 P.M. during winter months and 10 A.M. to 5 P.M. in summer months on non-holiday weekdays.[16] Reports on the initial response of primary (industrial and commercial) customers are available.[17]

While some time-of-day pricing options exist for residential and commercial customers, there is concern over the cost of meters and absence of knowledge of customer response.[18] Cost-effectiveness studies of load management options, load controls, and time-of-day rates are being conducted. Table 1 shows that residential time-of-day pricing is being tested through voluntary experiments by both major utilities. Experimental rates for small commercial customers and optional rates for farm and space-heating customers have been met with limited consumer enthusiasm. This lack of response is likely due to (1) recent availability of rates; (2) customer reluctance to take rates that are believed not to be advantageous (commercial customers may find it difficult to shift consumption when their operating hours are during peak periods); and (3) the need for unusual amounts of publicity and consumer education to increase awareness and acceptance of this new pricing method.

Time-of-day prices in Michigan have been based on fully allocated cost studies using embedded (average accounting) costs. In the final order in a

1976 Consumers Power rate case, the Michigan Public Service Commission rejected the use of marginal cost pricing in designing rates, stating, "[Marginal cost] determinations are replete with uncertainty and . . . can accomplish little more in the way of providing correct price signals to the consumer than the current . . . method."[19] The revenue reconciliation problem, adjusting rates based on marginal costs to equal the revenue requirement based on accounting costs, was a factor in this determination. The commission has directed the major utilities to submit marginal or incremental cost studies later this year (1979); thus, marginal costs may be used in the future.[20]

Ohio

In an important decision designed to deal with problems of rising electricity costs, earnings erosion, and increasing rates caused by rapid demand growth and declining system efficiency, the Ohio Public Utility Commission on 28 December 1978 authorized Dayton Power and Light Company (DP&L) to implement time-of-day pricing for all residential, commercial, and industrial customers. This action is significant among Ohio rate reform activities, which include generic hearings on the long-run incremental costs of electricity production[21] and authorization of optional residential time-of-day rates for Ohio Edison Company,[22] and experimental residential rates for forty Cleveland Electric Illuminating Company customers.[23]

In the Dayton Power and Light Company situation, the combination of two factors, declining load factor and high demand growth, resulted in a high construction rate without sales being sufficient to generate required revenues. This situation not only caused financial strain for the company but also caused electricity rates to increase rapidly. Consequently, rates based on marginal cost were proposed by the utility in order to promote more efficient use of its capacity.

Projections were made that time-of-day prices along with load control could *reduce* winter growth from 6.7 to 5.2 per cent and summer growth from 5.8 to 4.6 per cent (with time-of-day prices accounting for one third of the difference) and could increase the load factor by 1 per cent by 1987. The implementation cost was estimated to be $43 million over the next ten years. These costs would be offset by the estimated reduction in the need for 447 megawatts of additional capacity by 1987, with an associated savings of $110 million.[24]

Implementation is planned on a geographical basis, by substation or meter-reading route. It is hoped that 34 per cent of DP&L's customers can be on the new rates by 1987.

A final decision has not yet been reached on rate structure. Commission staff report that rate structures are currently derived using an embedded cost

Table 2
STATUS OF TIME-OF-DAY PRICING IN WISCONSIN

Utility	Customer Class	Affected Customers	Number	Estimated Percentage of Total Electric			PSC Docket And Order Date	Implementation
				Sales	Revenue	Peak Demand		
Wisconsin Power & Light Company	Large Industrial and Commercial (Cp-1 and Cp-3)	Demand Greater than 500 kW	144	31	22	20	2-U-8085 November 12, 1976	February 1977
	Large Industrial and Commercial (Cp-1 and Cp-3)	Demand Between 200 and 500 kW	275	8	6	6	6680-UR-5 (Interim) December 22, 1978	December 1978
Madison Gas & Electric	Large Industrial	Largest Demand	2	13	10	10	3270-UR-1 November 9, 1976	January 1977
	Large Industrial and Commercial (Cg-2)	Demand Greater than 300 kW	109	20	17	17	3270-ER-5 January 22, 1979	June 1979
	Industrial Commercial Residential (Cg-3 and Cg-4)	Optional Demand Less than 300 kW	300 Max.	Unavbl.	Unavbl.	Unavbl.	3270-ER-5 January 22, 1979	June 1979

Utility	Customer Class	Eligibility					Docket	Date
Wisconsin Public Service Corporation	Large Industrial (Cp-1)	Demand Greater than 1,000 kW	85	32	24	27	6690-ER-7 Pending	Pending
	Residential (Experimental)	Randomly Selected	501	Less than 1	Less than 1	Less than 1	6690-ER-5 February 18, 1977	May 1977 Until May 1980
Northern States Power Company	Large Industrial and Commercial	Demand Greater than 1,500 kW	34	22	16	13	4220-ER-8 March 12, 1979	April 1979
	Large Industrial and Commercial	Demand Between 500 and 1,500 kW	61	Unavbl.	Unavbl.	Unavbl.	4220-ER-10 Pending	Pending
Superior Water and Power Company	Large Industrial and Commercial	Demand Greater than 1,000 kW	13	60	53	41	5820-UR-3 Pending	Pending
Wisconsin Electric Power Company	Large Industrial (Cp-1, Rate Area 1)	Demand Generally Greater than 300 kW	452	31	23	24	6630-ER-2/5 January 5, 1978 (Interim) July 20, 1978 (Final)	January 1978
	Small Industrial and Commercial (Cg-3, Rate Area 1)	All Above 30,000 kWh/ Month	2,099	12	13	14	6630-ER-2/5 January 5, 1978 and July 20, 1978	All by July 1980

Table 2
STATUS OF TIME-OF-DAY PRICING IN WISCONSIN
(Continued)

Utility	Customer Class	Affected Customers	Number	Estimated Percentage of Total Electric			PSC Docket And Order Date	Implementation
				Sales	Revenue	Peak Demand		
	Residential (Rg-2, Rate Areas 2 and 3)	Largest Annual Consumption	577	Less than 1	Less than 1	Less than 1	6630-ER-2/5 Jan. 5, 1978, and July 20, 1978	July 1978
	Large Industrial (Cp-1, Rate Areas 2 and 3)	Demand Greater than 300 kW	39	6	4	3	6630-ER-8 March 6, 1979	March 1979
	Commercial (Cg-1, Rate Areas 2 and 3)	All Above 30,000 kWh/Month	209	2	2	2	6630-ER-8 March 6, 1979	July 1980
	Residential (Rg-1, Rate Areas 1 through 3)	Next Largest	3,000	Unavbl.	Unavbl.	Unavbl.	6630-ER-8 March 6, 1979	Schedule Pending

Source: Estimates provided by the various utilities listed.

allocation method in which the capacity costs are mainly allocated to peak periods on a peak responsibility basis. For DP&L, however, a marginal cost study has been prepared. Prices equal to marginal costs would produce revenues exceeding the revenue requirement because marginal costs exceed average costs. Therefore, a method for adjusting marginal prices downward will have to be used. Hearings for setting the final rates will be held soon. Commission staff expect some customer opposition to the new prices, especially from large industrial customers who may have difficulty shifting usage in response to the time-of-day prices.

Estimates of the DP&L residential customer response came from a Department of Energy-funded experiment involving about one hundred customers. The seventeen-month experiment with voluntary participation was designed to gain experience with time-of-day pricing and other load management technologies and to obtain insight concerning customer reaction. The final report has not yet been completed. However, preliminary analysis indicates that the demand of customers on time-of-day prices was less than the demand of control customers during peak-price periods. This reduction was maintained during the hour of system annual peak and on the hottest day of the year.[25]

The generic hearings on long-run incremental costs of electricity production require the electric utilities to prepare studies on estimating the total and incremental costs of providing electricity. In determining the long-run incremental and total costs for the next ten years, various load management alternatives (including time-of-day pricing) must be considered.

Wisconsin

The usefulness and importance of marginal cost pricing for reflecting economic costs associated with providing utility service was acknowledged by the Wisconsin Public Service Commission in August 1974.[26] Today, time-of-day electricity pricing, an outcome of the application of marginal cost-pricing principles, is widely implemented among large industrial and commercial firms. Implementation activities extend beyond the large firms, however, affecting household and small commercial customers as well. The commission has combined mandatory, optional, and experimental pricing in its time-of-day pricing programme. Table 2 summarizes the breadth of time-of-day pricing use in Wisconsin. Table 3 presents some current rate structures.

The first utility in Wisconsin to implement mandatory electric rates with time-varying energy and capacity charges is the Wisconsin Power and Light Company (WP&L). One hundred and thirty industrial and commercial customers with maximum demands of over 500 kilowatts began paying these prices in January 1977. Table 2 shows that the expansion of time-of-day pricing to lower-demand stratum customers (200-500 kW) was recently authorized.

Table 3
TIME-OF-DAY ELECTRICITY PRICES IN WISCONSIN[a]

Utility and PSC Order Number With Approved Rate	Customer Class	Summer Prices			Winter Prices			On-Peak Hours[b]	Fixed Charge $/Month
		On-Peak Energy $/kWh	Off-Peak Energy $/kWh	On-Peak Demand $/kW	On-Peak Energy $/kWh	Off-Peak Energy $/kWh	On-Peak Demand $/kW		
Wisconsin Power and Light Company 6680-UR-5 December 21, 1978	Large Industrial and Commercial (Cp-1 and Cp-3)	.0253	.0123	4.75[c]	.0253	.0123	4.75[c]	8 A.M.–10P.M. Mon.–Sat.	12.50
Madison Gas and Electric Company Sp-3 and 4 by 3270-UR-1, November 9, 1976	Largest Industrial (Sp-3 and Sp-4)	.0253	.0100	6.45[d]	.0200	.0100	5.10[d]	10 A.M.–9 P.M. Mon.–Fri.	—
	Large Industrial and Commercial (Cg-2)	.034	.0100	1.10+ 5.96[e]	.0268	.0100	.78+ 4.10[e]	10 A.M.–9 P.M. Mon.–Fri.	20.00
Cg-2, 3 & 4 by 3270-ER-4/5, January 21, 1979	Optional Industrial, Commercial and Residential (Cg-3)	.1125	.0150	—	.0782	.0150	—	10 A.M.–9 P.M. Mon. - Fri.	5.50 or 11.00
	Optional Industrial and Commercial (Cg-4)	.034	.0100	6.78	.0268	.0100	4.69	10 A.M.–9 P.M. Mon.–Fri.	7.50 or 15.00

Utility and PSC Order Number With Approved Rate	Customer Class	Summer Prices			Winter Prices				
		On-Peak Energy $/kWh	Off-Peak Energy $/kWh	On-Peak Demand $/kW	On-Peak Energy $/kWh	Off-Peak Energy $/kWh	On-Peak Demand $/kW	On-Peak Hours[b]	Fixed Charge $/Month
Wisconsin Electric Power Company 6680-ER-8, March 6, 1979	Large Industrial (Cp-1 Rate Areas 1, 2 and 3)	.0280	.0140	3.96[f]	.0280	.0140	2.87[f]	8 A.M.–8 P.M. C.S.T. Mon.–Fri.	645.00
	Small Industrial and Commercial (Cg-3, Rate Area 1 and Cg-1, Rates Areas 2 and 3)	.0300	.0150	6.38	.0300	.0111	3.52	9 A.M.–9 P.M. C.S.T. Mon.–Fri.	200.00
	Residential (Rg-2, Rate Area 1)	.0897	.0143	—	.0568	.0143	—	7 A.M.–7 P.M. C.S.T. Mon.–Fri.	5.00

a. Specific rates may vary depending upon power factor, fuel adjustment, voltage level, or other factors. Refer to approved rate sheets for detailed occurring information.
b. All hours are considered off-peak on holidays.
c. Billed demand is the larger of the highest on-peak demand or 50 per cent of the highest off-peak demand. Interruptible customers pay a demand charge of $3.50 per kW of interruptible load prorated for hours of interruption to hours in the billing period.
d. Demand charge slightly different between the two customers on this rate.
e. Total demand charge equals the sum of the first price times the maximum fifteen-minute demand plus the second price times the maximum two-hour demand during on-peak hours. Interruptible customers pay no on-peak demand charge for their interruptible loads.
f. Billed demand is the highest on-peak demand. The demand charge is reduced by 70 per cent for interruptible loads (Cp-2 rate).

Wisconsin Power and Light's initial time-of-day prices were determined by considering cost-of-service studies, load research, customer impact, and regulatory constraints. The final price design referenced data from five submitted cost-of-service studies, three using embedded costs and two using long-run incremental costs. Load research data collected before final rate design allowed allocation of costs to the industrial and commercial classes based on electrical usage patterns. Using magnetic tape-recording meters, load data were collected from a sample of affected customers before rate implementation. Load data not only played a part in the price design but also helped estimate customer impact of the new prices. To ensure that no customer would receive significant bill increases, prices were set so no sampled customer would receive more than a 7 per cent increase if they made no usage change (i.e., zero price elasticity). Final prices were also constrained by the revenue requirement derived for the customer class by the peak responsibility-embedded cost method.

Table 3 presents the current prices for WP&L customers. Energy prices were set close to marginal cost. The demand charge reflects a proration of marginal capacity costs to satisfy the revenue requirement based on accounting costs. The pricing periods were selected based on examination of hourly energy costs. The peak-period length of fourteen hours lessens the chance of a shifting peak problem, thus stabilizing the pricing periods for consumer and utility.

In a report issued on 17 April 1978, WP&L describes customer response to the time-of-day prices after one year. The company estimates that the affected customer class has shifted 23 megawatts (MW) of demand and about 76,000 MWH or 6.1 per cent of electricity usage from the on-peak to off-peak periods. A customer survey indicates that not all customers find it possible to shift usage, but those that do seem to utilize revised operating hours or energy management equipment (demand controllers) to reduce on-peak usage.[27] The economics of paying premiums for night shift work versus paying peak-period electricity prices enters into the response to time-of-day prices.

Another Wisconsin utility, Madison Gas and Electric Company (MG&E), also has mandatory time-of-day prices for its large industrial and commercial customers (in the case with demands greater than 300 kW), but it is unique in that it has an optional rate available to all other customers. The optional rate was authorized in January 1979, and customer acceptance has not been evaluated. Table 3 summarizes present MG&E rates. The MG&E's rate design incorporates data from studies of the marginal energy and capacity costs. The seasonal price differential is due to the relatively higher demand for electricity during the summer and the associated higher marginal energy and capacity costs.

The Wisconsin Public Service Corporation, co-operating with the U.S. Department of Energy and the Wisconsin Public Service Commission, has been sponsoring a scientifically designed residential pricing experiment involving about five hundred customers paying mandatory time-of-day prices.[28] A variety of peak lengths and on-peak to off-peak electricity price ratios were tested for a trial three-year period ending in May 1980. The experiment's goal was to determine how residential customers with different life-styles, appliance stocks, and levels of electricity usage respond to time-of-day prices. Net benefits from implementing such prices for the general residential population (or subsets of that population) can then be estimated.

The Wisconsin Electric Power Company (WEPCO), Wisconsin's largest electric utility, has *mandatory* time-of-day prices in effect for all large industrial and commercial customers and for portions of other customer classes, including residential. Table 2 shows that presently WEPCO has 491 large industrial customers (demand generally greater than 300 kW) and 577 residential customers on time-of-day prices. By July 1980, approximately 2300 small industrial and commercial customers with average consumptions greater than 30,000 kilowatt hours per month began paying time-of-day prices when special time-of-day meters were installed. Three thousand additional residential customers will be added to those residential customers already paying time-of-day prices.

The Wisconsin Public Service Commission's 5 January 1978 order authorizing the initial implementation of time-of-day rates for WEPCO describes the four steps by which the rates were designed:

1. Class revenue levels were set at adjusted current revenue levels.
2. The off-peak energy rate reflects four considerations:
    a. System lambdas (a measure of the operating cost of the next unit of production)
    b. Average fuel cost
    c. Operating characteristics
    d. Voltage losses
3. The on-peak energy rate is based on the cost differential indicated by the system lambdas.
4. The demand charge and the customer facilities charge were adjusted to equal the class revenue level. The demand charges were designed to reflect the seasonal differential of the cost of providing service.[29]

The demand charge mentioned above is assessed for the highest fifteen minutes of use in peak hours and has the impact of substantially raising the price for consumption in that period, putting the effective on-peak to off-peak price ratio between 4 and 6 to 1.[30]

Peak periods have the interesting characteristic of being slightly different for each customer class. Each is twelve hours long, and each begins at a different time. System characteristics, such as load duration curves, available capacity, and loss of load probabilities, were analysed for peak-hour selection. In addition, an attempt was made to set hours that would provide customers an opportunity to reduce or shift their on-peak consumption. Extensive load research was conducted before the rates were implemented.

Assessment of the potential impact of time-of-day prices on the bills of large industrial and commercial customers showed that 35 per cent would have higher electricity costs and 9 per cent would have annual increases exceeding 5 per cent, assuming no change in usage.[31] It was determined that many of the large consumption firms would experience increases not because of the time-varying prices, but because of losing the price advantage of declining block prices.

The Wisconsin Electric Power Company analysed the usage patterns of large industrial and commercial customers placed on time-of-day prices in January 1978 and reported no noticeable usage pattern change in response to the new pricing signals.[32] Additional analysis (including a customer survey) is planned. The inability to identify any price response is likely due to the short time the new prices have been in effect (long-run elasticity will be greater than short run) and to the inadequacy of summary statistics in isolating price response in an environment where stronger, more influential variables, such as economic activity or weather, are present. Also, the use since the early 1930s of a rate with no off-peak demand price (called the off-peak excess demand provision) probably mitigated the response to the new prices.

The Wisconsin Public Service Commission examined the potential environmental effects of time-of-day pricing implementation in a generic environmental impact statement.[33] For the WEPCO rates, a particular impact statement was prepared before granting final rate approval.[34] Several groups and individuals opposed to time-of-day price adoption appealed the commission's rate action on the grounds that the environmental impact analysis and statement were inadequate. However, their appeal was denied because it was found that ''the Commission clearly considered environmental factors in arriving at its decision.''[35]

PROPOSED FRAMEWORK FOR THE CONSIDERATION AND IMPLEMENTATION OF COST-BASED, TIME-OF-DAY ELECTRICITY RATES

The experience of utilities that have already implemented time-of-day rates for electricity provides important background and input for formulating a framework for considering and implementing cost-based, time-of-day rates. The proposed framework contains the following activities and steps: (1) load research, (2) rate design, (3) cost-benefit analysis, (4) information flow with customers, (5) actual application, and (6) reporting and analysis of results (actual experience). Figure 1 provides a diagram of these activities

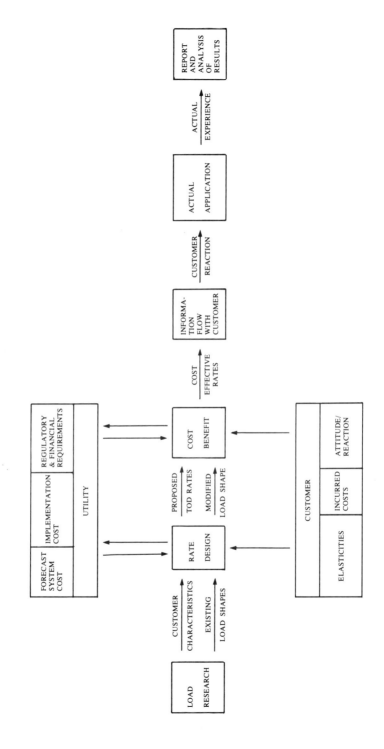

Figure 1
ACTIVITIES FOR CONSIDERATION AND IMPLEMENTATION OF TIME-OF-DAY RATES

and their interrelations. The PURPA addresses some of these activities, specifically load research, rate design, and cost-benefit analysis. This section of the paper discusses activities in the framework and provides references for additional information.

Load Research Activities

Load characteristics are one of the factors that affect every step in the framework. By providing time, place, classification, and form to an otherwise amorphous, aggregate usage profile, load research provides data concerning the potential savings by using load management, time-of-day rates, and load controls. Load research activities are of benefit to utilities in various ways; the data can be used for forecasting future demand for electricity usage, estimating growth patterns by customer, forecasting future revenues, and designing rates.

The following guide-lines concerning load research are proposed for regulatory commissions and electric utilities that are examining the feasibility of implementing a time-of-day pricing programme.
1. Clearly define the expected uses of the load research data, such as forecasting electric demand, formulating pricing periods, and providing billing information to customers.
2. Select customer groups or subgroups for a load research programme and establish relevant sampling techniques for each group. The following techniques are proposed.
   a. Assign meters to all large industrial and commercial customers that have a maximum demand (kW) greater than a specified level.
   b. Develop a statistical sampling procedure to allocate meters to a portion of the remaining commercial and industrial customers. Because this is such a heterogeneous group, it may not be sufficient to stratify the sample only by consumption levels. *The sampling procedure could also consider, in addition to usage, such variables as maximum demand or industry classification.*
   c. Formulate a statistical sampling procedure to allocate meters to residential customers. For this group, which is a relatively homogeneous class, sampling by annual (average) consumption levels or strata is probably sufficient to obtain a statistically valid sample.
   d. Determine the number of meters that should be allocated to each consumption stratum within a customer class. Because meter failures and other problems may decrease the amount of usable data, the number of meters allocated to each stratum should be substantially more than the minimum number necessary to obtain a statistically valid sample.
   e. Develop a load research programme with sufficient flexibility so that customers that significantly change their level of consumption can be assigned to a different stratum.

f. If resources for load research are very limited, the meters and other equipment should be allocated in such a way as to maximize the information needed to meet the objectives of the programme. For example, the meters could be allocated according to the customer groups or subgroup contribution to system peak demand.

3. Identify and obtain the computer equipment and programmes that are needed to analyse the complex and voluminous load data that will be generated. To be useful, equipment must be able to analyse data on a timely basis.

4. Select and install needed translators, readers, magnetic tape meters, and other related equipment. It is reasonable to assume that guide-lines 1 through 4 will take one year to implement.

5. Survey customers to determine the important similarities and differences among customers in a consumption stratum. Surveys should be taken both before and after the installation of meters and can be accomplished by written questionnaires, related personal interviews, and other similar techniques.

6. Twelve consecutive months of reliable load data are needed before time-of-day rates are implemented. These data are useful in providing important information to customers concerning the potential impacts on their electricity bills and in analysing changes in usage patterns caused by time-of-day pricing.

7. If adequate load data are currently available, the need for these preliminary activities is lessened. Therefore, the target date for actual implementation of time-of-day rates can be advanced.

The PURPA discusses the need to gather load research data.[36] The Rate Design Study has been conducting research in the area of load research for time-of-day rates.[37] The Rate Design Study is a national study on load management, time-of-day rates, and load controls that is being sponsored by the Electric Power Research Institute (EPRI), the Edison Electric Institute (EEI), the American Public Power Association (APPA), and the National Rural Electric Cooperative Association (NRECA) at the request of the National Association of Regulatory Utility Commissioners (NARUC).[38]

Rate Design Activities

After the load research programme has commenced, the next major activity necessary for considering and implementing time-of-day pricing is designing the rates. The process should begin during the time period (in the second year of the comprehensive programme) that relevant load data are being obtained. As more data are obtained and analysed, the rate design can be revised and refined. The following guide-lines for such rate design activity are suggested.

1. Separate the revenue requirement case from the time-of-day pricing case in order to minimize confusion about changes in customers' bills.
2. Estimate the rates based on an accepted measure of costs in order to formulate a defined quantitative starting point and compare these rates with present conditions. There are two basic measures of costs: accounting (average) costs and differential (marginal) costs.

    a. *Accounting costs*—Rates based on time-differentiated accounting costs require a cost-of-service study that differentiates usage by time of day. Alternative methodologies exist and are being refined for calculating rates based on accounting costs.[39]

    b. *Differential costs*—There exist alternative methods for estimating marginal cost (differential cost methods).[40] These methods differ most clearly in their handling of the marginal cost of generating capacity.

3. Specify the reasons (qualitatively or quantitatively) for deviations in rates from an accepted measure of costs. These reasons could include the following considerations: (a) revenue requirement, (b) rate simplicity, (c) revenue stability, (d) consumer impact, (e) historical practices, and (f) environmental impact. If strict cost-based rates are unacceptable for one or more of these reasons, then rates should move at least closer to costs.

4. *If necessary*, minimize adverse consumer bill impact in the short term by specifying a maximum allowable percentage increase in annual bill per customer (using the load data and assuming a zero price elasticity) during the initial stage of implementing time-of-day rates. However, if such an approach is adopted, customers should be provided with information and direction concerning probable future modifications in time-of-day rate design. This process could help customers in making their future investment decisions.

5. Examine (a) system and customer group demand (kW) patterns and (b) capacity availability, after including scheduled maintenance, in order to determine appropriate peak and off-peak time periods. Loss-of-load probability (supply side) and probability of contribution to system peak (demand side) provide useful quantitative measures for formulating the pricing periods, but considerations and adjustments for maintenance scheduling, pool versus company loads, and interconnections should be addressed. Practical judgement and experience should be used in formulating pricing periods. Therefore, consideration should be given to formulating time periods that enable customer groups to shift loads, that stabilize revenues for the utility, that reduce the need for capacity additions, and that improve system load factor.

Since PURPA specifies special rules concerning cost-of-service and time-of-day rates, rate-making activities assume *renewed* importance.[41] Research being conducted by the Rate Design Study on costs and rates should be helpful to regulatory commissions and utilities in considering PURPA.[42]

Cost-Benefit Analysis

Cost-benefit analysis provides a means of determining whether cost-based, time-of-day rates are cost effective. The following items should be considered in a cost-benefit analysis:

1. Select the criteria to be used in choosing among alternative rate structures. The criteria to measure cost of rates are numerous and varied. The cost-benefit analysis is not performed in a vacuum, and consequently the criteria chosen reflect the perspective of the analyst. A cost-benefit approach can be employed regardless of the criteria for cost-effectiveness chosen.

   a. The PURPA stipulates that long-run benefits to utilities and customers should exceed metering costs and other "associated costs" to be cost effective and specifies that this relationship must be achieved on a class-by-class basis.[43]

   b. Utility system planners and engineers use the criterion of whether a "sufficient" shift in load is achieved to evaluate whether rates are cost effective.

   c. Economists use the criterion of whether rates reflect costs at the margin to judge whether rates are efficient, and judge rates to be cost effective if they give consumers a correct price signal.

2. Develop a methodology that employs the criteria selected to evaluate alternative rate structures. Some cost-benefit methodologies have been developed to demonstrate cost advantages of load shifts under various time-of-day rate structures. Simulation methods based on generation expansion and production costing *and* a linear programming optimizing approach are two methodologies currently being developed and tested.[44]

3. Calculate the costs and benefits that will be inputs to the methodology. The shift in load shapes, which may take place when a price differential exists between peak and off-peak electricity rates, results in a cost savings to the utility due to more efficient use of existing capacity, deferral of new planned capacity, and substituting cheaper fuels for more expensive fuels. These savings do not come without associated increased costs, however. New metering equipment and maintenance costs, additional billing procedures, potential adverse customer relations, potential revenue erosion, and increased administrative costs are examples of additional utility costs. Customer inconvenience, additional investment in devices to use off-peak energy, costs associated with changing industrial processes, and possibly higher bills are examples of increased consumer costs.

4. Apply the methodology and select among alternative rate structures on the basis of the chosen criteria. If the benefits exceed the costs, the rates are cost effective. Determining what are benefits and what are costs, identifying and possibly selecting between long-run and short-run costs and benefits, specifying to whom the benefits and costs accrue, and

ensuring that the two net out at each level (customer, utility, society) are questions that must be answered within the criteria/methodology/selection framework.

5. Consider time-of-day rates as part of a total load management strategy. Rates cannot be implemented in a *vacuum* but rather interact with and support other forms of load management such as load controls and thermal storage devices. In evaluating the costs and benefits of time-of-day rates, it is important to consider associated costs of the other aspects of the complete load management strategy.

The basic concern in applying any cost-benefit methodology is that only *assumed* load shapes are available at the time the analysis takes place. In addition to the question of how much load shifting is reasonable to expect from time-of-day rates, several other unresolved issues remain in a cost-benefit analysis.

1. To what extent should rates reflect general efficiency and equity considerations and to what extent should they be judged solely from the costs and benefits accruing to the individual utility and its customers?
2. What is the impact on the transmission and distribution system?
3. Excessive shift of load from peak to off-peak periods and needle peaking could result in unstable rates.
4. Large industrial plants may relocate away from areas in need of jobs.
5. Modification of load shape (especially baseload) may adversely affect utility financial ''health.''

Current research and implementation experience address many of these issues, and answers or at least insights should be available as results are published.

Information Flow with Customers

Customers' understanding and acceptance of time-of-day rates and of the objectives of utilities' implementation programmes determine the success of such programmes. Load shifting and cost savings to the utility are dependent on understanding of customer class characteristics and customer attitudes. Research has been done to identify key variables in communicating utility objectives to customers and obtaining the feedback necessary to modify existing programmes and design new rate restructuring programmes.[45] Additional utility resources must be committed to the task of working with each customer class individually to accomplish implementation objectives and to address customer concerns and reservations about the new rate schedules.

In addition to explaining proposed time-of-day rates, the following activities may facilitate the information flow with customers:

1. Preparation of one year of sample bills based on the utility's proposed new rates, to be compared with bills computed under the old rates

2. Instruction on how to use these sample bills to experiment with different activities to reduce expenses
3. A guarantee of the longevity of the new rates that would encourage basing long-run investment decisions on the new rate structure.

Actual Application

When the regulatory commission approves and orders the implementation of the new schedules of time-of-day rates, the first month's application activities are key to the success of the implementation. Examples of such activities are the following:
1. Introduction of new billing periods and procedures
2. Modification of meter-reading techniques
3. Answering of additional administrative and operational questions.
Introduction of the new rates may be staggered on the basis of month of implementation, class or subclass, usage level, or any other appropriate basis. In order for the new rates to reflect accurately the additional costs of implementation, many of these application decisions must be made earlier during the rate design period. This process is *iterative*.

Reporting and Analysis of Results

The reporting of actual experience and the analysis of results are important in the implementation process of time-of-day rates. The utility's experience with the new rates should be reported to the customers, the stockholders, and the concerned public. The following items should be considered:
1. Before and after load profiles by customer class
2. Comparison of typical bills under the old and new rates
3. *Ex post* cost-benefit analysis
4. Customer survey of attitudes and reaction to new rates
5. Utility plans for future implementation.

SUMMARY

This paper has attempted to provide regulatory commissions, electric utilities, and other interested groups with information concerning recent legislative developments and implementation activities relating to time-of-day pricing of electricity. Instead of "reinventing the wheel," electric utilities and regulatory commissions should seriously examine and benefit from existing time-of-day pricing programmes that have made progress in formulating, applying, and analysing these innovative and potentially useful rate structures.[46]

168 / *Energy Crisis*

NOTES

* The authors wish to express appreciation to the following professionals who contributed to the preparation of this paper: Dr. Hasso Bhatia, Michigan Public Service Commission; Mr. Stefanos Enkara, Public Utilities Commission of Ohio; Ms. Jean Frazee, Illinois Commerce Commission; Mr. Terry Nicolai, Wisconsin Public Service Commission; and Mr. Robert Stemper, Wisconsin Public Service Commission. The authors, of course, assume responsibility for the contents of this paper. This paper was presented at the Midwest Economics Association annual meeting held in Chicago, April 1979.

[1] For information on regulatory activities concerning innovative rate designs for industrial, commercial, and residential customers, see J. Robert Malko, "Implementing Time-of-Use Pricing," paper presented at the Engineering Economy for Public Utilities Program, Stanford University, Palo Alto, California, July 1978, and *1977 Survey, State and Federal Regulatory Commissions, Electric Utility Rate Design and Load Management Activities*, prepared by Elrick and Lavidge for the Electric Utility Rate Design Study, Electric Power Research Institute, Palo Alto, California, October 1977.

[2] Madison Gas and Electric Co., (Wis. 1974) Docket No. 2-U-7423. For analysis of the major issues in this case, consult Richard D. Cudahy and J. Robert Malko, "Electric Peak-Load Pricing: Madison Gas and Beyond," *Wisconsin Law Review* (1976): 47-73.

[3] *Public Utility Regulatory Policies Act* of 1978, Public Law 95-617, Title I, Sections 111 and 115, November 1978.

[4] *The National Energy Act, (NEA) General Information*, prepared by the U.S. Department of Energy, Office of Public Affairs, November 1978, Section I, p. 2.

[5] *Ibid.*, Section I, p. 1

[6] *Public Utility Regulatory Policies Act* of 1978, Public Law 95-617, Title I, Sections 101 and 111, November 1978.

[7] *Public Utility Regulatory Policies Act* of 1978, Public Law 95-617, Title I, Section III (d), November 1978. For a detailed discussion of rate-making standards and related issues in PURPA, see *Reference Manual and Procedures for Implementing PURPA*, prepared for the National Association of Regulatory Utility Commissioners by the Electric Utility Rate Design Study, Electric Power Research Institute, Palo Alto, California, Final Version, March 1979.

[8] *Public Utility Regulatory Policies Act* of 1978, Public Law 95-617, Title I, Sections 115(a)-(c), November 1978.

[9] *Ibid.*, Section 112(b).

[10] *Ibid.*

[11] Illinois Commerce Commission, Order No. 76-0698, November 26, 1977.

[12] Illinois Commerce Commission, Docket 76-0698.

[13] Illinois Commerce Commission, Docket No. 76-0045.

[14] Michigan Public Service Commission, Case No. U-4807, dated March 30, 1976, p. 68.

[15] Michigan Public Service Commission, Case No. U-5845.

[16] Consumers Power rates are found in Michigan Public Service Commission, Consumers Power, Case U-5331, order dated July 31, 1978; Detroit Edison's rates found in Michigan Public Service Commission, Detroit Edison, Case U-5502, dated September 28, 1978.

[17] Consumers Power Company, Electric Department, *Summary of the Time-of-Day Study Ordered in MPSC Order U-4840*, and Detroit Edison, *Effect of Time-of-Use Rate on Load Characteristics for Primary Rate Customers 1975 – 1978*.

[18] Testimony by Dr. Hasso Bhatia, Director, Office of Tariff Analysis, Michigan Public Service Commission, in Case U-5845, pp. 5 and 6.

[19] Consumers Power, Michigan Public Service Commission, order in Case No. U-4840, p. 71.

[20] Detroit Edison Company, Michigan Public Service Commission, Case No. U-5502.

[21] Ohio Public Utilities Commission, Case No. 76-892-EL-COI.

[22] Ohio Public Utilities Commission, Case No. 77-1249-EL-AIR.

[23] Ohio Public Utilities Commission, Order No. 78-923-EL-ATA, 14 June 1978.

[24] Ohio Public Utilities Commission, Case No. 78-92-EL-AIR, 28 December 1978, p. 5.

[25] "DOE: TOD Rates Shift KWH and KW," *Electrical World* (15 November 1978), pp. 110-12.

[26] Madison Gas and Electric Co., Docket No. 2-U-7423. For analysis of the major issues in this case, consult Richard D. Cudahy and J. Robert Malko, "Electric Peak-Load Pricing: Madison Gas and Beyond," *Wisconsin Law Review* (1976): 47-73.

[27] Wisconsin Power and Light Company, *Report on Load Management and Time-of-Day Rates*, 17 April 1978.

[28] For additional information, see "Developing and Implementing a Peak-Load Pricing Experiment for Residential Electricity Customers: A Wisconsin Experience" by Dennis J. Ray, J. Stanley Black, and J. Robert Malko for the Midwest Economics Association Meeting, Chicago, 6 April 1978.

[29] Wisconsin Public Service Commission, Docket 6630-ER-2/5, order dated 5 January 1978.

[30] Wisconsin Public Service Commission, order No. 6630-ER-2/5, 5 January 1978, p. 26.

[31] State of Wisconsin, Public Service Commission, *Environmental Impact Statement for the Proposed Wisconsin Electric Power Company Tariffs for Electric Utility Service*, Docket 6630-ER-2/5, March 1978, pp. 45-45.

[32] Wisconsin Electric Power Company, *Load Management Report in Response to PSC Order in Docket 6630-ER-2/5*, 16 February 1979.

[33] Wisconsin Public Service Commission, *Generic Environmental Impact Statement on Electric Utility Tariffs*, Docket No. 1-AC-10, June 1977.

[34] State of Wisconsin Public Service Commission, *Environmental Impact Statement for the Proposed Wisconsin Electric Power Company Tariffs for Electric Utility Service* Docket 6630-ER-2/5, March 1978.

[35] State of Wisconsin, Circuit B. Court Branch IV, Dane County, *Wisconsin's Environmental Decade, Inc., et. al vs. Public Service Commission of Wisconsin*, Memorandum Decision dated February 20, 1979, p. 31.

[36] *Public Utility Regulatory Policies Act* of 1978, Public Law 95-617, Title I, Section 133(a), November 1978.

[37] *1978 Plan of Study*, Electric Utility Rate Design Study, Electric Power Research Institute, Palo Alto, California, December 1977, pp. 9-10 and RDS Report 74.

[38] The first phase, 1975–1977, of the Rate Design Study has been completed, and the second phase, 1978–1981, of research is currently being completed. A listing of available reports is presented by the *Guide to Electric Utility Rate Design Reports*, Electric Power Research Institute, Palo Alto, California.

[39] *Rate Design and Load Control: Issues and Directions, A Report to the National Association of Regulatory Utility Commissioners*, Electric Utility Rate Design Study, Electric Power Research Institute, Palo Alto, California, November 1977, pp. 26-32.

[40] The Rate Design Study has conducted research on comparing alternative methodologies for estimating marginal costs, and a final report is available. See *1978 Plan of Study*, Electric Utility Rate Design Study, Electric Power Research Institute, Palo Alto, California, December 1977, p. 6 and RDS Report 66 and 67.

[41] For a discussion of costing and rate-making issues in PURPA, see *Reference Manual and Procedures for Implementing PURPA*, prepared for NARUC by the Electric Utility Rate Design Study, Electric Power Research Institute, Palo Alto, California, Final Version, March 1979.

[42] *1978 Plan of Study*, Electric Utility Rate Design Study, Electric Power Research Insitute, Palo Alto, California, December 1977, pp. 3-10.

[43] *Public Utility Regulatory Policies Act* of 1978, Public Law 95-617, Title I, Section 115(b), November 1978.

[44] For a discussion of existing cost-benefit methodologies, see *Rate Design and Load Control: Issues and Directions*, A Report to NARUC, Electric Utility Rate Design Study, Electric Power Research Institute, Palo Alto, California, November 1977, Chapter 7. The Rate Design Study has conducted research to refine and develop two cost-benefit methodologies, and final reports are available. See RDS Reports 75 and 78.

[45] *Rate Design and Load Control: Issues and Directions*, A report to NARUC, Electric Utility Rate Design Study, Electric Power Research Institute, Palo Alto, California, November 1977, Chapter 6.

[46] Reports on the experiences of three utilities that have implemented time-of-use rates include the following: Richard A. Abdoo, Wisconsin Electric Power Company, "Does Time-of-Use Pricing Work?," presented at the 1979 Symposium of Problems of Regulated Industries, Kansas City, February 1979; Wisconsin Power and Light Company, "Report on Load Management and Time-of-Day Rates," April 1978; Pacific Gas and Electric Company, "Time of Use Rates for Very Large Customers," March 1978.

Chapter Nine

3230
6131

Tax Normalization, Regulation, and Economic Efficiency[1]

by
E. R. Berndt, J. R. Kesselman and *G. C. Watkins*

INTRODUCTION

The question of whether regulated utilities should be required to "flow through" the tax benefits of accelerated depreciation to their customers remains controversial. The principal alternative is some form of "normalization," by which the taxes charged for rate-setting purposes are those payable if straight-line rather than accelerated depreciation were required. The relative merits of these two regulatory approaches have generally been assessed in terms of accounting, financial, and "equity" considerations. This article deals with a neglected aspect of the issue: economic efficiency.

More specifically, the issue of how to treat tax liabilities arises when the rate of depreciation allowable for tax purposes—which hereafter we call the capital cost allowance—differs from the rate of depreciation chargeable for accounting purposes in defining net company income, which we shall call accounting depreciation. It is assumed that accounting depreciation is computed on a straight-line basis, the normal treatment for regulated utilities in both the United States and Canada.

Accelerated depreciation is no more than the term suggests. Tax is deferred, not reduced. Total chargeable depreciation over the life of an asset remains the same under either the accelerated or accounting schedules.[2] The accounting profession in Canada recommends that under these circumstances, a deferred tax reserve should be established to reflect the reduction in current tax payments afforded under accelerated depreciation.[3] The accountants argue that part of the value of an asset—its right to recover historical cost free of tax—is prematurely consumed, which should be recognized as a current cost, not an increase in earnings. When the capital cost allowance falls below the level of accounting depreciation, the deferred tax reserve would be disbursed to offset higher tax payments later in the life of an asset. This type of procedure would "normalize" taxes by treating taxation charges as a function of accounting depreciation rather than capital cost allowances.

Under present utility regulation, agencies normally require straight-line depreciation for accounting and rate-making purposes. The establishment of a deferred tax reserve is often not allowed.[4] Instead, any reduction in current tax payments is treated as an increase in earnings. The benefits of tax deferral "flow through" to customers. An argument occasionally made here is that as long as a company grows, tax postponement is tantamount to a tax reduction and should be enjoyed as a customer benefit. However, it is usually assumed that any deferred tax reserve established under normalization would carry an interest charge that would be credited—implicitly or explicitly—to customers of the utility. This assumption removes a distributional issue between the utility and its customers, which frequently arises in discussing normalization and flow-through.[5]

Advocates and opponents of tax normalization have advanced numerous arguments in support of their positions. Arguments that primarily concern financial and accounting aspects will not be discussed here. These include the propriety of providing for uncertain future expenditures; costs measured as the ability to generate tax deductions; and the permanence of tax deferral. In this paper, attention is confined to those issues actually or seemingly involving economic efficiency.

The contents of the article are as follows. The first section concerns tax normalization and the question of risk under regulation. The second section discusses the capital intensity of a regulated industry and tax normalization. Potential effects on the capital durability of a regulated industry are examined in the third section. Resource allocation within an economy and tax normalization are discussed in the fourth section. The fifth section deals with certain distributional and economic welfare aspects. Our conclusions are summarized in the sixth section. At various points in our discussion, the case of a natural gas pipeline is used by way of illustration.

RISK AND TAX NORMALIZATION

We begin examining the problem of an investment decision that involves a risky future return by considering a competitive non-regulated industry, characterized by assured sales this year and uncertain sales next year. The behaviour of a competitive industry is the conventional standard used by economists to analyse economic efficiency.

How would a perfectly competitive industry approach the related investment decision, assuming the industry's machines can be constructed economically only with two years' physical durability? Clearly, existing and prospective firms would calculate a higher necessary rate of return in the first year of production to offset the chance of a low return in the second year. Only a higher than risk-free market rate of return in the first year—when a two-year asset life is utilized—would yield the socially optimal, or efficient, amount of investment. Viewed differently, a market rate of return will give the socially ''correct'' solution only if machines were depreciated fully over a period of less than two years. The uncertainty over future sales makes the expected economic life of the machines shorter than their physical life.

The analogy for a regulated industry with a risky future return is straightforward. To obtain the efficient amount of investment in such an industry *vis-à-vis* an industry with relatively riskless future earnings, a higher rate of return must be allowed the former in the earlier life of projects. Thus, the proper decision by a regulatory agency requires some advance estimates of the degree and nature of risks involved in particular projects. The corresponding adjustments might take the form of raising the permitted rate of return or decreasing the accounting life over which the asset can be depreciated for rate-setting purposes.

It could be argued that normalization is an alternative way of providing the higher rate of return for risky projects. In this case, there might not be any necessity, as assumed earlier, to levy an interest charge on deferred tax contributions. However, this would appear to be an unsatisfactory approach to the problem. First, it is not known *a priori* whether normalization would provide sufficient benefits to the firm for equivalence to a risk-compensated rate of return. Second, projects differ in their degree of risk, and some finer calculations would seem in order. Whereas one project might warrant full normalization, a second might warrant half normalization, while yet a third might warrant more incentives than full normalization.

In short, a general policy of normalization appears to be a poor substitute for a more direct approach to risk compensation. This is not an argument against normalization *per se*; it is merely an argument that the presence of risk alone is an insufficient justification for normalization. Regulated rates of return may already reflect the average riskiness of regulated firms' investments.

Normalization permits the firm to recover the taxes on a particular project more quickly, against the risk that the project will not generate sufficient revenues in its later years. But the occurrence of this very outcome—low revenues and perhaps current losses from the project in later years—also implies a correspondingly lower tax liability. Thus, it is possible that the firm will have accumulated deferred tax reserves in the early years of a project that exceed the tax liabilities generated by the project in its later years. Of course, the firm would still benefit by applying these deferred tax reserves against the tax liabilities of other projects that are profitable in those years, but then the logical connection between the particular project and its deferred tax reserves has been weakened.

An important risk faced by a regulated utility is that its permissible rate of return would not be raised high enough to cover all of its revenue requirements. This danger would normally be associated with a shortage in one of the inputs needed in the utility's operation or a decreased demand for its service. Thus, even if the regulatory agency permits the regulated price to increase substantially, there may be a maximum feasible revenue that can be generated. Suppose the firm's maximum feasible revenue falls short of its full revenue requirements by an amount equal to its planned corporate tax liability. Then the return to owners of the firm will fall short of the regulated return by this amount, except that the loss of normal return will be partially cushioned by reductions in the current corporate tax liability itself. It might be argued that establishment of a deferred tax account in previous years would have protected the firm against this outcome. This is not a satisfactory argument against the proposition that differentially high project risk should be handled by differentially high regulated rates of return.

Another argument advanced in favour of tax normalization is its ability to reduce the cost of capital to the company even if full interest were charged to the deferred tax reserve. The reasoning here is that investors perceive a lesser risk associated with a company allowed to practice tax normalization rather than flow-through. Therefore, funds will be forthcoming at a lower interest cost than would otherwise be the case. There is some theoretical support for this position in the sense that, other things being equal, normalization would improve debt coverage ratios; there is also some econometric evidence of lower debt cost with tax normalization, but not for equity.[6]

CAPITAL INTENSITY OF THE REGULATED INDUSTRY

The provision of accelerated depreciation under the tax laws was primarily intended to induce greater capital formation. This objective was accomplished by lowering the price of capital relative to the prices of other inputs to production. Firms tended to substitute the input whose price had been depressed for other inputs.

In the non-regulated sector, ordinarily we would consider the distortion of the relative prices of inputs to be inefficient. An alternative approach is to hypothesize that the tax incentives favouring the demand for tangible capital are somehow offsetting, or are intended to offset other tax distortions that inhibit the supply of capital (mainly the personal and corporate income taxes). In this case, it might be argued that investment incentives promote the overall efficiency of the economy. Otherwise, arguments related to the goal of economic growth are needed to support tax incentives for investment. For present purposes, we take the legislative intent in stimulating investment to be justifiable—for efficiency or other economic objectives.

The question of immediate concern, then, is whether a regulated firm will respond to tax incentives for investment. Further, are these incentives enhanced by the flow-through or by the normalization approach? With flow-through, the tax savings in the early years of a project resulting from accelerated depreciation are passed along to the customers of the firm. This will leave unaffected the price of capital relative to the price of other inputs, so that the firm's capital intensity should be unaltered. Similarly, in the later years, the heavier tax burdens will be recovered from customers through adjustments to the price of services. Again, there is no incentive for the firm to change its capital intensity so long as there is no change in risk. The consideration of risk in the later period's earnings implies a differential impact on investment under flow-through with the introduction of accelerated depreciation. The regulated service price would need to be higher in the later years of the project, which might reduce demand. This could induce lower utilization during the later years of the project life, and even though there would be partially offsetting adjustments in the regulated service price, the forecast of project profitability could be lowered. Under normalization, this contingency would not arise, so that investment incentives would not be adversely affected by the use of accelerated depreciation.

With normalization, the introduction of accelerated depreciation in the place of straight-line depreciation may offer additional inducements to invest for a regulated firm. Since the firm's (maximum) rate of return is regulated, the question is whether the firm is able to increase its rate base through the creation of deferred tax reserves. Then it must be known whether the increased rate base carries any increase in the firm's tangible capital. The outcome depends upon whether the interest charged to the deferred tax reserves leaves the firm with any benefits. The fact that normalization improves the firm's ''coverage'' ratio between current revenues and interest expense will tend to improve the credit standing of the firm. Consequently, it may be able to borrow at a lower interest rate, with a resultant lowering in the price of capital relative to other inputs. But as discussed earlier, while such an effect might be plausible, empirical evidence of its existence is confined to debt interest costs.

To the extent that normalized firms enjoy lower borrowing costs than firms treated on a flow-through basis or are able to earn a higher return on equity through normalization, normalization would tend to transmit the investment incentives inherent in accelerated depreciation more effectively to regulated firms than would flow-through. However, such effects are likely to be rather small and are contingent on the stage of company development.

CAPITAL DURABILITY OF THE REGULATED INDUSTRY

In non-regulated industries, accelerated depreciation induces firms to choose less durable capital goods when expanding or replacing worn out capital goods. This can be explained by a simple example. Suppose first that only straight-line depreciation were permitted for tax deduction and further that it accurately measured true economic depreciation. The firm is facing the

choice between two capital goods, one more durable than the other, each of
which would provide the same amount of flow of services over its respective
life. The less (more) durable good has a capital cost of K_a (K_b) and a service
life of t_a (t_b).[7] We assume that the services provided by each machine would
be identical and would not involve differential maintenance costs. We also
assume that the firm would be indifferent between the two goods when only
straight-line depreciation were allowed, since both would give rise to the
same net-of-tax costs per period:

$$(1-\alpha)iK_a + (1-\alpha\beta) K_a/t_a = (1-\alpha)iK_b + (1-\alpha\beta) K_b/t_b$$

where i is the interest rate per period, α is the corporate marginal tax rate, and
β is a parameter reflecting the proportion of accounting depreciation
claimable for tax purposes. Thus, $(1-\alpha)iK$ would be the interest expense per
period for the respective good net-of-tax savings from its deductibility; the
term $(1-\alpha\beta)K/t$ would be the depreciation cost per period net-of-tax savings.
Since the more durable good would be more expensive initially ($K_b > K_a$),
the equation can hold only if $K_a/t_a > K_b/t_b$. That is, the economic
depreciation per period must be higher on the less durable good to offset its
lower interest cost and make the firm indifferent between the two goods.
Clearly, if the tax laws permit only the deduction of straight-line
depreciation, the parameter $\beta = 1$. The introduction of accelerated depreciation for tax purposes is like an increase in the value of β above unity in the
early years of the project.[8] This will swing the balance in favour of the less
durable good, since depreciation plays a larger role in its user cost as against
the more durable good.

The finding that tax allowances under accelerated depreciation induce
competitive firms to choose a less durable capital stock has no clear
efficiency interpretation. There is at least a mild presumption that the induced
change is in an inefficient direction, since efficiency is usually gauged by a
laissez-faire setting. However, there may be social benefits to an economy
from having a capital stock that is less durable. This might arise from the
development of technological innovations over time, which increase the
productivity of new capital goods, but which cannot be incorporated in
existing capital goods. An economy with less durable capital will have on
average a less obsolete capital stock and therefore a more productive one. It
might be argued that this "option" value of using less durable capital is
already recognized by entrepreneurs in a competitive economy and that any
further tax inducement in this direction is not efficient. However, some
technological innovations are "unanticipated," and a less durable capital
stock stands to benefit more. This issue has not been resolved by economists,
so we must leave open the efficiency evaluation of the inducement toward
less durability. To our knowledge, no empirical evidence has been
forthcoming on the quantitative effect of accelerated depreciation on capital
durability.[9]

Turning now to the regulated industry, we see a different set of incentives at play. The firm would prefer to increase its rate base on which the permissible rate of earnings is calculated by the regulatory authority. Depending on the regulations, this may result in an enlarged capital stock—the familiar Averch-Johnson effect[10]—as well as the possibility of a more durable capital stock. The latter outcome is one method of justifying more capital for a given rate of flow of capital services.

We now consider the incentives that are imposed on the regulated firm by the alternative treatments for accelerated tax provisions. Moving from straight-line to accelerated depreciation under the flow-through method would appear to have minimal effects on the firm's choice of durability. However, under normalization, the introduction of accelerated depreciation would induce the firm to choose less durable capital goods to the extent that it derives financial or psychic benefit from accumulating deferred tax reserves. This follows because the difference between accelerated and straight-line depreciation will be larger for the less durable goods, given their greater rate of depreciation per unit of capital service. In summary, normalization may cause the regulated industry to respond in a way analogous to the unregulated competitive industries. But the efficiency assessment of this move toward less durable capital does not have any generally accepted, unambiguous economic assessment.

ECONOMIC RESOURCE ALLOCATION AND TAX NORMALIZATION

Flow-through treatment of accelerated depreciation for a regulated firm leads to a pattern of service prices that varies over time, even though all input prices may be constant. If the firm had a single project, or a set of projects all begun at the same time and having the same service life, flow-through would generate low prices early in the life of the project, with the price rising over the life of the project as the depreciation tax deductions became exhausted. Of course, this distinct pattern could become blurred for a firm having many projects begun at different dates or having different service lives.[11] If again we abstract from general inflation, the use of normalization leads to a constant tax component in the service price over the life of a given project. For a firm with more than one project, the average service price could vary over time, but only as a result of changes in the nature of new projects or of capital replacements for fully depreciated projects.

To obtain an efficiency evaluation, we examine the non-regulated competitive sector of the economy. How would a firm in this sector behave with the introduction of accelerated depreciation for tax purposes? No competitive firm could afford to vary its product price over the life of a particular capital good to reflect the timing of benefits from accelerated depreciation. If it tried to do this early in the life of the project, it would set a low price. But since other firms could do the same thing when this particular firm's project exhausted most of its depreciation allowances, it could not successfully compete in later periods by charging a correspondingly high price. The forces of competition within the market for any particular commodity will compel all firms to charge the same price, regardless of how old their plant and equipment and regardless of how much depreciation is currently claimable. Thus, each firm will be forced to take a long-run view of the benefits of accelerated depreciation. To the extent that such tax provisions raise the rate of return in the industry above a ''normal'' return to capital, additional entry or expansion in the industry will be encouraged. This will bid down the market price of the goods or service and leave it at a stable level in the new industry position.

The competitive outcome relates something important about economic efficiency. A commodity price will equal its long-run marginal costs of production, after taking the corporate income tax as a cost to be added to the ''normal'' rate of return to capital, and after subtracting out the tax savings (properly discounted) from the deductibility of depreciation. Thus, accelerated depreciation enters as part of a long-run calculation, and its impact on commodity price is spread evenly over the capital service life. The analogy of tax treatment for the price setting of a regulated firm suggests that normalization is the closest counterpart to the competitive sector in terms of the tax component of rates. Flow-through yields variations in the price of output over time attributable to tax treatment, which would have no counterpart in a competitive industry.

We now examine the intertemporal pricing of a service where the nature of the service is essentially uniform over the facility life. For example, this property tends to characterize the transportation services provided by a utility such as a gas pipeline. Pipeline investments are principally long term. The eventual gas resources discovered and developed must be substantial to support a viable pipeline transportation network. Suppose pipeline systems were built to move gas on an even-flow basis with uniform operating costs. In this case, economic efficiency would not distinguish between costs assignable to gas transported at different times; unit costs allocated would be the same. Thus, we infer that normalization will be more consistent than flow-through with the uniform charge concept for services, which do not vary over time.[12]

The consideration of intertemporal efficiency can be extended to the sectors of the economy that use the services of the regulated sector. Arbitrary variations in the service prices of the regulated industry could induce further inefficiencies via the response of the non-regulated competitive users, involving investment commitments in non-regulated firms that over- or under-economize on use of the regulated service, and which themselves have long lives. Thus, normalization appears to be more consistent than flow-through with efficiency in the economy's intertemporal and intersectoral allocation of resources.

SOME DISTRIBUTIONAL AND ECONOMIC WELFARE ASPECTS

Two distributional issues are raised by the normalization versus flow-through controversy, neither of which is central to the assessment of efficiency. Normalization provides the regulated firm with funds in a deferred tax account. Unless there were an explicit or implicit credit to the firm's customers, interest-free funds would have been made available to the firm. In the case where there were an explicit interest charge, but not for the whole amount of the return on the funds, an increment to the approved rate of return apparently would be provided as remuneration to equity. This would redistribute funds from customers to the firm. Such an effect is analogous to an administrative or legislative decision to raise the rate of return that may be earned by the regulated utility.

Rather than leaving this question open, we assume that the regulated rate of return prior to the hypothetical move to normalization was already the proper one from the public policy standpoint. Thus, any move to normalization should be accompanied by an interest charge against the deferred tax account, with interest credited directly to customers or else offsetting the revenue requirement. This assumption also was made for most of the preceding analysis.

A second distributional issue has often been mentioned by proponents of normalization. This is the observation that flow-through treatment of the tax benefits from accelerated depreciation yields a lower utility service price for customers today as against customers later in the life of the facility. Such an outcome is claimed to be "inequitable" or "unjust" as between the two generations of customers.

The crucial observation is that the intertemporal pattern of utility service prices is fundamentally an efficiency question. If economic efficiency were to dictate a higher future price than the current price, *per se* this is neither "just" nor "unjust" for future customers. If public policy makers prefer to redistribute income to present customers by having differential service prices across time, then they should be conscious of the efficiency cost of distorting the efficient price pattern. Thus, the essential economic problem is reduced to one of determining whether normalization (which implies equal tax charges) or flow-through (which implies differential charges) is more efficient over time.

Implications for the economic welfare of Canadians relative to residents of other countries also may arise. Let us take the case of foreign buyers of utility services who enjoy lower prices through tax flow-through but do not attract higher prices when flow-through produces higher tax liabilities. For example, consider a pipeline exporting gas, when the price of the export gas is determined on the basis of an exogenous producer price plus pipeline tariffs (an "additive" mechanism). Suppose the bulk of gas exports were taken during the early part of accelerated tax provisions and export volumes were terminated before the end of the pipeline life. If pipeline tariffs were set under flow-through, Canadians would suffer a loss of economic welfare. The loss to Canadians imposed by flow-through would comprise the tax contributions that would be obtained from export customers under normalization. In effect, tax normalization protects against the possibility that foreign buyers of a utility company's services will enjoy the benefits of accelerated depreciation without incurring the deferred costs.

Other welfare costs can be incurred when "net-back" rather than "additive" pricing mechanisms prevail, that is, when market prices are exogenous and producer prices are the market price less the utility company charge. For instance, at the present time, producer prices for natural gas in Alberta are determined on a net-back basis from deemed prices in the domestic (Ontario) and export markets. A variation in utility pipeline transportation costs results in a corresponding variation in the opposite direction in well-head prices. Thus any incentive effects and distortion of efficiency will impact on the natural gas exploration and development sector. Assuming such net-back pricing continues over the long term, the pattern of service charges under tax normalization or flow-through has implications for the extent to which gas resources are economically recoverable and for the intertemporal pattern of exploration and development of gas reservoirs.[13]

Other things being equal, pipeline tariffs will be higher over the latter part of a pipeline's economic life under tax flow-through compared with tax normalization. Producer net-backs will be correspondingly lower during this latter period. Assuming the normalization "cross-over" point precedes reservoir exhaustion for the bulk of the reserves tied into the system, ultimate resource recovery will tend to be lower under tax flow-through, since reservoir recovery is positively related to eventual producer net revenue.[14] In short, reservoir abandonment may be premature.[15] Because tax charges are treated as transfer payments not involving the use of real resources, tax flow-through imposes a social loss on the economy. The effect is analogous to the impact of royalty schedules on resource recovery where such schedules are not truncated at the point where real marginal costs approach prices. Clearly, this outcome is economically inefficient; it would be possible under normalization to obtain more usable gas resources.

The use of tax flow-through methods has further effects on the intertemporal pattern of resource exploration and development. It may induce the premature development of some gas resources because the full tax benefits of accelerated depreciation are flowed through to producers. This consequence under tax flow-through was probably not intended by the framers of accelerated depreciation tax provisions. Their likely intentions were to stimulate the creation of additional tangible capital rather than the premature exploitation of natural resources.

CONCLUSIONS

We have focused on the relationship between tax flow-through, tax normalization, and economic efficiency in the context of a regulated utility. Our standard of efficiency is the behaviour of firms within a hypothetical competitive industry—as in conventional economic analysis.

Where future risk is prevalent, competitive markets would require a higher return in early years to compensate for future uncertainty. Normalization offers a way of providing such a higher return, but it would satisfy such objectives only by coincidence rather than by design. Normalization could lower the cost of capital for a regulated utility, but empirical evidence on this point is confined to debt costs.

The intent of accelerated depreciation was mainly to induce greater capital formation. Normalization will tend to transmit such investment incentives to regulated firms more effectively than would flow-through. Normalization will also tend to cause substitution of less durable for more durable capital goods, in so far as the nature of the utility's operation allows. The efficiency implications of such a change in the composition of the average capital stock in the regulated sector are not clear, although a less durable capital stock would be preferable if unanticipated technological innovations emerged.

More important is the question of the allocation of economic resources over time. A competitive system would tend to spread the impact of accelerated depreciation evenly over the service life of the capital asset. This suggests that tax normalization corresponds more closely to the competitive standard than does tax flow-through. Moreover, normalization will tend to avoid inducing inefficient behaviour elsewhere in the economy.

One special aspect is the efficiency of resource recovery where utility charges are a component in a net-back pricing mechanism. For example, in the case of a gas pipeline, tax flow-through could result in dead-weight economic losses from lower levels of resource recovery. Such losses could be avoided under normalization. National economic welfare can also be affected where additive rather than net-back pricing methods prevail. Here, foreign buyers of a commodity whose price includes utility charges could enjoy the benefits of tax normalization without incurring the deferred costs, if their later purchases were curtailed.

NOTES

[1] We acknowledge financial support for this work from Nova Corporation of Calgary, Alberta, but do not implicate it in any opinions expressed.

[2] The present value of the associated tax liabilities will be different, as will the real (constant) dollar values, given continuing inflation.

[3] See Canadian Institute of Chartered Accountants, *Accounting Recommendations*, paragraph 3470.13, p. 1924 (Toronto: September 1973). Provision is made for exempting regulated utilities in certain "rare situations." See *op. cit.* paragraph 3470.56 to 3470.58 inclusive.

[4] The use of normalization is limited although growing in Canada and the United States.

[5] See also discussion in the section "Some Distributional and Economic Welfare Aspects." For details on the analytics of tax normalization, see P. B. Linhart, "Some Analytical Results on Tax Depreciation," *Bell Journal of Economics and Management Science* 1 (Spring 1970): 82-112.

[6] For econometric evidence on debt costs using U.S. data, see E. R. Berndt, K. Sharp, and G. C. Watkins, "Utility Bond Rates and Tax Normalization," *The Journal of Finance* 34 (December 1979): 1211-20. A study on the cost of equity capital for electric utilities found a negative, though generally insignificant, relationship between flow-through and the cost of equity. See A. A. Robichek, R. C. Higgins, and M. Kinsman, "The Effect of Leverage on the Cost of Equity Capital of Electric Utility Firms," *The Journal of Finance* 28 (May 1973): 353-67.

[7] $t_b > t_a$.

[8] Discounting over the lives of the projects is ignored here, since the basic property is accurately portrayed by the first-period impact.

[9] For some analysis of this matter, see E. M. Sunley Jr., "The 1971 Depreciation Revision: Measures of Effectiveness," *National Tax Journal* 24 (March 1971): 19-30.

[10] See H. Averch and L. L. Johnson, "Behavior of the Firm Under Regulatory Constraint," *American Economic Review* 52 (December 1962): 1053-69.

[11] But the results of Linhart, *op. cit.* indicate that, in the long run, the distinct pattern would still re-emerge.

[12] This is analogous to the optimal pricing patterns required to minimize the distortion of consumer choices where marginal costs imply an upward adjustment to meet revenue requirements. Where demand for the utility company's services is not changing rapidly over time, a uniform percentage mark-up over marginal costs for all periods is preferable. See W. J. Baumol, "Optimal Depreciation Policy: Pricing the Products of Durable Assets," *Bell Journal of Economics and Management Science* 2 (Autumn 1971), p. 655.

[13] The analysis below is contingent upon a non-trivial divergence between the paths of service prices obtaining under flow-through and normalization respectively. This divergence grows with (i) the difference between accounting depreciation and the capital cost allowance; and

(ii) the share of depreciation in the total revenue requirements of the utility. The method of crediting the customer for the deferred tax account also affects the extent of the divergence over time.

[14] In 1972, the Alberta Energy Resources Conservation Board estimated that the recovery of gas from known reservoirs could increase by 5 TCF if gas prices rose by 10¢/MCF. See Energy Resources Conservation Board Report 72-E-OG, *Report on Field Pricing of Gas in Alberta* (Calgary: August 1972), Sections 6-14. For a general treatment of the effect of prices on petroleum recovery, see R. S. Uhler, ''Economic Concepts of Petroleum Energy Supply,'' in *Oil in the Seventies*, edited by G. C. Watkins and M. A. Walker (Vancouver: Fraser Institute, 1977).

[15] We cannot entirely ignore the possibility that some yet-to-be-discovered gas reserves would not be developed at all with lower net-backs that would obtain in the future under tax flow-through. However, as long as the price of gas continues to rise relative to the general price level, such an effect would likely defer rather than preclude exploitation. But we infer that any difference in the timing of development induced by tax flow-through is not socially beneficial, since it would be distortive.

Chapter Ten

Automatic Price Adjustment Clauses and Input Choice in Regulated Utilities

by
Scott E. Atkinson and *Robert Halvorsen**

INTRODUCTION

The requirement of regulatory approval of output price is pervasive in the energy, telecommunications, and transportation industries. Due to regulatory lags in the adjustment of output price, the high rates of inflation experienced in recent years have threatened the financial viability of major firms in these industries (Joskow and MacAvoy, 1975; Kendrick, 1975). One reaction to the new conditions has been the adoption of automatic price adjustment clauses, which allow automatic adjustments in output price in response to changes in specified cost elements.

Considerable concern has been expressed over the possible effects of automatic adjustment clauses on efficiency (C.R.S., 1974; N.A.R.U.C., 1974), but there has been almost no theoretical analysis of these effects. In this paper, we analyse the effects of fuel adjustment clauses on input choice in electric utilities.[1] The results are readily translated to other types of adjustment clauses and other industries.

The analysis indicates that the introduction of a fuel adjustment clause can result in either an increase or decrease in the use of fuel relative to other inputs. The direction of the effect depends on the elasticities of demand for output and for fuel as well as the ratio of total revenue to fuel costs. For the most plausible range of values for these arguments, the effect of fuel adjustment clauses in a period of rising fuel prices is to increase the use of fuel relative to other inputs. The net effect of regulation on input choice involves the effects of rate-of-return regulation as well as fuel adjustment clauses. Analysis of a model incorporating both types of effects indicates that they have opposite signs, and the net effect of regulation on fuel use cannot be determined *a priori*.

REGULATORY MODELS

We consider the effects of regulation on input choice under three sets of assumptions. In Case 1, it is assumed that output price is continuously adjusted to reflect the allowed rate of return on capital. The model used in this case is a modified version of the Averch-Johnson model. In Case 2, it is assumed that the only allowed changes in output prices occur in response to changes in the cost of fuel through the operation of the fuel adjustment clause. The firm is assumed to ignore the possibility that rate-of-return regulations will be enforced during a future regulatory review. The combined effects of rate-of-return regulation and fuel adjustment clauses are analysed in Case 3 by introducing the possibility of a future review into the model used for Case 2.

Case 1: Continuous Regulatory Review

In their path-breaking analysis of the effect of rate-of-return regulation on input choice, Averch and Johnson (1962) assumed that the rate-of-return constraint is continuously enforced. Given this assumption, fuel adjustment clauses are redundant, since output price would be continuously adjusted to yield the allowed rate of return given the prices of all inputs in each period. In this section, we review the effects of continuously enforced rate-of-return regulation using a dynamic version of the Averch-Johnson model for the case of three inputs.

In the continuous enforcement model, the firm maximizes the present value of profits subject to the constraint that the price of output in each period results in the firm earning the allowed rate of return on capital. Mathematically, the firm's problem is to maximize[2]

$$\int e^{-\delta t} \{P_t Q_t(P_t) - w_t L_t[Q_t(P_t), K_t, F_t] - r_t K_t - g_t F_t\} dt, \tag{1}$$

subject to

$$P_t = \frac{s_t K_t + w_t L_t[Q_t(P_t), K_t, F_t] + g_t F_t}{Q_t(P_t)}, \tag{2}$$

where P = output price
       Q = quantity sold
       L = labour input
       K = capital input
       F = fuel input
       w = price of labour
       r = cost of capital
       g = price of fuel
       s = allowed rate of return, $s > r$.

Omitting time subscripts, the relevant first order conditions are

$$-w \frac{\partial L}{\partial K} - r + Y(s + w \frac{\partial L}{\partial K}) = 0, \tag{3}$$

$$-w \frac{\partial L}{\partial F} - g + Y(g + w \frac{\partial L}{\partial F}) = 0, \tag{4}$$

where $Y = \lambda/Q$ is the shadow value per unit of output of a price increase. From (3), if $Y = 1$, $s = r$, which contradicts the assumption that $s > r$. Thus $Y \neq 1$. If the constraint is not binding, $Y = 0$. Therefore, assuming continuity, $0 < Y < 1$.[3]

From (3), the marginal rate of technical substitution between capital and labour is

$$-\frac{\partial L}{\partial K} = \frac{r}{w} - \frac{Y(s-r)}{(1-Y)w} . \tag{5}$$

From (4), the marginal rate of technical substitution between fuel and labour is

$$-\frac{\partial L}{\partial F} = \frac{g}{w} . \tag{6}$$

The marginal rate of technical substitution between capital and fuel is obtained by dividing (5) by (6),

$$-\frac{\partial F}{\partial K} = \frac{r}{g} - \frac{Y(s-r)}{(1-Y)g} . \tag{7}$$

Given $s > r$ and $0 < Y < 1$, equations (5) and (7) imply that the marginal rates of technical substitution between capital and labour and between capital and fuel are less than the corresponding price ratios, implying the use of too much capital relative to fuel and labour. However, the marginal rate of technical substitution between fuel and labour is equated to the ratio of their prices.

The continuous enforcement model is not consistent with the discontinuous nature of the regulatory process. Rather than enforcing the rate-of-return constraint on a continuous basis, regulators adjust prices to enforce the allowed rate of return only at intermittent regulatory reviews. Adjustments in output price between regulatory reviews can occur only in response to changes in certain cost elements. The fuel adjustment clause is the most important source of price adjustments between regulatory reviews. In Case 2, we assume that the firm considers this to be the only source of price changes and ignores the possibility of future regulatory reviews.

Case 2: Fuel Adjustment Clause With No Regulatory Review

A fuel adjustment clause results in the price of output being equal to the price set in the last regulatory review plus the increase in fuel expenditures per unit of output. Thus the price in period t is

$$P_t = \frac{s_0K_0 + w_0L_0 + g_0F_0}{Q_0} + \frac{g_tF_t}{Q_t} - \frac{g_0F_0}{Q_0} = \frac{s_0K_0 + w_0L_0}{Q_0} + \frac{g_tF_t}{Q_t}, \tag{8}$$

where the subscript 0 indicates the last regulatory review period.

The change in output price in period t is

$$\dot{P} = \frac{Q(\dot{g}F + \dot{F}g) - \dot{Q}gF}{Q^2}, \tag{9}$$

where the dots indicate time derivatives and the time subscripts are suppressed for simplicity. Assuming that the firm ignores future regulatory reviews, the firm's optimal control problem is to maximize the present value of profits, equation (1), subject to (9). The relevant first order conditions are

$$-w \frac{\partial L}{\partial K} - r = 0, \tag{10}$$

$$-w \frac{\partial L}{\partial F} - g + \frac{\lambda(Q\dot{g} - \dot{Q}g)}{Q^2} = 0. \tag{11}$$

Solving (10) and (11) for the marginal rates of technical substitution between capital and labour and between fuel and labour,

$$-\frac{\partial L}{\partial K} = \frac{r}{w}, \tag{12}$$

$$-\frac{\partial L}{\partial F} = \frac{g}{w} - \frac{\lambda(Q\dot{g} - \dot{Q}g)}{wQ^2}. \tag{13}$$

The marginal rate of technical substitution between fuel and capital is obtained by dividing (13) by (12),

$$-\frac{\partial K}{\partial F} = \frac{g}{r} - \frac{\lambda(Q\dot{g} - \dot{Q}g)}{rQ^2}. \tag{14}$$

Thus, the marginal rate of technical substitution between capital and labour is equated to the ratio of their prices.[4] However, the use of fuel relative to both capital and labour will not be that which minimizes costs, unless $Q\dot{g}$ happens to be equal to $\dot{Q}g$.

It is not clear from inspection of (13) and (14) whether too much or too little fuel will be used relative to capital and labour. Assuming that λ is positive,[5] the fuel adjustment clause will result in the use of more (less) than the cost-minimizing amount of fuel relative to capital and labour if $Q\dot{g} - \dot{Q}g$ is positive (negative). Thus

$$\text{sign} [FAC] = \text{sign} [Q\dot{g} - \dot{Q}g], \tag{15}$$

where sign [FAC] represents the sign of the effect of a fuel adjustment clause on the use of fuel relative to capital and labour.

The value of $\dot{Q}$ in (15) will be a function of $\dot{P}$, and thus of $\dot{F}$. In order to keep the analysis tractable, it will be assumed that[6]

$$\dot{Q} = Q_P\dot{P}, \tag{16}$$

and

$$\dot{F} = F_g\dot{g} + F_Q\dot{Q}, \tag{17}$$

where $Q_P = \partial Q/\partial P$, $F_g = \partial F/\partial g$, and $F_Q = \partial F/\partial Q$. Substituting in (15) from (16),

$$\text{sign} [FAC] = \text{sign} [Q\dot{g} - Q_P\dot{P}g]. \tag{18}$$

From (9), (16), and (17),

$$\dot{P} = \frac{Q[\dot{g}F + g(F_g\dot{g} + F_Q Q_P\dot{P})] - gFQ_P\dot{P}}{Q^2} = \frac{\dot{g}F + gF_g\dot{g}}{Q}$$

$$+ \frac{\dot{P}[QgF_Q Q_P - gFQ_P]}{Q^2} = \frac{\dot{g}F(1 + gF_g F^{-1})}{Q - gF_Q Q_P + gFQ^{-1}Q_P}. \tag{19}$$

Substituting in (18) from (19), and multiplying by $\dfrac{P}{P}$,

$$\text{sign} [FAC] = \text{sign} [Q\dot{g} - \frac{Q_P Pg\dot{g}F(1 + gF_g F^{-1})}{PQ - gF_Q Q_P P + gFQ^{-1}Q_P P}]. \tag{20}$$

The price elasticity of demand for output is defined as

$$\eta \equiv - \frac{Q_P P}{Q},$$

and the price elasticity of demand for fuel is defined as

$$\epsilon \equiv - \frac{F_g g}{F}.$$

Substituting in (20),

$$\text{sign} [FAC] = \text{sign} \{Q\dot{g}[1 + \frac{\eta gF(1 - \epsilon)}{PQ + g\eta(F_Q Q - F)}]\}.$$

Thus the sign of the effect of a fuel adjustment clause will be the same as the sign of $\dot{g}$ if

$$\frac{\eta g F(1-\epsilon)}{PQ + g\eta(F_Q Q - F)} > -1. \tag{21}$$

The denominator of the left-hand side will be positive if the term in parentheses is positive. This term will be positive if the firm is in the economic region of production, since

$$F_Q Q = \frac{Q}{\dfrac{\partial Q}{\partial F}} = \frac{Q}{MP} ,$$

where MP is the marginal product of fuel,

$$F = Q \cdot \frac{F}{Q} = \frac{Q}{AP},$$

where AP is the average product of fuel, and $MP < AP$ in the economic region of production. Since $\eta g F$ will also be positive, (21) can be rewritten

$$1 - \epsilon > -\frac{PQ + g\eta F_Q Q - g\eta F}{\eta g F} ,$$

or

$$\eta(\epsilon - D) < \frac{PQ}{gF}, \tag{22}$$

where $D = F_Q Q / F = AP/MP \geqslant 1$.

Therefore, the sign of the effect of a fuel adjustment clause on the use of fuel relative to other inputs is a function of the price elasticities of demand for output and for fuel, the ratio of average product to marginal product, and the ratio of total revenue to total expenditure on fuel. The introduction of a fuel adjustment clause in a period of increasing fuel prices would increase the use of fuel relative to other inputs if (22) is satisfied, but would decrease the use of fuel relative to other inputs if

$$\eta(\epsilon - D) > \frac{PQ}{gF}.$$

Thus the effects of a fuel adjustment clause cannot be determined *a priori*. However, the available empirical information on the arguments in (22) indicates that the introduction of fuel adjustment clauses in the recent period of rising fuel prices almost certainly resulted in an increase in the use of fuel relative to other inputs. Studies of the price elasticity of demand for electricity indicate that $\eta < 1.5$ (Halvorsen, 1978; Taylor, 1975). Less information is available on the price elasticity of fuel in electricity production, ϵ, but a reasonable upper bound is 2.0.[7] Since $D \geq 1$, the implied upper bound for the left-hand side of (22) is 1.5. This is substantially less than the ratio of total revenue to total expenditure on fuel, PQ/gF, which has been about 3.0 in recent years (U.S. F.P.C., various).

To summarize, analysis of Case 1 indicates that rate-of-return regulation continuously enforced would result in the use of more capital relative to fuel and to labour than would occur if the firm minimized costs. However, rate-of-return regulation would not affect the use of fuel relative to labour. Analysis of Case 2 indicates that fuel adjustment clauses affect the use of fuel relative to capital and to labour but do not affect the use of capital relative to labour. Under present conditions, fuel adjustment clauses would cause firms that ignored the possibility of future enforcement of rate-of-return regulation to use more fuel relative to capital and to labour than would occur under cost minimization.

Neither case indicates the net effect of regulation on input choice, which will involve elements of both Case 1 and Case 2. Even though rate-of-return regulation is not continuously enforced, it will affect input choice due to the possibility of a review during which it will be enforced. The possibility of a review, which was ignored in Case 2, is incorporated in Case 3 to provide a more complete model of the effects of fuel adjustment clauses on input choice.[8]

Case 3: Fuel Adjustment Clause With Regulatory Review

Assume for simplicity that there will be only one review, which will occur at date T. The rules to be applied by the regulatory agency are known to the firm but the timing of the review is not. The date T is treated as a random variable with exogenous probability density function ω_t,

$$\omega_t > 0, \quad \int_0^\infty \omega_t dt = 1.$$

The firm is assumed to maximize the expected present value of profits,

$$E \int_0^\infty e^{-\delta t} \{P_t Q_t(P_t) - w_t L_t[Q_t(P_t), K_t, F_t] - r_t K_t - g_t F_t\} dt$$

$$= E \int_0^\infty e^{-\delta t} \Pi_t(P_t, K_t, F_t) dt.$$

Define $W(P_T, K_T, F_T)$ as the constrained maximum of the present value of profits in the post-review period. Then

$$E \int_0^\infty e^{-\delta t} \Pi_t(P_t, K_t, F_t) dt$$

$$= \int_0^\infty \omega_T \left\{ \int_0^T e^{-\delta t} \Pi_t(P_t, K_t, F_t) dt + W(P_T, K_T, F_T) e^{-\delta t} \right\} dT. \quad (23)$$

Integrating the right-hand side of (23) by parts,

$$E \int_0^\infty e^{-\delta t} \Pi_t(P_t, K_t, F_t) dt$$

$$= \int_0^\infty e^{-\delta t} \left\{ \Pi(P_t, K_t, F_t) \Omega_t + \omega_t W(P_t, K_t, F_t) \right\} dt, \quad (24)$$

where

$$\Omega_t = \int_t^\infty \omega_\tau \, \tau$$

is the probability that the review does not occur prior to period t. The firm's optimal control problem is then to maximize (24) subject to (9). The relevant first order conditions are

$$\left(-w \frac{\partial L}{\partial K} - r \right) \Omega + \omega W_K = 0, \quad (25)$$

$$\left(-w \frac{\partial L}{\partial F} - g \right) \Omega + \omega W_F + \frac{\lambda(Q\dot{g} - \dot{Q}g)}{Q^2} = 0, \quad (26)$$

where time subscripts have been omitted for simplicity, $W_K = \partial W / \partial K$, and $W_F = \partial W / \partial F$.

Solving (25) and (26) for the marginal rates of technical substitution between capital and labour and between fuel and labour,

$$-\frac{\partial L}{\partial K} = \frac{r}{w} - \frac{1}{w} \frac{\omega}{\Omega} W_K, \quad (27)$$

$$-\frac{\partial L}{\partial F} = \frac{g}{w} - \frac{1}{w} \frac{\omega}{\Omega} W_F - \frac{\lambda(Q\dot{g} - \dot{Q}g)}{w \Omega Q^2}. \quad (28)$$

The marginal rate of technical substitution between fuel and capital is obtained by dividing (28) by (27),

$$-\frac{\partial K}{\partial F} = \frac{g}{r} + \frac{-\dfrac{1}{w}\dfrac{\omega}{\Omega}W_F - \dfrac{\lambda(Q\dot{g} - \dot{Q}g)}{w\Omega Q^2} + \dfrac{g}{w}\dfrac{1}{r}\dfrac{\omega}{\Omega}W_K}{\dfrac{r}{w} - \dfrac{1}{w}\dfrac{\omega}{\Omega}W_K}. \qquad (29)$$

The assumption in Case 2 that the firm ignores the possibility of future regulatory review is equivalent to the assumption that $\omega_t = 0$ and $\Omega_t = 1$ for all t. In this case, equations (27), (28), and (29) reduce to equations (12), (13), and (14) respectively. In the more general case in which the firm does not ignore the possibility of future review, alterations in input choice due to fuel adjustment clauses are modified by consideration of their effects on post-review profits.

When the review occurs, the regulatory agency will adjust output price to equate profits to sK, where s is the allowed rate of return. As discussed in Case 1, profit maximization subject to this constraint involves the use of more than the cost-minimizing quantity of capital relative to other inputs. If the firm were maximizing profits at time T subject to this constraint, both W_K and W_F would equal zero. However, the existence of fuel adjustment clauses, combined with the uncertainty of the time of the review, will result in the firm not using the bundle of inputs at time T which would result in $W_K = W_F = 0$. As discussed in Case 2, the probable effect of fuel adjustment clauses is a decrease in the use of capital relative to fuel. Therefore, W_K can be assumed to be positive and W_F can be assumed to be negative.

Given $W_K > 0$, equation (27) indicates that the possibility of a review results in the use of more than the cost-minimizing amount of capital relative to labour. This is in contrast to Case 2, in which the marginal rate of substitution between capital and labour is equated to the ratio of their prices. Note that the fraction ω_t/Ω_t in (27) is equal to the conditional probability of a review in period t given that it has not occurred earlier. *Ceteris paribus*, the higher the probability of a review in a given period, the greater the effect on the use of capital relative to labour.

As shown in (28), allowing for the possibility of a review introduces a term involving W_F in the equation for the marginal rate of technical substitution between fuel and labour. For $W_F < 0$, the sign of the new term is opposite to that of the term involving $\dot{g}$. Therefore, the possibility of a review offsets the tendency of fuel adjustment clauses to result in the use of too much fuel relative to labour. Similarly, allowing for the possibility of a review introduces terms in W_K and W_F in equation (29), which offset the tendency of fuel adjustment clauses to result in the use of too much fuel relative to capital. In both cases, the extent to which the effect of fuel adjustment clauses is offset by the possibility of a review will be greater the greater the conditional probability of a review, ω_t / Ω_t.

CONCLUDING COMMENTS

The net effect of regulation on input choice includes both the effects of regulation of the rate of return on capital and the effects of fuel adjustment clauses. In the absence of a fuel adjustment clause, rate-of-return regulation results in the use of more than the cost-minimizing amount of capital relative to labour and fuel. In the absence of future regulatory reviews, a fuel adjustment clause can be expected to result in the use of more than the cost-minimizing amount of fuel relative to capital and labour in periods of increasing fuel prices. Thus the fuel adjustment clause introduces a new distortion with respect to the marginal rate of technical substitution between labour and fuel, but has an effect on the marginal rate of technical substitution between capital and fuel that is opposite in sign to the distortion from rate-of-return regulation. Therefore, it is not clear whether inefficiency in the choice of inputs under regulation is increased or decreased by the introduction of fuel adjustment clauses.[9]

NOTES

* Research for this paper was supported by the Department of Energy. Views expressed in the paper are ours and do not reflect the policies or views of the Department of Energy. We are grateful to Richard Hartman for helpful comments on an earlier draft of this paper.

[1] In recent years, fuel adjustment clauses have accounted for about two thirds of the total increase in electric utility rates.

[2] The results for the first model are equivalent to those obtained by Petersen (1975) for a static three-input model.

[3] See Averch and Johnson (1962) and Takayama (1969). Alternative derivations of the result 0 $<$ Y $<$ 1 are also available, see Baumol and Klevorick (1970) and Zajac (1972).

[4] As discussed below, allowing for the effect of future regulatory reviews would result in the use of more than the cost-minimizing amount of capital relative to labour.

[5] The sign of λ indicates the effect on profits of an increase in output price. Since a main purpose of regulation is to reduce price below that which maximizes profit, λ can be assumed to be positive at the time of the regulatory review, see Klevorick (1973). However, the effect of the fuel adjustment clause on output price could eventually cause λ to become negative.

[6] Assumption (16) will be satisfied if the effects of all other determinants of $\dot{Q}$, such as changes in income and the prices of substitutes, just offset each other. Assumption (17) will be satisfied if the prices of capital and labour remain constant over time or if their effects just offset each other.

[7] Atkinson and Halvorsen (1976) estimated elasticities of demand for coal, oil, and gas in steam-electric generation. The mean of the estimated elasticities for the individual fuels was 1.0.

[8] The analysis of Case 3 is adapted from the treatment of uncertain technological change in Dasgupta and Heal (1974).

[9] It can be shown that use of total expenditure as the rate base, which would be equivalent to having automatic price adjustment clauses with respect to all inputs, would result in the elimination of inefficiency in input choice, see Appelbaum and Halvorsen (1977) and Bailey and Malone (1970).

REFERENCES

Appelbaum, Elie and Halvorsen, Robert. (1977) "Rate of Return Regulation Without Input Distortion." Discussion Paper 77-2. Seattle: University of Washington, Institute for Economic Research.

Atkinson, Scott E. and Halvorsen, Robert. (1976) "Interfuel Substitution in Steam Electric Power Generation." *Journal of Political Economy* 84 (October): 959-78.

Averch, Harvey and Johnson, Leland L. (1962) "Behavior of the Firm Under Regulatory Constraint." *American Economic Review* 52 (December): 1052-69.

Bailey, Elizabeth E. and Malone, John C. (1970) "Resource Allocation and the Regulated Firm." *Bell Journal of Economics and Management Science* 1 (Spring): 129-42.

Baumol, William J. and Klevorick, Alvin K. (1970) "Input Choices and Rate-of-Return Regulation: An Overview of the Discussion." *Bell Journal of Economics and Management Science* 1 (Autumn): 162-90.

Congressional Research Service. (1975) "Electric and Gas Utility Rate and Fuel Adjustment Clause Increases, 1974." Washington, D.C.

Dasgupta, Partha and Heal, Geoffrey. (1974) "The Optimal Depletion of Exhaustible Resources." *Review of Economic Studies, Symposium on the Economics of Exhaustible Resources*, pp. 3-28.

Halvorsen, Robert. (1978) *Econometric Models of U.S. Energy Demand*. Lexington, Mass.: Lexington Books.

Joskow, Paul L. and MacAvoy, Paul W. (1975) "Regulation and the Financial Condition of the Electric Power Companies in the 1970's." *American Economic Review* 65 (May): 295-301.

Kendrick, John W. (1975) "Efficiency Incentives and Cost Factors in Public Utility Automatic Revenue Adjustment Clauses." *Bell Journal of Economics* 6 (Spring): 299-313.

Klevorick, Alvin K. (1973) "The Behavior of a Firm Subject to Stochastic Regulatory Review." *Bell Journal of Economics* 4 (Spring): 57-88.

National Association of Regulatory Utility Commissioners. (1974) "Automatic Adjustment Clauses Revisited." Washington, D.C.

Petersen, H. Craig. (1975) "An Empirical Test of Regulatory Effects." *Bell Journal of Economics* 6 (Spring): 111-26.

Takayama, Akira. (1969) "Behavior of the Firm Under Regulatory Constraint." *American Economic Review* 59 (June): 225-60.

Taylor, Lester D. (1975) "The Demand for Electricity: A Survey." *Bell Journal of Economics* 6 (Spring): 74-110.

U.S. Federal Power Commission. (various) *Statistics of Privately Owned Electric Utilities in the United States*. Washington, D.C.: Government Printing Office.

Zajac, Edward E. (1972) "Lagrange Multiplier Values at Constrained Optima." *Journal of Economic Theory* 4 (April): 125-31.

Chapter Eleven

S220
6352
7230

Optimal Investments in Geothermal Electricity Facilities: A Theoretic Note

by
Peter Blair, Thomas Cassel, and *Robert Edelstein*

INTRODUCTION AND MOTIVATION

This paper, a theoretic-oriented presentation, examines optimal, multi-period, investment decision making for electric generation facilities. In principle, the analysis is relevant to many other energy-investment situations; however, it is easiest to imagine our discussion in the context of electric generation plants that utilize geothermal resources, especially liquid- and vapour-dominated (hydrothermal) resources. Utilizing a time-state preference framework, this study analyses optimal investment decision making under conditions of perfect intertemporal foresight as well as uncertain demand for and costs of production of electricity.

For hydrothermal resources, the amount of usable resource over the life of the electric plant may depend upon the time-path of electric generation and resource extraction. In turn, the rate of energy extraction may affect both fluid replacement and thermal recharge of the reservoir, and hence the temperature and the pressure of the resource. Given the technology of electric generation using hydrothermal resources, the design and operation of the electric generation facilities are sensitive to and dependent upon the time-path of available energy resources (Blair *et al.*, 1980).

Our presentation develops a set of analytic models. Commencing with a basic model, we then proceed, by changing one key assumption at a time, to alter the basic framework to refine the analysis. In this light, the subsequent discussion is separated into five sections. The second section casts the basic model under conditions of certainty. Also, the key analytic assumptions are outlined in this section. The third section introduces and explores the effects of temporally interdependent cost functions and demand functions on optimal investment choices. The fourth section focuses on how uncertainty in demand for electricity can alter optimal investment decisions. In the fifth section, "exogenous" uncertainty (such as a change in public sector energy policy) is introduced into the basic model. Finally, the sixth section demonstrates how opportunities for temporally continuous investment adjustments will relate to the optimal time-path for capital investments.

THE BASIC MODEL

For both simplicity and realism, the hydrothermal-powered electricity generation facilities are owned independently from the electric utility company. The electricity producer in this scenario delivers electric power to the electric utility grid-network under contract at pre-specified prices and quantities. This separation of ownership—the power producer and the electricity distributor—is for analytic convenience, but characterizes the way geothermal-based electricity is likely to be produced and sold in the foreseeable future. It probably reflects both the institutional features of the electric utility company and the intrinsic risks of geothermal resources and energy technology.

The hydrothermal-powered electricity facilities consist of four integral elements: the well-field system for drilling, collecting, and reinjecting hydrothermal energy resources; a network of pipes, tanks, and pumps for delivering and returning the hydrothermal power to and from the electric generation plant and the well-field; the electricity generation plant, which converts hydrothermal energy to electrical energy; and a system for electricity distribution from the generation plant to the electric utility grid-network. For our purposes, these four elements are treated as a single, profit-oriented entity. Hence, the objective function for the electricity production (taken as a whole) will be to maximize net expected discounted revenues over the life of the facilities.

It is important to emphasize, therefore, that the well-field is assumed to be "unitized" and is operated for the overall benefit of the electricity power project. We do *not* concern ourselves with how the net profits are distributed between well-field owners and generation plant owners, and so forth.

The analysis also assumes that the financing arrangements and taxing benefits (e.g., depreciation allowances, depletion considerations, etc.) are predetermined and already accounted for in all data.[1] Finally, the analysis presumes that key sub-optimization problems are understood beforehand. In particular, for each time path of electrical production over the life of the plant, the optimal facility design may differ and is assumed to be contained within cost-investment functional relationships (Blair *et al.*, 1980).

Our basic model is an investment decision analysis. The criterion for investing in the electricity facilities, as discussed above, is a net revenue analysis. The unique feature of the model is the explicit introduction of a binding capacity constraint on the quantity of electricity produced per year.[2] This is an important, real-world feature in most electrical power generation plants. For the moment, it is assumed that all data are available and known with certainty. Our notation is as follows:

D(t) = the riskless time discount factor for period t; that is, the present value of a dollar for certain t periods; hence, it will be the appropriate discount factor for both costs and revenues

X(t) = the quantity of electricity generated and sold in period t

V(X(t), t) = the "demand function" as seen by the electricity producers for producing X(t) in period t

C(X(t), K, t) = the operating and maintenance costs for X(t) in a plant with capacity K

I(K) = the initial investment needed to construct a plant with a capacity X=K per period; implicitly, the cost of capital for the project is assumed to be the investors' time preference interest rate.

The objective function is

$$\underset{X(t),\,K}{\text{Max}}\; P = \sum_{t=1}^{n} D(t) \left[\int_{0}^{X(t)} V(Y(t),\,t)dY - C(X(t),\,K,\,t) \right]$$

$$- I(K) - \sum_{t=1}^{n} \lambda(t)\,(X(t)\text{-}K), \tag{1}$$

where n is the relevant life of the project under consideration. The marginal conditions of optimization are

$$D(t) \left[V(X(t),t) - \frac{\partial C(X(t),\,K,\,t)}{\partial X(t)} \right] = \lambda(t);\; t = 1, \ldots, n \tag{2}$$

$$\sum_{t=1}^{n} D(t) \frac{\partial C(X(t),\,K,\,t)}{\partial K} + I'(K) = \sum_{t=1}^{n} \lambda(t) \tag{3}$$

$$\lambda(t) = 0 \text{ or } X(t) - K = 0;\; X(t) \geqslant 0;\; K \geqslant 0;\; t = 1, \ldots, n. \tag{4}$$

For the subsequent analysis, it will be assumed that for at least one K > 0, P > 0, and unless otherwise specified, second-order conditions for maximization are satisfied. (Obviously, if P ≤ 0 for all K > 0, the electrical plant should be rejected on "efficiency" grounds.)

This solution is easy to interpret. Note that λ(t) will be zero for periods when X(t) < K. In such a case, equation (2) suggests that if the electrical plant is operating at less than capacity, it is desirable to equate discounted marginal revenues with discounted marginal operating costs. This is analogous to the theory of the firm where, in the short run, it does not consider its fixed costs when equating discounted marginal costs and revenue in order to maximize the present value of profits. The equilibrium condition, represented by equation (3), states that the difference between savings on discounted operating costs and the marginal costs of the investment induced by an expansion in capacity is equal to the sum of the shadow prices. Hence, using equation (2), the sum of the λ's equals the sum of the *net* discounted marginal revenues for all periods when the capacity constraint is binding (the sum equals zero if capacity is not binding in any period). In other words, the sum of the shadow prices represents the discounted net value for a marginal expansion of capacity and, therefore, suggests that capacity will be expanded until the discounted marginal revenues are equal to the marginal costs.

Graphically, the multi-period model is illustrated in Figure 1. The $\lambda(t)$ terms represent the values consistent with equation (2) above. Therefore, $\sum_t \lambda(t)$ is the net discounted marginal benefits, which is consistent at the margin with varying the capacity level K. The upward sloping curve, though not necessary for the analysis, is assumed and is derived from the left-hand side of equation (3). It is the net discounted marginal cost associated with each capacity level K. The marginal conditions for equilibrium occur where the two functions intersect, implying that the net discounted marginal revenues and marginal costs are equal and that no further capacity expansion is profitable. The implications for the optimal level of electricity production

Figure 1
THE MULTI-PERIOD MODEL

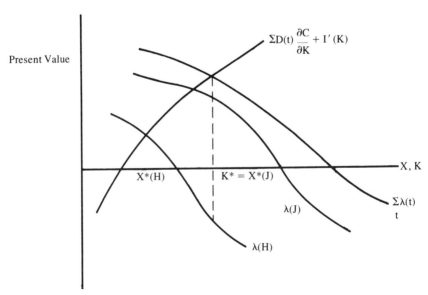

Note: $\lambda(t) = D(t) \quad V(X(t), t) - \dfrac{\partial C(X(t), t, K)}{\partial X(t)}$

for two particular time periods are illustrated in Figure 1. In time-period J, capacity is binding, and K* = X*(J), with the λ*(J) > 0. In period H, capacity is "optimally" underused for X*(H) < K* and λ*(H) = 0.

Of course, there may not be any time-period for which capacity is binding, implying that all λ(t) will be zero. This situation can occur when additional capacity dramatically reduces operating costs relative to the cost of the additional capital. This could occur if there existed significant economies of scale in geothermally powered electricity production. That is, if for all periods the discounted marginal revenues are increasing less rapidly than the discounted marginal costs, including capacity *and* operations costs, the local extrema will be a maximum with λ(t) = 0 for all t. In microeconomic parlance, the maximization occurs if the discounted marginal cost curve intersects the discounted marginal revenue function from below. Therefore, this does not rule out the possibility of economies of scale (or another relatively plausible possibility in cases of our electrical plant analysis, that the demand function for electricity faced by plant owners may be upward sloping in certain ranges due to contractual arrangements).

EXTENSIONS OF THE BASIC CERTAINTY MODEL: THE CASES OF TEMPORAL INTERDEPENDENCE

Our strategy now is to demonstrate the flexibility of this framework for doing analysis under various sets of altered assumptions. Space limitations do not permit us to examine all of the interesting possible cases. However, two very important cases will be illustrated:

A. Temporally interdependent demand
B. Temporally interdependent costs

To this point, the demand and cost functions have been assumed to be temporally independent. However, it is easy to conceive of situations where these functions are *not* temporally independent. In the case of hydrothermal energy, the costs incurred in the production of electricity are likely to be dependent-upon earlier period electricity production levels. This is true because hydrothermal resource temperature declines are related to past rates of energy extraction (Bechtel, 1977). Similarly, the level and value of electricity sales may depend upon earlier period production levels. This may be true because of contractual obligations or because of the growth rate in an area's demand for electricity, and so forth. For whatever reason, in our patois, the current period demand function for electric power would be dependent upon all X(t)s that occurred before the present period.

A. Temporally Interdependent Demand for Electricity

The following demand function is an illustration of the effects of this type of temporal interdependency. $V(h) = V(\sum_{j=1}^{h} X(j))$ where h is the current period. This demand function signifies that previous electricity production, no matter when it occurred in the past, creates identical increases in demands according to the sum total of electricity production throughout all previous time periods.[3] (Also, we assume that V and C are stable functions, independent of time.)

The objective function for the electric power producers now will become

$$
\underset{X(t), K}{\text{Max } P} = \sum_{t=1}^{n} D(t) \left[\int_{0}^{Q} V(Y(t))dY - C(X(t), K) \right] \quad (5)
$$

$$
- I(K) - \sum_{t=1}^{n} \lambda(t)(X(t)-K),
$$

where

$$
Q = \sum_{j=1}^{t} X(j) \text{ and } t = 1, \ldots, n
$$

The marginal conditions for optimizing behaviour will be

$$
\sum_{t=h}^{n} D(t) V\left(\sum_{j=1}^{t} X(j) \right) - D(h) \frac{\partial C(X(h), K)}{\partial X(h)} = \lambda(h) \quad (6)
$$

$$
h = 1, \ldots, n
$$

$$
\sum_{t=1}^{n} D(t) \frac{\partial C(X(t), K)}{\partial K} + I'(K) = \sum_{t=1}^{n} \lambda(t). \quad (7)
$$

Equation (6) demonstrates that the true marginal value of electricity production in the hth period must take into account the discounted revenues created (i.e., gain or loss) for all future periods. In equilibrium, the discounted stream of marginal revenues for the current and future periods less the discounted marginal cost will be the shadow price for capacity for the present period.

The joint solution of equations (6) and (7) will yield the optimal investment of capacity, shown as equation (8).

$$
\sum_{t=1}^{n} D(t) \left[t \cdot V\left(\sum_{j=1}^{t} X(j) \right) - \frac{\partial C(X(t), K)}{\partial X(t)} \right] = \quad (8)
$$

$$
\sum_{t=1}^{n} D(t) \frac{\partial C(X(t), K)}{\partial K} + I'(K).
$$

The interpretation of equation (8) is straightforward. The right-hand side is, as discussed earlier, the net discounted marginal costs for creating additional electricity generation capacity. The left-hand side is the net discounted marginal revenues, including "intertemporal externalities," for electricity production for optimal capacity utilization. In equilibrium, these discounted marginal revenues and costs must be identical.

B. Temporally Interdependent Cost Functions

The cost function for producing electricity is likely to vary in accordance with the amount of electricity generated in earlier periods. As already observed, and as explained in detail in Blair *et al.* (1980), electrical production rates from hydrothermal resources follow a time-path relating to the resource extraction-flow-temperature gradient. For example, assume that the cost function is $C(h) = C(\sum_{j=1}^{h} X(j), K, h)$; this formulation is similar to the revenue function time-dependence analysis above. Then, the objective function will become

$$
\begin{array}{ll}
\text{Max P} & = \sum_{t=1}^{n} D(t) \left[\int_{0}^{X(t)} V(Y(t), t)\, dY - C(F(h), K, t) \right] \\
X(t), K
\end{array}
$$

$$
- I(K) - \sum_{t=1}^{n} \lambda(t) (X(t)\text{-}K), \tag{9}
$$

where

$$
F(h) = \sum_{j=1}^{h} X(j); t = 1, \ldots, n; \text{ and } h = 1, \ldots, n.
$$

The marginal conditions will be equations (10) and (11).

$$
D(h)V(X(h), h) - \sum_{t=h}^{n} D(t) \frac{\partial C(\sum_{j=1}^{t} X(j), K, t)}{\partial X(t)} = \lambda(h)
$$

$$
\text{for } h = 1, \ldots, n \tag{10}
$$

$$
\sum_{t=1}^{n} D(t) \frac{\partial C(\sum_{j=1}^{t} X(j), K, t)}{\partial K} + I'(K) = \sum_{t=1}^{n} \lambda(t). \tag{11}
$$

Equation (10) signifies that the discounted expected marginal revenue in each period is equated with the discounted marginal operating costs and the appropriate period shadow price, λ. However, the discounted marginal operating cost functions are altered in *all* future periods because of the assumed temporal interdependence. That is, more production in an earlier period, subsuming normal cost function shapes, increases marginal costs in all subsequent periods and, therefore, should be taken into account in the current period production decisions. Equation (11) shows how capacity choice, K, should be taken into account in the intertemporal costs analysis.

$$\sum_{t=1}^{n} D(t) \left[V(X(t), t) - t \cdot \frac{\partial C(\sum_{j=1}^{t} X(j), K, t)}{\partial X(t)} \right] =$$

$$\sum_{t=1}^{n} D(t) \frac{\partial C(\sum_{j=1}^{t} X(j), K, t)}{\partial K} + I'(K). \tag{12}$$

Equation (12) combines equations (10) and (11). The left-hand side of equation (12) is the net discounted marginal revenue, including intertemporal production cost "externalities," from increasing electricity production, given the optimal capacity. The right-hand side of equation (12) is for an optimal time-path of electricity production, the discounted *reduction* in the intertemporal costs of production of electricity associated with capacity increases plus the added costs of increasing that capacity. In equilibrium, as before, these discounted quantities (i.e., left-hand and right-hand sides of equation (12)), are equal to the sum of the shadow prices for capacity expansion and, hence, are equal to each other.

UNCERTAINTY IN THE VALUE OF THE ELECTRICITY

In general, the precise magnitudes of the future demand price and quantity of electricity, the operating costs of the facilities, and the cost of the initial investment for the project are not known. For example, the costs of operating and maintaining the facilities may be affected by unanticipated changes in the rate of inflation for input factors or by changes in the physical characteristics of the geothermal resource, both of which are uncertain. Furthermore, the construction outlays (i.e., the initial investment), as is true with most large-scale construction projects, may not be known with certainty. At best, these values may be known in terms of likelihood or probability distributions. In this section, we will disregard the uncertainties involved in the cost structure of the electric plant, and will instead focus upon the difficulties of choosing the optimal size of the power plant created solely by the uncertainty about the demand function for future electricity.

In the case of hydrothermal power plants, the causes for this type of uncertainty are real and stem from various sources. First, if the total quantity of electricity generated in future periods were known, the value (revenue) attributed to it might depend on exogenous factors. The value of revenue paid to the producer for X kilowatts of electricity might depend upon the "state" of the world or general energy policies. For example, if the price paid by the electric utility is determined contractually by the utility's cost of producing electricity from alternative fuels, such as oil, then the world price for oil, an uncertain event, will determine the future streams of revenue for the geothermal electric producer. Second, and similarly, the quantity of alternative fuel available for the utility may be uncertain and may affect the demand for the amount of electricity, X, that can be sold to the utility in future periods. Third, the need for electricity in an area may be linked to the technology employed by industry, which changes in an uncertain fashion over time. Furthermore, the need for electricity generation may be related to the level of particular economic activities, which themselves are uncertain. That is, the level of production over time is contingent on, and determined by, an "uncertain" derived demand for electrical power. Fourth, the future technology of alternative electrical generation systems is uncertain and will affect the selling position of the geothermal power plant. The effects of these elements on the electric utility are dependent upon, as well as determine, the electric utility's ability to create other sources of electric generation. This, in turn, will affect the amount and price of electricity that the geothermal plant will be able to sell.

With these considerations in mind, theoretically it is straightforward to extend our analysis of the optimal, geothermal power project for conditions of uncertainty by extending our time-state preference model. The ability to do this analysis requires an *a priori* estimate of the conditional probability distributions for the appropriate time discount rate, and for revenues and cost functions for each future period. That is, uncertainty about conditions in future periods must be described by specifying a set of possible "states of the world," only one of which will actually occur in each future period. The realized values of the revenues, costs, and the discount rate are also contingent on which state actually occurs.

For example, if there were only one future time period with two mutually exclusive, but exhaustive, possible states of the world, the appropriate estimates of the discount factors are, say, .50 and .30 for state 1 and state 2 in time period 1 respectively. That is, if one were to pay $0.50 now, he would receive $1.00, *contingent upon* state 1 occurring in the future period; similarly $0.30 now would yield $1.00 in the future period, contingent upon state 2. In essence, the discount factors take into account the probability of each state occurring *and* the time preference interest rate that will be relevant if that state were to occur. (This also can be related to the estimate of the risk-free rate. In this example, an expenditure of \$0.80 now (i.e., \$0.50 + \$0.30) will *guarantee* a dollar in the next period. $D(t=1, s=1)$ + $D(t=1, s=2) = .80 = \dfrac{1}{1+r}$ where r is the risk-free rate. Therefore, in this example the risk-free time preference rate is 25 per cent.)

If our would-be investors are assumed to possess the relevant *a priori* data estimates, the objective function of the net expected discounted profits from the project, constrained by capacity, will be

$$\operatorname*{Max\,P}_{X(s,t),K} = \sum_{t=1}^{n} \sum_{s=1}^{m} D(t,s) \left[\int_{0}^{X(t,s)} V(Y(t,s), t,s)\,dY - C(X(t,s), t, s, K) \right]$$

$$- I(K) - \sum_{t=1}^{n} \sum_{s=1}^{m} \lambda(t,s,) \left(X(t,s) - K \right), \qquad (13)$$

where n   = the relevant future life of the project

m   = the mutually exclusive, but exhaustive, possible states of the world in each future period

$D(t,s)$ = the discount factor for period t if state s obtains

$X(t,s)$ = the quantity of electricity produced in period t contingent upon state s

$V(X(t,s),t,s)$ = the revenue function for producing the *last* unit, $X(t,s)$, in period t if state s occurs; it is the "demand function" as seen by the producer

$C(X(t,s,),t,s,K)$ = the operating and maintenance costs associated with producing $X(t,s)$ in a plant with capacity K if state s obtains in period t.

The marginal conditions for optimization will be

$$D(t,s) \left[V(X(t,s), t,s) - \frac{\partial C}{\partial X(t,s)} \right] - \lambda(t,s) = 0 \quad \begin{Bmatrix} t = 1, \ldots, n \\ s = 1, \ldots, m \end{Bmatrix} \qquad (14)$$

$$- \sum_{t=1}^{n} \sum_{s=1}^{m} \left(D(t,s) \frac{\partial C}{\partial K} - \lambda(t,s) \right) - I'(K) = 0 \qquad (15)$$

$$\lambda(t,s) = 0 \text{ or } X(t,s) - K = 0; \ X(t,s) \geqslant 0; \ K \geqslant 0 \text{ for all } t,s \qquad (16)$$

The constrained objective function (13) and the marginal conditions (14), (15), and (16) are similar to equations (1)-(4), except that we have introduced states of the world for each time period. The interpretation of these marginal conditions is, as before, easy. In any time period t, for each state of the world, s, the $\lambda(t,s)$ is a shadow price that equals the appropriately discounted (i.e., present value) difference between marginal expected revenues and marginal expected costs. If $X(t,s)$ is less than capacity K, $\lambda(t,s)$ will be zero with the net revenues being zero. Analogous to the certainty model, capacity need not limit the *expected* electricity production for any state of any future time period. That is, it is possible that no state will or could occur that actually utilizes this capacity. Moreover, the sum of the shadow prices $\lambda(t,s)$ over all states and time periods represents the expected present value of net marginal revenues from expanding capacity; and capacity should be increased until the marginal expected costs incurred through expansion equal the marginal expected revenues over all periods and states. Hence, the introduction of uncertain future states offers no new analytic and interpretative difficulties.

THE EFFECTS OF UNCERTAIN "EXOGENOUS" PUBLIC SECTOR ENERGY POLICY

To this point, the analysis implicitly has treated public sector energy policies over the relevant planning time period as exogenous and known. It is more realistic to assume that future general public sector energy-related policy such as fuel tax charges and levels of pollution standards for electrical generation and enforcement practices are uncertain. Ideally, the planner-investor should incorporate his expectations about the risks of changes in public sector policies into his decisions for the electric plant's capacity and anticipated scale of operations.

As an illustration of this issue, we will examine the case for designing the hydrothermal electric plant with one future period, consisting of two potential states in that future period (depending on exogenous public policy, affecting the prices for fuels used in alternative systems for generating electricity). In particular, assume an alternative fuel tax. As the tax decreases, the amount of electricity that will be required from our plant will diminish and/or the value to the producer of the X units of electricity will fall. Let Z be the policy parameter for measuring alternative fuel tax policy, with $Z=0$ representing the existing situation. If the alternative fuel tax rates were to change by a *small* quantum, represented by $\delta Z > 0$, it will be assumed that the value function for electricity produced by the plant will be shifted according to the functional rule f(Z). (For expositional convenience, it is assumed that changes of Z affect only the gross revenue value function.) According to an economist-engineer's approximation of f(Z), it is known that

$$f(Z + \delta Z) = f(Z) [1 - \rho \cdot \delta Z], \qquad (17)$$

where $\rho > 0$ is a linear approximation of the effects of small δZ on f(Z).

Using (17), the definition of the total derivative, the following expression can be created:

$$\frac{df(Z)}{dZ} = \lim_{\delta Z \to 0} \left(\frac{f(Z+\delta Z) - f(Z)}{\delta Z} \right) = -\rho f(Z).$$

Therefore, the f-function is the form $f(Z) = e^{-\rho Z}$ with $\rho > 0$ and $f(0) = 1$, as required from our initial conditions with respect to $Z = 0$. Hence, the revenues achieved by the electricity plant will be for $\delta Z = 0$ and some other $\delta Z > 0$

$$f(Z) \cdot V(X) \text{ such that}$$

a. For $(S=1, t=1)$, original exogenous policy state in the future period,

$f(Z) = f(0) = 1$ implies that the
$$f(Z)V(X) = V(X). \tag{18}$$

b. For $(S=2, t=1)$, the altered exogenous policy state in the future period,

$f(Z) = f(0 + \delta Z) = 1 - \rho Z'$, with $\delta Z = Z' > 0$, implies that
$$f(Z)V(X) = (1-\rho Z')V(X). \tag{19}$$

The marginal condition for the optimizing solution is equation (20), and is found by applying the conditions derived in equations (18) and (19) with the objective function for one future period with two mutually exclusive exhaustive states:

$$Z' \cdot \rho \cdot D(1,2) \cdot V(X(1,2)) =$$

$$\sum_{s=1}^{2} D(1,S)(V(X,(1,S)) - \frac{\partial C}{\partial X} - \frac{\partial C}{\partial K} - I'(K). \tag{20}$$

An examination of equation (20) indicates that "uncertain" exogenous public sector policies normally will be expected to reduce either the planned capacity K and/or the levels of electrical energy generation X, conditional on the states of world. First, note that if $Z' = 0$, the case for *no* anticipated changes for exogenous public policy, the left-hand side of equation (20) vanishes, and the solution of the right-hand side is identical with the complete certainty case. If, however, $Z' > 0$, under the normal assumptions about the shapes of the revenue and costs functions and second-order conditions for maximization, decreases either in K or in the expected level of operations of the plant, X, in each state of the world will be required to engender the required positive value for the right-hand side of equation (20).

OPTIMAL POLLUTION CONTROL WITH CONTINUOUS INVESTMENT OPPORTUNITIES

We shall now briefly demonstrate how we can modify our model to incorporate *continuous* investment activities in the plant as well as *continuous* depreciation of the existing plant. For the sake of convenience, we will return to a model with perfect foresight for cost and demand functions over an infinite time horizon. Also, it will be necessary to modify the time functions from discrete to continuous forms. $X(t)$, $I(t)$, and $K(t)$ are the amounts of electricity produced at time t, the value of expenditures on capital for increasing capacity at time t, and the total electricity capacity at time t, respectively. X and I are considered control variables and K is a *state* variable. We define $R(K, X, I)$ as the net revenue function and, at any instant in time t, it is evaluated to be consistent with our prior models, as

$$R(K, X, I) = \int_0^X V(Y)\,dY - C(X, K) - IK.$$

We will utilize an instantaneous discount rate r such that $D(t) = e^{-rt}$. We continue to constrain the control variable X by the state variable K such that

$$K(t) - X(t) \geq 0 \text{ at each t.}$$

Finally, we need to know the state transition or motion differential equation for the state variable:

$$K = K + f(I) - \delta K,$$

where $f(I)$ = the amount of capacity that can be installed for an expenditure of I, and δ = the instantaneous depreciation rate for the current capacity level (i.e., capital stock).

In this model, the objective function we wish to maximize is

$$\int_0^\infty R(K, X, I)e^{-rt}dt, \text{ subject to}$$

$$K = K + f(I) - \delta K \text{ and } K - X \geq 0.$$

The problem can be solved by defining the Hamiltonian, H:

$$H = e^{-rt}\left(R(K, X, I) + \lambda(K-X) + u(K + f(I) - \delta K)\right), \qquad (21)$$

where $\lambda(t)$ is the shadow price for the capacity constraint in each state, and $u(t)$ is the shadow price for the rate of return on capacity growth at any point in time.

Our analysis, being illustrative, will only delineate the first-order conditions for the maximal, *interior* time-path solution. Applying Pontryagin's maximum principle, we know that a maximum for the problem is achieved if we maximize the Hamiltonian and if the shadow price u satisfies (assuming appropriate initial conditions):

$$\frac{du}{dt} = u \cdot r - \frac{\partial H}{\partial K} = -u(1\text{-}r\text{-}\delta) - \lambda - \frac{\partial R}{\partial K} \qquad (22)$$

$$\lim_{t \to \infty} \left(e^{-rt} u(t) \right) \geq 0 \text{ and } \lim_{t \to \infty} \left(e^{-rt} u(t) K(t) \right) = 0.$$

Also, maximization requires that

$$\frac{\partial H}{\partial I} = 0 = \frac{\partial R}{\partial I} + uf'(I) \qquad (23)$$

$$\frac{\partial H}{\partial X} = 0 = \frac{\partial R}{\partial X} - \lambda. \qquad (24)$$

Equation (23) signifies that, in equilibrium, the net marginal profitability from investment expenditures ($\frac{\partial R}{\partial I}$) must equal the investment in capacity times the marginal rate of return for capacity (u). This is completely analogous to our earlier models. Equation (24) generates a marginal condition that also mirrors our earlier models' developments. If capacity is binding at time t, it implies that $-\lambda(t) \geq 0$ and that the net marginal profitability for electricity production ($\frac{\partial R}{\partial X}$) will equal λ. If $-\lambda = 0$, which occurs if capacity is a slack variable, the net marginal profits for additional capacity will be zero.

Equation (22) appears to be a quantum change from our previous models, but really is not. It is used to connect through time the change in the rate of return on capacity growth (i.e., the intertemporal investment incentive), taking into account the time-discount factor, the degree of capacity utilization (K-X), the depreciation rate of the current capacity, and the change in the short-term cost structure with respect to changing K (i.e., $\frac{\partial R}{\partial K}$).

While we will not go through the ramifications of steady-state equilibria, smooth paths to equilibria, or boundary solutions, this type of problem form has been analysed, and yields solution counterparts to the simpler models presented earlier in this paper (Arrow, 1968).

A FINAL COMMENT

This paper develops a set of theoretical models for analysing optimal investment-production decisions for hydrothermal-powered electricity facilities. The theoretical thrust of the presentation is directed toward incorporating various assumptions about the world, including uncertainty, into the analytic structure of the problem-solving procedure. At present, the concepts developed here are being applied and operationalized in research being conducted under the auspices of the United States Department of Energy. The model being used is a computerized gradient search-stochastic-programming model. At a later time, we hope to be able to present our empirical results.

NOTES

[1] This issue, as well as the choice of optimal investment financing of the project and the optimal pricing and use of the plant once it is built, are important, but beyond the scope of this paper.

[2] The capacity constraint reflects the total volume of electricity that can be generated per period. In fact, this is a modification of the usual technology capacity constraints faced in the operations of hydrothermal plants. More commonly, capacity is in terms of the total volume of fluid that can be used per period for certain ranges of fluid temperature, conditional on several factors, such as salinity.

[3] This functional form is meant to be illustrative, though it is likely that contractually, revenue for the hydrothermal facilities will be related to past deliveries of power.

REFERENCES

Arrow, K.J. (1968) "Applications of Control Theory to Economic Growth." In *Mathematics of the Decision Sciences, Part 2*, edited by G.B. Dantzig and A.F. Veinott Jr., pp. 85-119. Providence, R.I.: American Mathematical Society.

Bechtel International, Inc. (1977) "Advanced Design annd Considerations for Commercial Geothermal Power Plants at Heber and Niland, California." San Francisco.

Blair, P. and Cassel, T. (1978) "Optimal Geothermal Resource Extraction for Electric Power Applications." In *Proceedings of the Joint Automatic Control Conference* held in Philadelphia. Pittsburgh: Instrument Society of America.

Blair, P.; Cassel, T.; and Ervolini, M. (1979) "Analysis of Resource Pricing for Geothermal Electric Power Applications." In *Modeling and Simulation*, Vol. 10, edited by W.G. Vogt and M.H. Mickle. Pittsburgh: Instrument Society of America.

Blair, P.; Cassel, T.; Edelstein, R.; and Paik, I. (1980) "The Economics of Optimal Geothermal Resources Extraction for Electric Power." *Journal of Energy and Policy*.

Cassel, T. and Edelstein, R.H. (1978) "Forecasting Capital Investment Behavior for Geothermal Electric Power Industry." Paper presented at the Western Economic Association, 21 June.

Cassel, T.; Edelstein, R.H.; Blair, P.; and Amundsen, C. (1979) "Geothermal Investment Analysis." Final Report No. C00-4713-1, prepared by the University of Pennsylvania and Technecon Analytic Research, Inc. for the U.S. Department of Energy under contract no. ET-78-S-01-4713.

Electric Power Research Institute. (1976) "Utilization of U.S. Geothermal Resources." Palo Alto, Calif.

Kuller, R.G. and Cummings, R.G. (1974)) "An Economic Model of Production and Investment for Petroleum Reservoirs." *American Economic Review* 64 (March): 66-79.

Tinbergen, J. (1956) *Economic Policy: Principles and Design*. Amsterdam: North Holland.

Turvey, R. (1963) "On Divergences Between Social Cost and Private Cost." *Economica* 30 (August): 309-13.

Chapter Twelve

6(3(
72 30
63$2

Interprovincial Electrical Energy Transfers: The Constitutional Background

by
Leo Barry

PART ONE—INTRODUCTION

Since the Organization of Petroleum Exporting Countries (OPEC) began flexing its muscles and increasing dramatically the price of petroleum, oil importing countries have been attempting to reduce their dependency upon foreign supplies. In Canada, there have been a number of steps taken to improve the efficiency of internal electrical energy transmission. In 1976, the ten provincial premiers set up a committee of public employees, the Interprovincial Advisory Council on Energy (IPACE), to investigate the net benefits that might accrue from expanding the existing interprovincial electrical network into an integrated national energy grid, which might permit the complete co-ordination of all provincial electrical energy systems.[1] The Provinces of Nova Scotia, New Brunswick, Prince Edward Island, and the federal government negotiated towards the establishment of a Maritime Energy Corporation.[2] The Province of Newfoundland has been promoting the Gull Island hydroelectric site on the Lower Churchill River in Labrador as a possible source of energy for transmission to the Island of Newfoundland by an underwater cable or tunnel.[3] The four western premiers have retained consultants to examine a preliminary report that suggests that $4.3 billion could be saved between 1979 and 1988, through deferring new generating stations and developing a better regional interconnection at a cost of $600 million.[4] The purpose of this paper is to outline the ways in which Canada's constitutional structure may affect schemes to give Canada improved electrical planning, generation, and transmission on a regional and national basis.

It will be seen that the main question here, as in so many other areas of Canadian constitutional law, concerns the division of jurisdiction between the federal and provincial governments. Do the provinces by themselves have sufficient authority to develop an electrical transmission system adequate for Canada's needs? At what point, if any, does the federal government have jurisdiction over electrical transmission systems, and to what extent will provincial jurisdiction be displaced by that of the federal government?

In attempting to answer these questions, it will be necessary to identify the provisions of the *British North America Act*[5] *(B.N.A. Act)*, which may contribute to either federal or provincial jurisdiction over electrical transmission systems. The main heads of federal authority (Peace, Order and good Government; Trade and Commerce; Works and Undertakings connecting Provinces; and the Federal Declaratory Power) will be examined together with the main heads of provincial authority (Management of Public Lands belonging to the Province; Property and Civil Rights in the Province; and Matters of a merely local or private Nature in the Province). This paper will then briefly relate the relevant heads of jurisdiction to the various functions involved in an electrical energy system (planning, allocation of rights, construction, operation, and marketing). Judicial policy considerations, which arise in determining jurisdiction over interprovincial electrical interconnections, will be discussed. Certain conclusions will be drawn that indicate a role here for both federal and provincial governments but identify areas of constitutional uncertainty and possible drastic shifts in the federal-provincial balance of power if co-operative arrangements are not developed between the two levels of government. Federal jurisdiction, although not yet asserted, almost definitely exists over those electrical works crossing provincial or international borders. The major problem lies in trying to predict how far an extraprovincial electrical connection will carry federal authority into areas that are under provincial jurisdiction before the connection.

PART TWO—RELEVANT PROVISIONS OF THE *B.N.A. ACT*

Basis of Federal Authority

The main heads of federal authority with respect to electrical power installations are Peace, Order and good Government (s. 91, preamble and conclusion); Public Property (s. 91(1A)); Trade and Commerce (s. 91(2)); Defence (ss. 91(7) and 117); Navigation and Shipping (s. 91(10)); Sea Coast and Inland Fisheries (s. 91(12)); Indian Lands (s. 91(24)); Criminal Law (s. 91(27)); Extraprovincial Works and Undertakings (ss. 91(29) and 92(10)); Works declared for the general Advantage of Canada (ss. 91(29) and 92(10)); Agriculture (s. 95); and Empire Treaties (s. 132).

Although worded as a residual head of legislative authority, the "Peace, Order and good Government" clause has been narrowly interpreted by the courts, and its scope has been severely restricted by the courts' interpretation of the provincial authority over "Property and Civil Rights in the Province" and over "Matters of a merely local or private Nature in the Province." As will be seen below, the "Peace, Order and good Government" clause may give the federal Parliament certain authority over international and interprovincial river improvements.

The effect of jurisdiction over "Public Property" of the Crown in right of Canada is to give the federal government certain legislative and executive authority in addition to the authority specifically allocated by the *B.N.A. Act*. In Canada, the Crown is entitled to ownership of all ungranted lands. Before Confederation, Crown lands, including mineral resources, had become the public property of the provinces in which they were situate.[6] Part VIII of the *B.N.A. Act* distributed resources between the Crown in right of Canada and the Crown in right of the provinces. Section 108 assigned property specified in the Third Schedule of the act to Canada. Sections 109 and 117 confirmed that the residue remained with the provinces, including the beds of rivers, even navigable rivers.[7] The resources, ownership of which are relevant in determining authority over electrical power systems, consist of (*a*) Crown lands that may be the sites of generating plants or transmission lines, and (*b*) energy sources, primarily water, petroleum, coal, and uranium. A government can deal with the property it owns just as can a private owner, subject to any restrictions imposed by the common law or by validly enacted federal or provincial legislation. Thus a government can, as a right of ownership, attach to licences relating to the use of its property conditions dealing with matters that would ordinarily fall within the jurisdiction of the other level of government.[8] Of course the latter government can override the condition by proper legislation.

Jurisdiction over public property permits the federal government to enter into businesses otherwise under provincial jurisdiction. These federal businesses may be free of provincial regulatory and taxing powers.[9]

One important consequence flowing from authority over its public property has not yet been mentioned—the manner in which the federal authority, through its spending power, may significantly influence action in areas under provincial jurisdiction. Money is public property and can be used by the federal government to promote the development of, for example, interprovincial electrical transmission lines, through grants to the provinces or to private organizations. Conditions attached to the grants may influence the course of the development. But this combined use of the federal taxing and property powers must stop short of "regulating" matters under provincial jurisdiction.[10]

The effect on electrical power developments of federal authority over "Trade and Commerce" will be discussed below. Although for a time severely limited in scope by the courts' restrictive interpretation of it, this head of authority has recently been imbued with new life by the courts and now probably holds considerable potential for the exercise of federal jurisdiction over interprovincial energy transfers.

The "Defence" power, while unlimited in time of war, would appear to be of little consequence for electrical power networks during peace-time.

Federal authority over "Navigation and Shipping" means that federal permission is necessary before a province is entitled to interfere with navigation on a navigable river by the erection of a power dam. It is not settled whether the federal Parliament has jurisdiction over hydroelectrical potential arising from a dam constructed for purposes of improving navigation. A federal statute, which has as its "pith and substance" or main purpose a matter within federal jurisdiction, is valid, even though it may "incidentally affect" a matter normally within provincial jurisdiction. But doubt has been expressed whether federal jurisdiction over a hydroelectric plant would be justified merely because it is associated with a navigation project.[11]

The "Sea Coast and Inland Fisheries" authority of the federal government means that provinces, in constructing transmission lines or hydroelectric works, must consider the impact upon fish in the watercourse. For example, if the damming of a river would so drastically affect the flow of water as to wipe out a salmon run, the federal government could require that measures be taken so as to avoid this impact upon the fish.[12] But the federal regulatory power over fisheries must not be used to completely deprive the provinces of proprietary rights.[13] Riparian rights and rights to ownership of the bed of the river are incidents of provincial property.[14] Accordingly, co-operation is necessary between provinces and the federal authorities to ensure optimal water use.

With respect to "Indian Lands," it appears that transmission lines could not be constructed without federal-provincial co-operation. The underlying title to most reserves belongs to the provinces, and federal authority to deal with the property is therefore limited. And because the usufructuary title of the Indians (not yet precisely defined) falls within federal jurisdiction, the provinces cannot interfere with this by electrical developments. There are, however, certain reserves where the underlying title has been transferred from the provinces to the Dominion and the federal Parliament has complete control over these lands, including the right to use them for power development.[15]

Authority over "Criminal Law" gives the federal Parliament the right to prohibit certain activity for the protection of the public health, safety, and morals.[16] This may be relevant to hydro development in certain cases, for example, if the flushing action of a river in removing sewage or industrial waste were to be decreased by the installation of a power dam to the extent that health hazards might be created in areas adjoining the river. (It would, of course, make more sense to stop the pollution at its source, but this is not always accepted as feasible).

One of the most relevant heads of federal authority for an interprovincial electrical network is that over "Extraprovincial Works and Undertakings."

Hydro developments and transmission lines extending across provincial boundaries may fall under federal jurisdiction through this authority. This is discussed more fully below, as is the authority over "Works declared for the general Advantage of Canada."

Federal authority over "Agriculture" may become relevant in hydro developments should there arise a conflict between use of water for transmission line construction and use for federally legislated agricultural purposes.

The federal authority over "Empire Treaties" has been held inapplicable by the courts to those treaties negotiated since Canada acquired the capacity as a sovereign state to negotiate its own treaties apart from Great Britain. The federal government now has the authority to negotiate and execute all treaties[17] but cannot legislatively implement, without provincial agreement, those treaties covering matters under provincial jurisdiction.[18]

Some types of treaties may be implemented by executive action and do not require legislative implementation.[19] These generally relate to incidents of boundaries,[20] including water boundaries.[21] They may be dealt with by the federal government on its own authority. Generally, however, federal-provincial co-operation will be required for the implementation of treaties affecting provincial waters.

There are still "Empire Treaties" in effect that give the federal government authority to make certain legislation affecting water within provinces. The most important is probably the Boundary Waters Treaty with the United States.[22]

Although the provinces have no power to execute treaties,[23] there is probably no constitutional barrier to provinces entering into agreements with foreign states. However, doubts have been raised whether these can be binding in an international or domestic sense.[24] But if the making of such agreements is a prerogative of ownership, as an agreement concerning water rights or electrical power probably is, then there seems to be nothing to prevent these agreements being binding. A province would not be infringing upon federal treaty-making power in making an electrical power agreement with a U.S. state, but merely acting as owner of the power development. Of course, the provincial agreement could be abrogated or modified or prohibited by federal legislation under a valid head of federal jurisdiction.[25]

Basis of Provincial Authority

The main sources of provincial authority over electrical power installations are "Public Lands belonging to the Province" (s. 92(5)); "Local Works and Undertakings" (s. 92(10)); "Property and Civil Rights in the Province" (s. 92(13)); and "Matters of a merely local or private Nature in the Province" (s. 92(16)).

As previously discussed, the provinces, through their authority over the management and sale of ''Public Lands belonging to the Province,'' may deal with their lands as would a private person.[26] They may issue water rights subject to conditions that deal with matters ordinarily under federal jurisdiction—for example, on condition that power generated shall not be exported.[27]

Little provincial authority has been attributed to the head of ''Local Works and Undertakings,'' since the courts have tended to rely primarily upon authority over ''Property and Civil Rights in the Province'' and ''Matters of a merely local or private Nature in the Province'' to find provincial jurisdiction over water supply, pollution, power development, water conservation, flood control, recreation, and pollution.[28] Provincial authority under sections 92(13) and 92(16) has been used by the courts to restrict the scope of federal authority under ''Peace, Order, and good Government'' and ''Trade and Commerce.'' This is discussed more fully below.

Restrictions upon Federal and Provincial Authority

Section 121 of the *B.N.A. Act* was originally interpreted as forbidding only customs barriers between provinces.[29] It reads, ''All Articles of the Growth, Produce, or Manufacture of any one of the Provinces shall, from and after the Union, be admitted free into each of the other Provinces.'' This wording was held to permit provincial retail sales taxes to be levied on goods brought into the province where the tax also applied to goods purchased in the province.[30]

Subsequent statements by Mr. Justice Rand indicated a possible larger role for section 121.[31] He stated:

'Free' in s. 121, means without impediment related to the traversing of a provincial boundary. If, for example, Parliament attempted to equalize the competitive position of a local grower of grain in British Columbia with that of one in Saskatchewan by imposing a charge on the shipment from the latter representing the difference in production costs, its validity would call for critical examination. That result would seem also to follow if Parliament, for the same purpose, purported to fix the price at which grain grown in Saskatchewan could be sold in or for delivery in British Columbia. But burdens for equalizing competition in that manner differ basically from charges for services rendered in an administration of commodity distribution. The latter are items in selling costs and can be challenged only if the scheme itself is challengeable.[32]

In upholding the right of the Canadian Wheat Board to impose charges upon producers shipping wheat from one province to another, Mr. Justice Rand concluded:

I take s. 121, apart from Customs duties, to be aimed against trade regulation which is designed to place fetters upon or raise impediments to or otherwise restrict or limit the free flow of commerce across the Dominion as if provincial boundaries did not exist. That it does not create a level of trade activity divested of all regulation I have no doubt; what is preserved is a free flow of trade regulated in subsidiary features which are or have come to be looked upon as incidents of trade. What is forbidden is a trade regulation that in essence and purpose is related to a provincial boundary.[33]

Mr. Justice Rand's comments indicate that possibly a provincial government may not be entitled to impose a tax upon sales outside the province under a contract entered into between a provincial utility and a customer in another province. But one must conclude that the extent to which section 121 bars impediments to interprovincial trade has not been finally settled.

Section 125 provides "No Lands or Property belonging to Canada or any Province should be liable to Taxation." This section did not prevent the federal government from collecting customs duties on a shipment of liquor imported by British Columbia for sale through one of its liquor stores.[34] Nor did it relieve the Quebec government from liability to pay a federal government 6 per cent tax on long-distance telephone messages where federal legislation had authorized a telephone company to pass on such charges to its customers.[35] One writer suggests that in determining the constitutionality of a particular fiscal provision one must determine:

Whether it is a tax imposed solely for the raising of a revenue, a financial imposition for some broader regulatory purpose of a non-taxing character, or a fee for services incorporated in a supplier's overall charges, rather than a straight burden on public "lands" or "property". Although each case will be decided on its own facts, probably the first example would be unconstitutional and the other two would be valid.[36]

A current dispute between Alberta and the federal government could eventually lead to the courts taking a further look at the meaning to be given to section 125. At present, federal legislation prevents a lessee under a provincial oil and gas lease from deducting from federal corporate income tax the royalty payments due to the provincial Crown. It has been suggested that this is the same thing as the federal Parliament attempting to tax the provincial Crown itself for that Crown's income from royalties.[37] The matter has not yet been brought to the courts and may be resolved in the political arena instead.

PART THREE—RECONCILING FEDERAL AND PROVINCIAL JURISDICTION

Peace, Order, and Good Government

The authority of the federal Parliament to legislate for the "Peace, Order and good Government" of Canada is considered by some the main legislative authority of the federal Parliament, with the specified heads such as "Trade and Commerce" being merely examples of this broader authority. The position adopted by the courts, however, is that the "Peace, Order, and good Government" authority applies only to those matters *not* covered by the *enumerated* federal or provincial heads of authority.[38]

Where the legislation in question deals with a subject matter that could not have been in the contemplation of the United Kingdom Parliament at the time of the passing of the *B.N.A. Act* because that subject matter had not been invented (for example, aeronautics), the courts are inclined to find that the subject matter does not fall within any of the specified provincial heads of authority should there be a national dimension to the subject matter or should the subject dealt with be of concern to the country as a whole. Instead, the courts will conclude that the subject matter falls under the "Peace, Order and good Government" authority.[39]

Where, however, an aspect of the subject matter has been dealt with by the provinces for some time and was visible at the time of Confederation, the courts will probably want to be shown that there exists an emergency or crisis (or at least a rational basis for the federal government concluding one exists) before the courts will be prepared to find that there is an overriding federal aspect to the subject matter that justifies the exclusion of provincial jurisdiction and the application of the "Peace, Order and good Government" authority of the federal Parliament.[40]

It appears that whenever they are faced with federal legislation dealing with subject matter that has been previously dealt with from a different aspect by the provinces, the courts have the option of classifying the primary purpose or "matter" or "pith and substance" of the legislation with a name that may not appear to fall within any of the enumerated heads of legislative authority, but instead may conveniently be found to fall within the "Peace, Order, and good Government" authority.[41]

Whether engaged in finding a national concern, or in classifying the primary purpose of legislation as non-provincial, the courts will apparently be influenced by a conclusion that the problem being tackled "is beyond the power of the provinces to deal with."[42]

The courts will be more inclined to find federal intervention justified on the basis of national concern (even though no emergency or "crisis" exists), if the subject matters dealt with are limited and specific, than they will be where the category of the subject matter is so sweeping and general as to pose the danger of severe erosion of customary provincial authority.[43]

The Russian electrical engineer, Paul Jablochkov, had lit up the boulevards of Paris with arc lights as early as 1867.[44] It can therefore be argued that the United Kingdom Parliament would have been aware of this new technology of generating and distributing electrical power and would have contemplated its inclusion within one of the enumerated heads of authority. On the other hand, the first electric power station was not constructed until 1882 when Thomas Edison's Pearl Street Station went into operation in New York City. That same year, a small hydroelectric plant commenced operations in Appleton, Wisconsin.[45] The courts would therefore probably accept the argument that the technology of electrical generation and distribution was not sufficiently developed in 1867 as to lead to a presumption that the United Kingdom Parliament had any particular intentions concerning legislative authority over this new technology.

But just because electrical generation and distribution is a ''new'' subject matter (having developed after 1867) does not mean that laws concerning it automatically come within the ''Peace, Order and good Government'' authority. It may instead fall within ''Property and Civil Rights in the Province'' (s. 92(13)) or ''Matters of a merely local or private Nature in the Province'' (s. 92(16)).[46] If the subject matter is merely of local concern, then it will probably be held to fall within provincial jurisdiction under sections 92(13) or 92(16). If, however, the subject matter is changed (such as by existing transmission lines within a province being connected and operationally integrated with lines in another province) so as to become a matter that may have impact upon the nation as a whole, then the courts may conclude that the subject matter falls within the ''Peace, Order, and good Government'' authority.[47]

Trade and Commerce

Generally it has been accepted by the courts that intraprovincial trade and commerce (that is, transactions completed within a province) falls within provincial power under ''Property and Civil Rights in the Province'' or ''Matters of a merely local or private Nature in the Province,'' the federal ''Trade and Commerce'' power being restricted to interprovincial or international trade and commerce and ''the general regulation of trade affecting the whole Dominion.''[48]

Recently the Supreme Court of Canada has upheld federal regulation of intraprovincial transactions in oil on the ground that such regulation was incidental to the main object of regulating the interprovincial flow of that commodity.[49]

It is not yet clear what degree of interprovincial connection there must be to support federal regulation of activity that may be completed within a province. Up until now, Canadian courts have not been prepared to permit the size of the market to determine whether there was a sufficient interprovincial element to justify federal regulation. But recent cases striking down provincial marketing schemes have indicated the Supreme Court's recognition of the notion of a national market requiring regulation on a national basis.[50]

The willingness of Canadian courts to uphold federal regulation of the trade in grain and oil, even though purely local transactions were thereby regulated, appears to be a significant expansion of the federal ''Trade and

Commerce'' power.[51] The recent *Ontario Egg* case[52] establishes, however, that there are still limitations on federal power to intrude into local trade.[53] The Supreme Court of Canada there held that the federal government could not legislate to authorize adjustment levies on intraprovincial trade.

Although there are cases where the courts have permitted provincial regulation where there was some effect upon interprovincial trade,[54] on the ground that any such effects were merely incidental to an essentially intraprovincial scheme, the most recent cases indicate a good possibility of the Supreme Court of Canada permitting the federal Parliament to exclusively regulate, under the ''Trade and Commerce'' power, any market that extends beyond any one province.[55] Then, in the event of conflict with federal legislation, any provincial regulation with effect outside the province would be overridden (assuming that such provincial regulation was not declared *ultra vires* on the basis of the *Manitoba Egg* and *Burns Food* cases).

The extent of federal authority to legislate for the ''general regulation of trade affecting the whole Dominion'' is not clear. It had not been applied as a head of federal power in recent years until the Supreme Court of Canada in 1976, while striking down a section of the federal *Trade Marks Act*[56] (which prohibited and provided a civil remedy against business practices ''contrary to honest industrial or commercial usage in Canada''), which had been upheld by the Federal Court of Appeal as within the general category of ''Trade and Commerce,'' indicated the section might have been valid if it had been part of a ''regulatory scheme'' administered by a ''federally-appointed agency.''[57] Whether this presages additional scope for the federal ''Trade and Commerce'' power is still too early to say.

Extraprovincial Works and Undertakings[58]

The federal Parliament has jurisdiction over ''Lines of Steam or other Ships, Railways, Canals, Telegraphs, and other Works and Undertakings connecting the Province with any other or others of the Provinces, or extending beyond the Limits of the Province'' (s. 92(10)(a)).

The courts have interpreted the general phrase ''Works and Undertakings'' as referring only to those works and undertakings ''*ejusdem generis*,'' or of the same type as the specific examples preceding the phrase. In other words, section 92(10)(a) confers a power that is confined to works and undertakings involved in transportation or communication.[59]

In *Hewson* v. *Ontario Power Co.*,[60] two of five judges of the Supreme Court of Canada concluded that electrical power systems are included in the phrase ''other Works and Undertakings.'' Doubts have been raised, however, in other cases, as to whether such inclusion is correct.[61] It should not be unrealistic to view an electrical transmission system as transporting a flow of electrons in a manner analogous to a canal's carrying of water, and this should be sufficient to satisfy the criterion of ''transportation,''[62] even if one were not prepared to draw a comparison to telegraph lines. An interprovincial oil pipeline has been held to be a ''Work or Undertaking'' within the meaning of section 92(10)(a),[63] and it seems unlikely that the Supreme Court would treat the movement of electrical energy any differently from the movement of petroleum.

Mere physical connection of a local work with an interprovincial undertaking, or with an undertaking that has been declared to be for the general advantage of Canada, will not be sufficient to extend federal jurisdiction over the local work. There must be more than just a link—there must be some degree of operational integration between the undertakings to make them ''connecting'' within the meaning of section 92(10)(a).[64] It is not clear from the cases what degree of integration the courts will require. Various factors have been considered relevant in determining whether an undertaking within a province, physically linked with an extraprovincial undertaking, becomes part of the latter and under federal jurisdiction—for example, the type of corporate organization, the nature of the physical connection, the extent of the organizational interconnection, the purpose of the local undertaking, and the purpose of the connection.[65]

There is some possibility that the courts will find that the incorporation of a company with the object of operating a work or undertaking falling within section 92(10)(a) can only be carried out by the federal government. The argument here flows from the interpretation of section 92(11), which gives the provinces authority over ''the Incorporation of Companies with Provincial Objects.'' There are two meanings given to the expression ''with Provincial Objects'':

(a) since the object is non-Provincial in the sense that it envisages territorial operation outside a single Province, the incorporation of this type of company falls under the federal residual power, or
(b) since the object is non-Provincial in the functional sense, in that the proposed activity does not fall to be regulated under any Provincial head of legislative jurisdiction (which may be, though not necessarily, for territorial reasons), then the incorporation of this type of company falls under the federal residual power.[66]

Some would argue that the incorporation of a company with the object of operating a work or undertaking within section 92(10)(a) is jurisdictionally a matter for the federal government on either of the above interpretations.[67] But it seems the better view is that while there is clearly a territorial limitation on the provincial power to incorporate companies,[68] this limitation in itself would not necessarily prevent a provincially incorporated company from dealing with an extraprovincial work or undertaking. This is because it is just as clearly established that a province may give its companies *the capacity to accept* extraprovincial powers even though the province cannot create a corporation with a legal existence outside the province.[69] If a province grants its own company the power to deal with that portion of an extraprovincial work or undertaking within its boundaries, and if the province also grants this company the capacity to accept powers and rights from the province in which the other part of the work or undertaking is located, then the provincially incorporated company will have all the power necessary for dealing with the complete extraprovincial work or undertaking. And since the arguments against any functional limitation on the provincial incorporation power appear to outweigh the arguments in favour of such limitation, I submit that the courts will most likely conclude there is no problem with a provincially incorporated company constructing and operating an extraprovincial work or undertaking.[70]

The Federal Declaratory Power

Section 92(10)(c) combined with section 91(29) brings within federal authority "Such Works as, although wholly situate within the Province, are before or after their Execution declared by the Parliament of Canada to be for the general Advantage of Canada or for the Advantage of Two or more of the Provinces." Although the reference is only to "works" and not to "undertakings" in this subsection, the terms "works" and "undertakings" have at times been used interchangeably by the courts.[71]

The section applies only to works "wholly situate within the Province." In *Toronto* v. *Bell Telephone Co.*,[72] a declaration concerning the Bell Telephone Company was held to be "unmeaning" because the planned "works" would not be wholly situate within the province.

The effect of a declaration over a work is to withdraw from provincial and bring under federal jurisdiction "not only the physical shell of the activity but also the integrated activity carried out in connection therewith."[73]

The courts will not normally question Parliament's decision that a particular work is for the general advantage of Canada. But conceivably the courts could step in, in the unlikely event that it could be shown that Parliament's decision was colourable in that its *only* purpose was to intrude upon matters within provincial jurisdiction.[74]

For political reasons, the federal Parliament would probably be very reluctant today to use the declaratory power—the power has not been used since 1961.

PART FOUR—FUNCTIONAL ANALYSIS OF AUTHORITY OVER ELECTRICAL SYSTEMS

Planning

Planning authority for electrical systems would flow to the provinces from their position as owner of riverbeds and of Crown land necessary for generation and transmission lines. This authority would be reinforced by the enumerated subheads of section 92 of the *B.N.A. Act*, giving the provinces legislative authority over ''the Management and Sale of the Public Lands belonging to the Province,'' ''Property and Civil Rights in the Province,'' and ''Matters of a merely local or private Nature in the Province.''

Federal planning authority could arise under the headings of section 91 dealing with ''the Public Debt and Property,'' ''the Regulation of Trade and Commerce,'' ''Subjects as are expressly excepted in the . . . Subjects . . . assigned exclusively to the Legislatures of the Provinces,'' as well as under the ''Peace, Order and good Government'' provision.

Questions may be raised concerning the spending of provincial money on matters that may not fall clearly within provincial jurisdiction—for example, spending on a survey to determine the amount of undeveloped hydroelectric potential in another province, which might be the subject of a joint development between two provinces. As was previously indicated with respect to the federal spending power, the better view appears to be that the provinces also may spend on other matters in addition to those within their jurisdiction under the *B.N.A. Act*. Their authority for this would flow either from the provincial prerogative power—that residue of power left to the Crown—or from the province's power over property and civil rights in the province.[75]

An interprovincial mechanism for collaborative planning between provinces will be a factor that the courts will look at in determining whether there is such an integration and operational interdependence of local and interprovincial electrical systems as to bring the local system under federal jurisdiction as part of an interprovincial work or undertaking. In the *Luscar Collieries Ltd. Case*,[76] the court found that the Luscar Railway Line was ''a part of a continuous system of railways operated together by the Canadian National Railway Company, and connecting the Province of Alberta with other provinces of the Dominion,'' so as to fall within section 92(10)(a). This is an example of where common operation was a factor influencing the court to conclude that the local line, because of its connection with an interprovincial line, fell under federal jurisdiction. As will be discussed more fully below, a common operation in itself is probably not sufficient to determine where jurisdiction lies, but it is a factor of some importance. Co-ordinated planning between the utilities of various provinces will not, in my opinion, in itself, be enough to cause the courts to conclude that the individual provincial electrical networks have become merged into an interprovincial network under federal jurisdiction. Combined, however, with joint development of generating and distributing systems, and with joint financing, the situation may be different.

Allocations

The authority to allocate rights with respect to resources also flows from ownership of these resources, complemented in the case of the provinces by jurisdiction over (*a*) "the Management and Sale of the Public Lands belonging to the Province," (*b*) "Property and Civil Rights in the Province," and (*c*) "Matters of a merely local or private Nature in the Province."

If the resource is owned by the federal Crown, its authority as owner is reinforced by jurisdiction over (*a*) "the Public Debt and Property," and (*b*) "Indians and Lands reserved for the Indians," should the resource be on Indian lands. Ownership of "Canals, with Lands and Water Power connected therewith," "Public Harbours," and "Rivers and Lake Improvements" were transferred to the federal government by the Third Schedule of the *B.N.A. Act*.

The question arises of how far a province can go in imposing conditions as owner upon its lease of water rights—for example, its lease to a corporation, having several provinces as shareholders, which proposes to export electrical power from the province. Could the province, for example, provide that the leases automatically terminate if the corporation should export the power to another province at less than a specified price or without a permit granted by the province? One would think that this would be an acceptable exercise of ownership rights.[77] Certain statements in the recent *CIGOL Case*[78] raise some doubts, however, about this. In that case, Mr. Justice Martland, while rendering the majority opinion, stated:

> Provincial legislative authority does not extend to fixing the price to be charged or received in respect of the sale of goods in the export market. It involves a regulation of interprovincial trade and trenches upon subs. 91(2) of the *British North America Act*.[79]

Mr. Justice Martland quoted from an earlier decision of Chief Justice Kerwin in the *Reference re The Farm Products Marketing Act*,[80] where the Chief Justice said:

> Once a statute aims at "regulation of trade in matters of interprovincial concern" it is beyond the competence of a provincial Legislature.
>   . . . The concept of trade and commerce, the regulation of which is confided to Parliament, is entirely separate and distinct from the regulation of mere sale and purchase agreements. Once an article enters into the flow of interprovincial or external trade, the subject-matter and all its attendant circumstances cease to be a mere matter of local concern.[81]

Mr. Justice Martland concluded that in the *CIGOL Case* the legislation was "directly aimed at the production of oil destined for export and has the effect of regulating the export price, since the producer is effectively compelled to obtain that price on the sale of his product." He therefore concluded that the mineral income tax and the royalty surcharge imposed by the Province of Saskatchewan was *ultra vires* that legislature and invalid.

Mr. Justice Martland's conclusions conflict with earlier authorities. For example, although the federal Parliament has legislative authority over aliens, still the Province of British Columbia was held to have jurisdiction to issue forest permits on the condition that they would be cancelled should certain aliens be employed.[82] Also, the Ontario Court of Appeal upheld a condition in Ontario timber leases requiring timber cut thereunder to be processed in Canada, although this condition had been challenged on the ground that it dealt with a matter under federal jurisdiction as relating to trade and commerce. The Ontario court concluded that the Ontario legislature was merely acting in its capacity as property owner, and since the condition was not prohibited by any existing law, therefore it was valid.[83] These cases may possibly be distinguished on the grounds that the leases in the *CIGOL Case* had been issued prior to the imposition of the tax and royalty surcharge. It could be argued that property had passed from the Crown, and the tax and royalty surcharge could not therefore fall within the ''Management and Sale of the Public Lands belonging to the Province.''[84] A case with which it is more difficult to reconcile *CIGOL* is the *Carnation Case*.[85] In that case, certain provincial regulations were upheld although their effect was to increase the cost of milk purchased in Quebec, which milk was mostly for sale outside the province. Mr. Justice Martland distinguished the *Carnation Case* on the grounds that their regulations were only ''indirectly'' related to export prices. He did not explain, however, the meaning of ''indirectly,'' nor did he explain how one determines what a statute is ''directly aimed at.'' The *CIGOL Case* and the *Carnation Case* are both examples of the unfortunate tendency of our Supreme Court to arrive at decisions (presumably based on hidden value judgements or policies) by a process of labelling the statute in question as either ''directly'' or ''indirectly,'' ''relating to'' or ''aimed at'' various subject matters. This approach to decision making will be considered further in Part Five below.

If a province were in the process of planning a hydro development and were concerned about the possibility of the federal government declaring the works to be for the general advantage of Canada, the province could possibly maintain some control over this eventuality by making it a condition of its lease of water rights to the private or Crown corporation responsible for the development, that the water rights would be determinable should the federal declaration be made. Then, although the federal government might obtain, by declaration, control over a generating plant, it is possible it might not be entitled to the use of the water.[86]

The manner in which conditions are attached to water rights may be important. Michael Crommelin has dealt with this point in connection with Alberta petroleum and natural gas leases. He states:

> First, the clause is drafted in the form of a determinable limitation rather than a condition subsequent. It does not prohibit the removal of gas from Alberta without approval or give the Alberta government a cause of action for breach. Instead, it provides that the rights of the lessee under the Crown lease shall determine upon the happening of such an event, without the necessity of any further action on the part of the Alberta government. The courts have long drawn a distinction between the effects

of a determinable limitation and a condition subsequent. In particular, the results are different if the conditions are void. A grant subject to a condition subsequent, which is judged to be void, becomes absolute, since the grant is properly limited even though the condition fails. However, a grant subject to a determinable limitation which is void wholly fails, since the grant has no proper limitation. Thus, even if the clause effecting the limitation of the lessee's interest was held to be inoperative as going beyond the powers of the Alberta government, the lease would determine, with the consequence that the future removal of gas from the province would be prevented.[87]

With respect to the allocation of rights in the territorial sea (which might be relevant, for example, for the proposed Bay of Fundy Tidal Power Project), the Supreme Court decided in the *B. C. Offshore Minerals Reference*[88] that the residual power of section 91 was one reason for concluding that the federal government had legislative authority over the territorial sea and Continental Shelf off British Columbia. The reasoning of the Supreme Court in this case has been severely criticized and it is questionable whether it will be applied to the eastern provinces, which in any event, with their different historical development, are probably not in the same position as British Columbia.[89]

Construction

Provincial ownership of the sites of generating stations and transmission lines will give the provinces authority over such works, which authority will be complemented by jurisdiction over (*a*) "Local Works and Undertakings" in addition to (*b*) "Management and Sale of the Public Lands belonging to the Province," (*c*) "Property and Civil Rights in the Province," and (*d*) "Matters of a merely local or private Nature in the Province."

In addition to the federal powers relevant for allocation of rights (which apply here as well), some federal jurisdiction will arise from authority over "Navigation and Shipping" (should interference with navigation result), "Sea Coast and Inland Fisheries" (should fish be affected), "Agriculture" (should irrigation projects be affected), "Criminal Law" (should the matter of water pollution arise), and from legislation presently in effect implementing international treaties, such as the International Boundary Waters Treaty, 1909, or the Migratory Birds Convention (should waters crossing or forming the border with the United States, or should waterfowl, be affected). Federal-provincial co-operation will be needed to permit developments where these matters arise.

Electric power generating and transmission systems have been at times declared works for the general advantage of Canada or for the advantage of two or more of the provinces.[90] A declaration could probably be made in terms of "all electric power generating stations, transmission lines and distribution lines, whether heretofore constructed or hereafter to be constructed."[91] By the *Atomic Energy Control Act*,[92] all works and undertakings for the production, use, and application of atomic energy are declared to be works for the general advantage of Canada. Even without this declaration, the provisions of this act are valid under the ''Peace, Order and good Government'' authority of the federal government.[93]

As previously indicated, the effect of a declaration is to apply federal jurisdiction not only to the physical shell or facility but also to the integrated activity connected with it. Thus declarations with respect to grain elevators have been effective to authorize federal regulation of the grain trade generally.[94] Up to now, federal jurisdiction has not been asserted over the transmission of electrical power from nuclear stations. However, as more nuclear stations are integrated into provincial electrical networks, the chances increase of federal jurisdiction becoming applicable over the entire system.

In addition to federal jurisdiction over generation from nuclear plants, federal authority will also probably arise should the generating source be a river that extends beyond the boundaries of a province. Where a physical plant extends beyond the limits of a province (for example, if part of a generating station or reservoir was in one province and part in another, or in another country), the Supreme Court has indicated that section 92(10)(a) will apply to make federal jurisdiction effective.[95] This is difficult to reconcile with the view that the works and undertakings referred to in that sectiom must be *ejusdem generis* with the specific works enumerated in section 92(10), which works have been classified as those relating to transportation and communication. Looking at the reference to ''Canals,'' however, it may be sufficient to show that the ''work'' in question ''transports water'' (a test that presumably a generating station would meet).

Where there have been no works or undertakings constructed or organized, that is, where an interprovincial or international river is still in its natural state, section 92(10)(a) would not apply because that section envisages something fabricated. Therefore, the federal government could not appropriate a river bed to construct its own generating plant. It is possible, however, that federal jurisdiction exists in respect of interprovincial and international rivers by virtue of the authority to legislate for the ''Peace, Order and good Government'' of Canada. An argument can be made that rights relating to interprovincial or international waters are neither matters in relation to ''Property and Civil Rights in the Province,'' nor ''Matters of a merely local or private Nature in the Province.'' And if there is, therefore, no provincial authority to deal with such matters, the authority must lie with the federal government under the ''Peace, Order, and good Government'' residual authority.[96]

If a province's activity on an interprovincial river should have an impact upon riparian owners in another province, other questions will arise, namely, what are the legal principles to be applied in determining the rights of riparian provinces and in what forum will these rights be determined. Considering the applicable principles, there seems to be a consensus among commentators that rights respecting the use of waters in interprovincial rivers are probably covered by the common law.[97] The principles of the common law could be brought in by Canadian courts adopting the doctrine of equitable apportionment, as did the Supreme Court of the United States concerning interstate disputes over rivers crossing their boundaries. This doctrine recognizes that, contrary to the Harmon Doctrine, an upstream state may not dispose of water in such rivers in any manner it chooses without regard for the impact on the downstream state. "Each state has an interest in the water that must be respected and reconciled."[98] Assuming the applicability of common law principles, there appears to be some confusion as to what these principles consist of. Some writers submit that common law riparian rights entitle an upstream owner to divert water for irrigation and manufacturing uses as well as for purely domestic or personal uses.[99] Others contend that only the latter uses are authorized.[100] It is my opinion that the former "reasonable use" doctrine is more likely to be applied by the courts than the latter "natural flow" doctrine.[101]

On the matter of the appropriate forum, because Canada lacks a constitutional provision for the judicial determination of interprovincial disputes such as there is in the United States, many questions cannot be judicially determined without the agreement of the provinces.[102] This does not apply, however, to constitutional disputes. If, therefore, a province exercises executive or legislative authority, which may be constitutionally defective (because, for example, it interferes with civil rights outside of the province[103]), its authority may be challenged by another province in the first province's courts and ultimately in the Supreme Court of Canada. In fact, such unconstitutional executive or legislative action by a province could be challenged by an individual as well, either as a member of the general public[104] or as someone particularly affected by the unconstitutional action (for example, should a question arise as to whether federal or provincial labour relations laws apply to a project on an interprovincial river[105]). Most interprovincial disputes concerning interprovincial rivers would probably be worked out at the political level. If, however, they were brought before the courts, either by a province or by an individual, there is a good possibility that the courts would conclude that the matter falls within the federal government's "Peace, Order, and good Government" authority. This is because, as previously discussed, the problem is beyond the power of the provinces to deal with, and the failure of one province to act may be harmful to the residents of another province.

In the event of an acute energy crisis in Canada, the courts would probably find that the federal government had the authority under the emergency power based on the "Peace, Order, and good Government" provision to construct generating plants wherever it decided to do so. A rational basis for concluding that an emergency existed would have to be established for the courts should the federal action be challenged.[106] Once the courts concluded that the federal government had a rational basis for concluding an emergency existed, federal involvement in generating plant construction would be justified on grounds of relieving dependence upon foreign energy sources through utilization of undeveloped Canadian resources. It is highly improbable that the courts will set aside what the federal Parliament concludes is necessary in order to cope with what Parliament (but not necessarily the court) believes to be a national emergency.[107]

Operation and Marketing

The provinces will normally have authority over transmission lines by virtue of their ownership of lands on which these lines will be placed and by their authority over ''the Management and Sale of the Public Lands belonging to the Province,'' ''Property and Civil Rights in the Province,'' ''Local Works and Undertakings,'' and ''Matters of a merely local or private Nature in the Province.''

Federal jurisdiction will normally be restricted to lines passing over federally owned land (''the Public Debt and Property'') or over ''Lands reserved for the Indians.'' However, where power is transmitted across provincial or international boundaries, the following heads of federal jurisdiction become relevant: (*a*) ''Extraprovincial Works and Undertakings''; (*b*) ''Trade and Commerce''; and (*c*) ''Peace, Order, and good Government.''

Extraprovincial works and undertakings have already been discussed generally under Part Three above. It has been indicated that electrical power systems are probably included in the phrase ''Other Works and Undertakings.'' The difficulty lies in determining when a transmission line within a province takes on the character of a work or undertaking ''connecting the Province with any other or others of the Provinces, or extending beyond the Limits of the Province.'' There are really two questions involved here: (*a*) whether the extraprovincial connection or extension exists, and (*b*) if it does exist, how far does the extraprovincial system intrude into the province. It is necessary to look at the factors that the courts in the past have considered relevant in order to arrive at any conclusions respecting these questions. The cases indicate that the following factors are most important in determining the existence and extent of an extraprovincial connection: (*a*) corporate organization, (*b*) physical and operational connection, and (*c*) purpose of the connection.[108]

In the *Empress Hotel Case*,[109] the Privy Council decided that the British Columbia *Hours of Work Act* applied to employees of the Canadian Pacific Railway Hotel in Victoria. The court, after concluding that a company could be authorized to carry on more than one undertaking, looked at the corporate powers of the C.P.R., found that the company had authority to enter into the hotel business generally, concluded that the actual operation of the Empress Hotel was little different from any non-railway hotel and competed with other hotel keepers for general hotel business, determined that the hotel was not built on land adjoining the railway line, and decided that the hotel was not part of the same work and undertaking as the railway.

The Privy Council in the *Empress Hotel Case* concluded that while the hotel business was obviously useful to the company's railway business, that did not prevent the two from being separate undertakings. This conclusion makes the decision in the *B.C. Power Case*[110] somewhat questionable. In the latter case, after deciding (for reasons discussed more fully below) that the company's bus and electrical systems fell within section 92(10)(a), Chief Justice Lett of the Supreme Court of British Columbia decided that the company's railway and gas business were all part of the one interprovincial undertaking—although he had concluded that the railway system standing by itself would have been within provincial jurisdiction, and although the only connection of the gas business to the other portions of the business was that one of the company's thermal-electric plants was supplied by the gas division of the company. The *B.C. Power Case* has been criticized on a number of different grounds and is probably of doubtful value as a precedent for future cases.[111]

The British Columbia Electric Company carried on its activities through the divisions of a single company. But this was not found to be a conclusive factor in *Retail, Wholesale and Department Store Union* v. *Reitmier Truck Lines Ltd.*[112] where an interprovincial motor carrier had taken over all the outstanding shares and the operation of an intraprovincial carrier, preserving the corporate identity of the latter but operating it solely with persons who were employees of the parent company. Mr. Justice Munroe, as an alternative reason for his conclusion that the local company was not subject to federal labour relations law, found that the local company had not become so integrated with the interprovincial carrier as to bring it within federal jurisdiction. In another case, Northern Electric Company, a subsidiary of the Bell Telephone Company of Canada, which manufactures the telephone equipment used by Bell, was held to fall within provincial jurisdiction for matters of labour relations.[113] This decision has been questioned because of the lack of evidence before the court concerning the degree of integration and interdependence of the two companies.

Generally, the cases indicate that corporate structure is not a factor that will have a conclusive effect upon the determination of whether a particular business entity is part of an interprovincial work or undertaking. In the *Stevedoring Reference,*[114] a stevedoring operation was found to be so integrated with interprovincial shipping that it fell within federal jurisdiction, even though the stevedoring was carried out through a corporation independent of the steamship companies involved.

Physical and operational connection has been an important factor in determining whether an extraprovincial work or undertaking exists and its extent. But the cases are not all consistent with each other. Thus, one railway company, which operated local lines, was found not to fall under the authority of the federal Board of Railway Commissioners, although it connected at several points with a railway that had been declared to be for the ''general Advantage of Canada,'' and although arrangements were made for the transporting of passengers from a point on one system to that on another.[115] The court concluded that since arrangements could probably be made with local railway companies so as to permit the federal company to carry out its federally imposed obligations, it was not necessary to find that federal jurisdiction extended to ''through traffic'' on local lines. On the other hand, another local railway, which connected with the Canadian National Railway, and was by agreement operated by the C.N.R., was ''a part of a continuous system of railways operated together by the Canadian National Railway Company, and connecting the Province of Alberta with other Provinces of the Dominion'' and under federal jurisdiction within section 92(10)(a) of the *B.N.A. Act*.[116] Still another local railway, which connected the lines of the C.N.R. with a third railway, which had been declared to be a work for the ''general Advantage of Canada'' under section 92(10)(c), and was also operated by the owners of the local railway under an agreement with the C.P.R., was held not to ''connect'' the province with another.[117] This and the *Montreal Street Railway Case* can be distinguished from the *Luscar Collieries* case because, unlike *Luscar*, the railways were not operated as a single unit, and the connections were with railway systems that normally would not have been under federal jurisdiction.[118] More recently, a commuter train service operated by the government of Ontario completely within the province, using its own rolling stock, but operating on C.N.R. tracks and using C.N.R. rail crews by agreement, was held to be part of the overall interprovincial undertaking of C.N.R. and within federal jurisdiction for the purpose of fixing tolls.[119]

In some lower court decisions, judges have refused to use a percentage computation of the amount of extraprovincial work as compared to the intraprovincial work of a company in order to determine whether the undertaking is one falling under section 92(10)(a). In one case, a transport company operating primarily in Ontario, with only 6 per cent of its trips outside that province, was held to be an ''undertaking'' within the section.[120] Yet in another case, where interprovincial business accounted for about 5.5 per cent of the company's growth revenue, the operations were held not to fall within section 92(10)(a).[121] In the latter case, the transport company operated its interprovincial work on an intermittent basis as requested by customers. This last case indicates that the extraprovincial connection must be to a degree ''regular and continuous'' in order for it to be significant in determining the question of jurisdiction. This view is supported by the Supreme Court of Canada decision, which refused to find federal jurisdiction over vessels that transported cargo between ports on the St. Lawrence River within the province of Quebec although they had, on three exceptional circumstances, travelled outside the limits of the province, twice to Toronto in 1964 and once to Nova Scotia in 1965.[122]

In the *B.C. Power Case*,[123] Chief Justice Lett had to determine whether the B.C. Electric Co., a provincially incorporated company, constituted an intraprovincial undertaking under section 92(10)(a) where it carried on business involving the distribution of natural gas, the generation and distribution of electricity, and railway and bus transportation. He decided that although the railway had links with the C.P.R. transcontinental system, it did not connect the province with other provinces within the meaning of section 92(10)(a). Also, he concluded that the business in natural gas, linked by a meter with pipelines going outside the province to the United States and Alberta, did not ''connect'' other provinces and the United States within the meaning of section 92(10)(a). But he did find that the electric system ''connected'' the province with the United States by reason of the electrical interconnection at the international boundary and because of the existence of the company's cables in U.S. territorial waters. The company's electrical system, he concluded, was an integral part of the Northwest Power Pool, a grid linking British Columbia and certain states of the United States, intended to supply mutual assistance during emergencies and for exchange of surplus energy. Chief Justice Lett also found that there was a continuous flow of a minimal amount of energy (''deviation'') back and forth between the B.C. system and the other parts of the Northwest Power Pool. Also, small amounts of energy were transmitted regularly and continuously from the B.C. company to an American company by delivery at the international border for sale to Point Roberts in the United States. The company's bus operations, although operating primarily in British Columbia, had, as a small portion of its business, sightseeing tours to the United States. Chief Justice Lett concluded that both the bus and electrical systems extended beyond the province and therefore the entire operations—gas system, railways, buses, and electrical system—fell within section 92(10)(a) as a single undertaking.

The *B.C. Power Case* has received much criticism.[124] As one writer has stated:

> . . . On the facts in the *British Columbia Power* case, there is much to be said for the view that the electric company there was an intraprovincial one. It generated and supplied electricity only in the province. Cooperation with other electric companies by means of a grid for balancing different peak periods and for emergencies is not, it is suggested, sufficient to alter the substantial character of the electric operations. Otherwise the interconnections in power systems that exist in the Maritimes would

probably make all the electric companies and provincial commissions subject to federal control in relation to their intraprovincial operations. The courts in other fields have strongly acted to prevent such easy federal passage into the provincial domain. The minor extensions of the electric company's works in American territory for more convenient operations in Canada should not be permitted to alter the matter either. The Privy Council was willing to ignore such minor extensions outside provincial territory in connection with colorable attempts to evade provincial responsibility. There seems to be no reason why it could not equally be ignored in determining whether in pith and substance an undertaking is provincial or interprovincial in scope. Moreover, provincially incorporated companies may be permitted by other authority to function outside the province; so territorial incompetence should be no problem. The grid system, itself, however may be a different matter.[125]

The courts, however, have shown a marked reluctance to divide up a company's operations into two undertakings, one interprovincial and one intraprovincial. In addition to the *B.C. Power Case*, we see the same inclination to find a single undertaking in the *Winner* case.[126] There Mr. Winner, an American citizen of Maine, ran a bus line between Boston and Halifax through New Brunswick. He had been granted a licence under the New Brunswick *Motor Carrier Act* for the operation of his buses, subject to the restriction that no passengers could be set off or taken on in the province. Mr. Winner disregarded that limitation and the question in the case was whether the New Brunswick act applied to his operations. In concluding that it did not, the Privy Council stated:

> The question is not what portions of the undertaking can be stripped from it without interfering with the activity altogether; it is rather what is the undertaking which is in fact being carried on. Is there one undertaking, and as part of that one undertaking does the respondent carry passengers between two points both within the province, or are there two?[127]

The *Winner* case has been criticized for failing to recognize the feasibility of separating the local bus service from the interprovincial and international service, for regulatory purposes.[128] The case has been recently applied by the Supreme Court of Canada in the *Capital Cities Communications Case*,[129] where it refused to find two undertakings in the operations of a cable television company that intercepted American television signals by antennae, deleted U.S. ads, and retransmitted within the province by cable. The court concluded that there was one single indivisible undertaking falling within federal jurisdiction.

The purpose of the connection between the local work and the extraprovincial line was one of the factors applied by the federal Board of Transport Commissioners in the *Westspur* case[130] to find that gathering lines located wholly within a province and connected to interprovincial trunk-lines were an integral part of the interprovincial undertaking, the purpose of the gathering lines being to feed the trunk-lines as well as to operate for the benefit of local producers. This, in itself, may not be a very helpful factor, however, in determining jurisdiction over intraprovincial electrical transmission lines that are connected to an interprovincial grid, since it seems that in all cases there will be regional or national aspects as well as a local aspect in the purpose of the connection.

The federal "Trade and Commerce" power is probably one of the most likely avenues available for the taking of jurisdiction by the federal government over extraprovincial sales of electricity. The transmission of electricity can be considered as the conveyance of a commodity, and when transmission occurs over provincial or international boundaries, this is probably interprovincial or external trade and commerce, falling under federal jurisdiction.[131] In the *Ottawa Valley Power Company* case, Mr. Justice Fisher stated:

> The transmission of such commodities as electricity, oil or gas is clearly the conveyance of a commodity and the Dominion has no regulatory power over such interprovincial exchange of commodities.[132]

The reference to electricity and oil or gas being a commodity is probably correct, but the assumption that the Dominion would have no regulatory power over the interprovincial exchange of oil and gas has been proven incorrect,[133] and interprovincial exchanges of electricity would also probably be found to fall within federal jurisdiction should the federal Parliament decide to regulate it.[134] As discussed in Part Three above, the most recent cases of the Supreme Court of Canada indicate that the court will tend to permit the federal Parliament to regulate, under the "Trade and Commerce" power, any market that extends beyond any one province. The Supreme Court has also indicated that federal regulation of interprovincial trade may extend to purely local transactions.

The transmission and sale of electricity between provinces probably also falls under federal jurisdiction through the "Peace, Order and good Government" power. Being an activity that has developed since Confederation, the courts will have to assign it to either the federal or provincial residuary powers. They will look at the significance of interprovincial electrical transmission for the nation as a whole. They will probably be influenced by the fact that the intransigence of one province in refusing, for example, to agree to a regional power development could mean the loss of a significant benefit for all should the matter be placed within provincial jurisdiction. Also, the subject matter of interprovincial transmission of electricity is limited and specific enough to avoid any significant erosion of provincial authority in other areas should the federal government be given jurisdiction.

PART FIVE—JUDICIAL POLICY CONSIDERATIONS IN DETERMINING JURISDICTION OVER INTERPROVINCIAL ELECTRICAL INTERCONNECTIONS

Should there come before the courts the question of whether the federal or provincial goverments have jurisdiction over interprovincial electrical interconnections, the matter will involve weighing federal authority over "Trade and Commerce," "Extraprovincial Works," and "Peace, Order and good Government" against provincial authority over "Property and Civil Rights in the Province," "Local Works and Undertakings," and "Matters of a merely local or private Nature in the Province." As mentioned earlier,[135] Canadian courts have shown an unfortunate tendency to resolve such issues by a question-begging process. They will state, often without further explanation, that a statute "aims at" or has as its "pith and substance" or "primary purpose" such and such. The such and such is so described by the court as to relate to one of the heads of federal or provincial authority set out in sections 91 and 92 of the *B.N.A. Act*. Naturally enough, the court is then able to find that the such and such is a matter exclusively within the jurisdiction of the government having the particular head of authority.

The conclusion is inescapable that the courts are engaging in hidden value judgements when they engage in such labelling techniques to arrive at particular decisions. Presumably, each decision depends upon the judge's view of the nature of Canadian federalism and the effect that a decision one way or the other will have upon the operations of the federation.

As discussed earlier,[136] cases dealing with the "Peace, Order and good Government" authority indicate that judges will want to be shown that a new (since 1867) subject-matter is of national concern or that an old matter of provincial responsibility has been overtaken by an emergency or crisis, before they will find federal jurisdiction under this head of authority. Federal authority under "Trade and Commerce" arises with respect to interprovincial or international trade and commerce or "the general regulation of trade affecting the whole Dominion"—transactions completed within a province will fall under federal "Trade and Commerce" jurisdiction only if this is necessary for the effective implementation of an otherwise legitimate national regulatory scheme (and, presumably, only if it does not involve too great an encroachment upon the normal provincial sphere of authority).[137] As for "Extraprovincial Works and Undertakings," federal authority will apparently only arise if there is a significant regular and continuous extraprovincial connection between works or undertakings within a province and those outside it.[138]

In deciding whether federal jurisdiction over a proposed scheme arises under any of the above heads of authority, it becomes relevant to consider the purpose and effect of the scheme. To what extent will the undertaking established under the scheme be a matter of national concern? Do all parts of the undertaking within the provinces form an essential part of a nationally integrated system? Are the potential benefits for the nation as a whole so great as to require that individual provinces not be permitted to frustrate the worthwhile endeavour? Is central decision making necessary for the successful operation of the undertaking? Questions such as these will probably arise in the mind of a judge considering jurisdiction over a scheme for strengthening interprovincial electrical interconnection. It should, therefore, be helpful to identify the aspects of such a scheme that may influence the judge in choosing between federal and provincial jurisdiction. In particular, it is useful to specify the, not necessarily compatible, federal and provincial interests that may be at stake.

The main aims of a scheme to strengthen interprovincial electrical interconnections will normally be as follows:
1. Minimize use of scarce resources
2. Maximize use of low-cost energy
3. Facilitate regional development
4. Improve efficiency of energy development
5. Increase flexibility of planning and operation
6. Minimize environmental damage.

The respective federal and provincial interests may be better identified by summarizing the methods through which some of the above aims could be met.
1. Scarce resources could be conserved by
   - Construction of transmission systems permitting development or usage of resources that would not otherwise be utilized (*federal interest*— usage of Canadian resources to reduce dependence upon foreign energy supplies; *provincial interest*—earliest development of provincial resources with resultant employment, revenues, and general economic development)
   - Interconnections allowing displacement of a scarce exhaustible fuel by a more abundant or renewable one (*federal interest*—national self-reliance in energy; *provincial interest*—conservation of a depleting resource or development of a renewable one)
   - Interconnections allowing diversity interchanges, permitting provinces to take advantage of differences in peak periods, both time zone and seasonal, and thereby maximizing use of renewable energy sources and reducing ultimate fuel consumption (*federal interest*—conservation of national resources and reduced dependency upon foreign fuel supplies, with improved balance of payments; *provincial interest*—increased revenue from otherwise surplus energy and savings through reduced fuel costs).
2. The use of low-cost energy could be maximized by
   - Interconnections permitting operation on ''economic dispatch'' on a regional or national basis with maximum production from plants with

lowest fuel costs and minimum production from plants with highest fuel costs (*federal interest*—stimulation of the national economy through lower energy costs; *provincial interest*—stimulation of the provincial economy; revenue flows to allow recovery of capital investment in plants whether they have high or low fuel costs).

3. Regional development could be facilitated by
   - Co-operation in planning so as to avoid construction of surplus regional generating capacity (*federal interest*—decreased pressure on the Canadian dollar from reduced capital requirements; *provincial interest*—deferred borrowing charges)
   - Acceleration of development in one province by installing capacity in advance of that province's needs for temporary use by another province or provinces (*federal interest*—possible improvement in the economic position of disadvantaged areas, possible deferral of increased reliance upon foreign energy supply; *provincial interest*—a diminished financial burden in construction of capacity for future needs; accelerated employment; revenue and general economic development in the resource-owning province as well as possibly lower-cost power for a time to the consuming province).

4. The efficiency of energy development and flexibility in planning and operating provincial utilities may be increased by
   - Reducing the reserve requirements of several systems by interconnecting and permitting the sharing of reserves and the benefits of diversity (*federal interest*—reduced capital requirements for the nation; *provincial interest*—reduced capital requirements and reduced energy costs)
   - Interconnections permitting the installation of larger generating units than an isolated system could safely absorb (*federal interest*—greater reliability throughout the nation; *provincial interest*—financial benefits through economies of scale).

5. Environmental damage might be minimized by
   - Interconnections reducing the number of development sites needed, for example, allowing one large development instead of many smaller ones (*federal interest*—possibly less hazard to health and safety; *provincial interest*—less impact in some provinces but more in others)
   - Interconnections permitting developments in remote instead of populated areas (*federal interest*—possibly less risk to health and safety of large numbers of people but probably greater interference with interests of native peoples; *provincial interest*—degree of impact within the province).

The interests of the federal and provincial governments will not always coincide. The federal government may decide to promote on a national basis the development of one form of generating plant, for example, nuclear, while a province may wish to develop another form. Or the federal government may decide that a certain uniformity in energy costs is desirable across the nation while a province will wish to obtain the best price for the sale of its resources. In determining jurisdiction, a judge decides which interest shall prevail.

To change from the existing provincial to federal control over electrical utilities would have far-reaching consequences. As one writer has pointed out:

> The raising of a revenue is not the sole reason that public property is of fundamental importance to the provinces. It also provides them with a powerful instrument for the control of their economic and political destinies. By requiring that resources for public property be processed within its boundaries, a province can materially contribute towards the establishment of secondary industries there, and prevent the export of raw material to other countries. It can also encourage colonization and the development of industry by judicious grants of such property. It can further shape the economy by public ownership of industry and thereby dictate what class will control the economic sphere of society.[139]

Changes in the type of generating facility can affect the extent of the mining, manufacturing, and construction activity that takes place in a province. The adoption of national energy pricing policies could have the effect of transferring benefits from energy-producing provinces to consuming provinces. These are only two of many areas where courts will have to recognize that tilting the balance in favour of the federal interest and granting greater federal control will significantly affect areas traditionally dealt with by the provinces.

It has already been pointed out[140] that the courts have been reluctant to find that an arrangement involving extraprovincial works and undertakings can be divided up into two undertakings, one under federal and one under provincial jurisdiction. In the case of interprovincial electrical interconnections, while it seems that federal jurisdiction will apply to the interprovincial operations, it is my opinion that despite their reluctance to find a divided jurisdiction, the courts will seek to reduce the extent to which federal jurisdiction, arising because of an interconnection, applies to operations existing within the province before the interconnection. Any other approach will mean too great a shift in the present balance of power between the provinces and the central government. The courts will probably restrict federal jurisdiction to those activities necessary for the interprovincial undertaking. But it will be difficult to distinguish between operations essential for interprovincial transmission and those directed at meeting provincial needs. Organizational arrangements may assist in this process of differentiation and help ensure that any federal jurisdiction imposed does not intrude too far into areas presently the sole responsibility of the province.

Consider the situation where two or more provinces decide there may be benefits gained from some form of co-operative effort in the electrical energy field. They decide to create a central electrical authority. This central body could function on a voluntary, informal, and non-contractual basis or it could, for either planning, or scheduling, or both, function in accordance with formal undertakings, contractual agreements, dealing with such matters as reserve sharing and pool operating procedures. There is an entire spectrum of possible arrangements moving from the completely informal on one hand to the completely formal and contractual on the other. The scheme of implementation adopted will influence the courts in determining whether legislative jurisdiction over it can be divided. The greater the degree of integration and interdependence of the electrical operations of the different provinces, the more likely will it be that federal jurisdiction will arise over the entire interconnected system. If one accepts that federal jurisdiction will be recognized with respect to interprovincial transmission lines, should this be asserted by the federal government or by a third party in the courts, then the question arises whether the provinces can limit the extent of the federal jurisdiction by the manner in which they organize their co-operative undertaking. It is submitted that they very likely can. For example, they could separate the ownership of facilities in the province to meet export needs from the ownership of facilities required for normal utility development. The latter would be owned by the provincial public utility, but the former could be owned by a separate corporation or title remain with the Crown. This might establish a sufficient dividing line for the courts to grasp in deciding where to define the limits of federal jurisdiction. Unfortunately, a certain loss of efficiency will probably be the cost of any organizational structure adopted for the purpose of limiting the extent of federal jurisdiction.

PART SIX—CONCLUSIONS

To avoid *all* jurisdictional uncertainty, any proposed interprovincial transmission scheme should operate with authority delegated from both provincial and federal governments. The arrangement could still be challenged in the courts, but if properly set up so as to draw its authority from both levels of government, the arrangement would remain legally valid even if one government were found to have no jurisdiction.[141]

By the use of its power to declare works to be for the "general Advantage of Canada" or for the advantage of two or more provinces, the federal government could at any time obtain jurisdiction over an interprovincial network, but political considerations might prevent this.

By issuing water rights to their Crown corporations on the basis that these rights are to be terminated in the event of the exercise of the federal declaratory power, the provinces might be able to ensure that regulation is not taken out of their hands by the federal government.

Since electricity will probably be treated as a commodity in interprovincial trade, the province might not have the *legislative* authority under the *B.N.A. Act* to regulate the price of energy transmitted interprovincially, but this is probably of no great significance, since the province, *as owner* of the electricity, can set its own prices so long as there is no federal legislative intervention.

The greater the degree of integration and interdependence of the electrical systems of different provinces, the more likely will it be that the federal Parliament could exercise jurisdiction on the basis that there is an extraprovincial work or undertaking involved in the connection of the various provincial systems.

The extent to which operations between various provincial systems are co-ordinated will be the main factor that the courts will look at to find the existence of a work or undertaking within federal jurisdiction. And since the very purpose of interprovincial networks will be to achieve significant co-ordination in operations, it appears inevitable that some degree of federal jurisdiction will be found, should this be asserted in the courts. It is not clear, however, to what extent federal jurisdiction will apply, in that event, to normal utility operations within a province. The better view seems to be that the courts will attempt to distinguish between interprovincial operations under federal jurisdiction and normal operations under provincial jurisdiction.

If federal jurisdiction is upheld, since the provinces will be owners of the generating plants and transmission systems, they will still have authority to make interprovincial arrangements until the federal Parliament by statute provides otherwise.

An interprovincial electrical network would probably be regarded by the courts as a subject matter of a national rather than local concern, and therefore to some extent within federal jurisdiction under its "Peace, Order and good Government" power. The provinces could still exercise concurrent jurisdiction by virtue of ownership until federal legislation prevents this. Normal utility operations within the province would probably not be affected.

Generally, the danger of interference with provincial jurisdiction is no greater than the risk of federal intervention through its declaratory power. To oust provincial jurisdiction by virtue of its "Trade and Commerce," "Extraprovincial Workings and Undertakings," or "Peace, Order and good Government" powers, the federal government will have to make a political decision to introduce legislation just as it would have if it wished to make a declaration.

The most likely area where provincial jurisdiction will be challenged, at least initially, will be with respect to those provincial statutes affecting the rights of individuals—for example, provincial jurisdiction to impose labour relations or workmen's compensation regulations upon workmen employed on an interprovincial transmission line or to impose mechanics liens might be challenged by individuals in the courts. Should provincial jurisdiction over, for example, an interprovincial transmission line be successfully challenged in the courts by individuals, then it might be easier for the federal government subsequently to politically justify intervening by legislation to exclusively regulate the interprovincial network. There does not appear to be any way, apart from federal-provincial co-operative arrangements, whereby jurisdictional uncertainty here may be avoided.

Current proposals before Parliament for amendment to the *B.N.A. Act* would remove some of the uncertainty concerning provincial jurisdiction over extraprovincial electrical energy transfers. A proposed new section 92A would in part read:

> 92A. (1) In each province, the legislature may exclusively make laws in relation to
>
> . . . . . .
>
> (c) development, conservation and management of sites and facilities in the province for the generation and production of electrical energy.
>
> (2) In each province, the legislature may make laws in relation to the export from the province to another part of Canada of the . . . production from facilities in the province for the generation of electrical energy, but such laws may not authorize or provide for discrimination in prices or in supplies exported to another part of Canada.[142]

It is still too early to say whether the proposed amendments to our constitution will be adopted.

APPENDIX—EXTRACTS FROM THE *BRITISH NORTH AMERICA ACT*

91. It shall be lawful for the Queen, by and with the Advice and Consent of the Senate and House of Commons, to make Laws for the Peace, Order and good Government of Canada, in relation to all Matters not coming within the Classes of Subjects by this Act assigned exclusively to the Legislatures of the Provinces; and for greater Certainty, but not so as to restrict the Generality of the foregoing Terms of this Section, it is hereby declared that (notwithstanding anything in this Act) the exclusive Legislative Authority of the Parliament of Canada extends to all Matters coming within the Classes of Subjects next herein-after enumerated; that is to say,—

. . . . . .

1A. The Public Debt and Property.

. . . . . .

2. The Regulation of Trade and Commerce.

. . . . . .

10. Navigation and Shipping.

. . . . . .

12. Sea Coast and Inland Fisheries.

. . . . . .

24. Indians, and Lands reserved for the Indians.

. . . . . .

27. The Criminal Law, except the Constitution of Courts of Criminal Jurisdiction, but including the Procedure in Criminal Matters.

. . . . . .

29. Such Classes of Subjects as are expressly excepted in the Enumeration of the Classes of Subjects by this Act assigned exclusively to the Legislatures of the Provinces.

And any Matter coming within any of the Classes of Subjects enumerated in this Section shall not be deemed to come within the Class of Matters of a local or private Nature comprised in the Enumeration of the Classes of Subjects by this Act assigned exclusively to the Legislatures of the Provinces.

92. In each Province the Legislature may exclusively make Laws in relation to Matters coming within the Classes of Subjects next herein-after enumerated; that is to say,—

. . . . . .

5. The Management and Sale of the Public Lands belonging to the Province and of the Timber and Wood thereon.

. . . . . .

8. Municipal Institutions in the Province.

. . . . . .

10. Local Works and Undertakings other than such as are of the following Classes:—
    (a) Lines of Steam or other Ships, Railways, Canals, Telegraphs, and other Works and Undertakings connecting the Province with any other or others of the Provinces, or extending beyond the Limits of the Province:
    (b) Lines of Steam Ships between the Province and any British or Foreign Country:
    (c) Such Works as, although wholly situate within the Province, are before or after their Execution declared by the Parliament of Canada to be for the general Advantage of Canada or for the Advantage of Two or more of the Provinces.

11. The Incorporation of Companies with Provincial Objects.

. . . . . .

13. Property and Civil Rights in the Province.

. . . . . .

16. Generally all Matters of a merely local or private Nature in the Province.

. . . . . .

95. In each Province the Legislature may make Laws in relation to Agriculture in the Province . . . ; and it is hereby declared that the Agriculture in all or any of the Provinces . . . ; and any Law of the

Legislature of a Province relative to Agriculture . . . shall have effect in and for the Province as long and as far only as it is not repugnant to any Act of the Parliament of Canada.

. . . . . .

108. The Public Works and Property of each Province, enumerated in the Third Schedule to this Act, shall be the Property of Canada.

109. All Lands, Mines, Minerals, and Royalties belonging to the several Provinces of Canada, Nova Scotia, and New Brunswick at the Union, and all Sums then due or payable for such Lands, Mines, Minerals, or Royalties, shall belong to the several Provinces of Ontario, Quebec, Nova Scotia, and New Brunswick in which the same are situate or arise, subject to any Trusts existing in respect thereof, and to any Interest other than that of the Province in the same.

. . . . . .

117. The several Provinces shall retain all their respective Public Property not otherwise disposed of in this Act, subject to the Right of Canada to assume any Lands or Public Property required for Fortifications or for the Defence of the Country.

. . . . . .

121. All Articles, of the Growth, Produce, or Manufacture of any one of the Provinces shall, from and after the Union, be admitted free into each of the other Provinces.

. . . . . .

125. No Lands or Property belonging to Canada or any Province shall be liable to Taxation.

. . . . . .

132. The Parliament and Government of Canada shall have all Powers necessary or proper for performing the Obligations of Canada or of any Province thereof, as Part of the British Empire, towards Foreign Countries, arising under Treaties between the Empire and such Foreign Countries.

. . . . . .

THE THIRD SCHEDULE

Provincial Public Works and Property to be the Property of Canada
  1. Canals, with Lands and Water Power connected therewith.
  2. Public Harbours.

. . . . . .

  5. Rivers said Lake Improvements.

. . . . . .

  10. Armouries, Drill Sheds, Military Clothing, and Munitions of War, and Lands set apart for general Public Purposes.

NOTES

[1] Report of IPACE Networks Study Group, October 1978. The author initially became involved in the subject-matter of this paper through preparing a report for IPACE on the topic.

[2] Subsequently replaced by a new body. See the Mid-Year Report of the Council of Maritime Premiers, 15 December 1980.

[3] *Energy Update* (Ottawa: Department of Energy, Mines and Resources, 1977), p. 30.

[4] *The Halifax Chronicle Herald* (28 March 1979), p. 1.

[5] (1867), 30 & 31 Vict. (U.K.), c. 3, as amended; R.S.C. 1970, App. II, No. 5. See the Appendix of this chapter for relevant provisions.

[6] Gerard V. LaForest, *Natural Resources and Public Property under the Canadian Constitution* (Toronto: University of Toronto Press, 1969), pp. 3-14.

[7] *A.-G. Can. v. A.-G. Ont. et al (the Fisheries Case)*, [1898] A.C. 700 (P.C.); *Montreal v. Montreal Harbour Commissioners*, [1926] A.C. 299 (P.C.).

[8] *Smylie v. The Queen* (1900), 27 O.A.R. 172 (Ont. C.A.); see also Michael Crommelin, "Jurisdiction Over Onshore Oil and Gas in Canada," *University of British Columbia Law Review* 10 (1975), pp. 92-95 and 102-6; G.V. LaForest and Associates, *Water Law in Canada—The Atlantic Provinces* (Ottawa: Information Canada, 1973); and R. Thompson and H.R. Eddy, "Jurisdictional Problems in Natural Resource Management in Canada," in *Essays on Aspects of Resource Policy*, Science Council of Canada Special Study No. 27 (Ottawa: Information Canada, 1973), p. 74.

[9] LaForest and Associates, *op. cit.*, pp. 18-19 and authorities cited therein.

[10] *Ibid.*, pp. 8 and 20-21; see also D. Gibson, "The Constitutional Context of Canadian Water Planning," *Alberta Law Review* 7 (1969), pp. 84-85.

[11] *Booth v. Lowery* (1917), 54 S.C.R. 421, per FitzPatrick, C.J., p. 424, and Duff, J., p. 429 and see Laskin, *Resources for Tomorrow Background Papers* (1961), p. 211.

[12] This seems to be no different from the authority to prevent pollution affecting fish, which is generally accepted as lying with the federal Parliament. See Laskin, *op. cit.*, p. 218; LaForest and Associates, *op. cit.*, p. 15.

[13] *A.-G. Can. v. A.-G. Ont. et al (The Fisheries Case), op. cit.*

[14] *Burrard Power Co. Ltd. v. R.*, [1911] A.C. 87.

[15] LaForest and Associates, *op. cit.*, pp. 42-45.

[16] *Reference re Validity of Section 5(a) of the Dairy Industry Act (The Margarine Reference)*, [1949] S.C.R. 1.

[17] This authority as it relates to provincial matters has been challenged in recent years. See G.L. Morris, "The Treaty Making Power: A Canadian Dilemma," *Canadian Bar Review* 45 (1967): 478; and A.E. Gotlieb, *Canadian Treaty-Making* (Toronto: Butterworths, 1968).

[18] *Reference re the Weekly Rest in Industrial Undertakings Act, The Minimum Wages Act, and The Limitation of Hours of Work Act, (The Labour Conventions Case)*, [1936] S.C.R. 461.

[19] *Francis* v. *The Queen*, [1956] S.C.R. 618, at p. 625, per Rand, J.

[20] *Ibid.*

[21] See the discussion in LaForest and Associates, *op. cit.*, p. 67.

[22] *International Boundary Waters Treaty Act*, R.S.C. 1970, c. I-20.

[23] This principle is not accepted by everyone; see Morris, *op. cit.*, and Gotlieb, *op. cit.*

[24] *A.-G. Ont.* v. *Scot et al*, [1956] S.C.R. 137; and see Laskin, *Resources for Tomorrow, op. cit.*, p. 220; LaForest and Associates, *op. cit.*, p. 68; and Gibson, *op. cit.*, p. 89.

[25] For example, the *National Energy Board Act*, R.S.C. 1970, c. N-6, s. 81 of which prohibits the export of electrical power except under licence from the Board.

[26] ''Lands'' in the common law sense means ''lands and waters.''

[27] *Smylie* v. *The Queen, op. cit.*, see also *Brooks-Bidlake and Whittall, Ltd.* v. *A.-G. B.C.*, [1923] A.C. 450 (P.C.). Some doubts have been raised, however, by the *CIGOL Case* discussed below at note 78.

[28] Laskin, *op. cit.*, pp. 215-20 and LaForest and Associates, *op. cit.*, p. 13, discuss the impact of these provincial powers upon water management.

[29] *Gold Seal Ltd.* v. *Dominion Express Co. and A.-G. Alta.* (1921), 62 S.C.R. 424.

[30] *Atlantic Smoke Shops, Ltd.* v. *Conlon*, [1943] A.C. 550.

[31] *Murphy* v. *C.P.R. and A.-G. Can.*, [1958] S.C.R. 626.

[32] *Ibid.*, pp. 638-39.

[33] *Ibid.*, p. 153. See, however, the comment of Mr. Justice Laskin in *A.-G. Man.* v. *Manitoba Egg and Poultry Ass'n.* (1971), 19 D.L.R. (3d) 169, pp. 189-90 where he indicated that federal authority existed to set up provincial barriers to commodities.

[34] *A.-G. B.C.* v. *A.-G. Can.*, [1924] A.C. 222.

[35] *R.* v. *Bell Telephone Co.* (1935), 59 Que. K.B. 205.

[36] W.H. McConnell, *Commentary on the British North America Act* (Toronto: Macmillan, 1977), p. 367. A view adopted by the Alberta Court of Appeal in the *Reference Concerning the Federal Excise Tax on Natural Gas* (1981), a case decided after this paper was written and as yet unreported, where the Court held that section 125 prevented the tax from applying to gas owned by the provincial government, since the statute imposing the tax was a ''taxing'' one and could not be characterized as ''regulatory'' under the ''Federal Trade and Commerce'' power.

[37] W.R. Lederman, ''The Constitution: A Basis for Bargaining,'' in *Natural Resource Revenues: A Test of Federalism*, edited by A. Scott (Vancouver: UBC Press, 1976), pp. 54-57.

[38] Discussed in Peter W. Hogg, *Constitutional Law of Canada* (Toronto: Carswell, 1977), pp. 243-44.

[39] *Johannesson* v. *West St. Paul*, [1952] 1 S.C.R. 292. *Re Regulation and Control of Radio Communication*, [1932] A.C. 304.

[40] *Reference re Anti-Inflation Act* (1976), 68 D.L.R. (3d) 452 (S.C.C.).

[41] *Munro* v. *National Capital Commission*, [1966] S.C.R. 663.

[42] D. Gibson, "Measuring 'National Dimensions'," *Manitoba Law Journal* 7 (1976), p. 33; and discussion in Hogg, *op. cit.*, p. 260.

[43] W.R. Lederman, "Unity and Diversity in Canadian Federalism: Ideals and Methods of Moderation," *Canadian Bar Review* 53 (1975): 597-620.

[44] *The World Book Encyclopedia*, vol. 6, 1974, p. 53.

[45] *Ibid.*, p. 140.

[46] Hogg, *op. cit.*, p. 246. Confirmed since writing by *Fulton v. Energy Resources Conservation Board et al.* (1981, S.C.C.), as yet unreported.

[47] *B.C. Power Corporation Ltd.* v. *A.-G. B.C. et al.* (1963), 47 D.L.R. (2d) 633 (B.C.S.C.).

[48] *Citizens Insurance Co.* v. *Parsons* (1881), 7 App. Cas. 96 (P.C.).

[49] *Caloil Inc.* v. *A.-G. Can.*, [1971] S.C.R. 543.

[50] *A.-G. Man.* v. *Manitoba Egg and Poultry Ass'n.*, *op. cit.*; *Burns Foods Ltd.* v. *A.-G. Man.*, [1975] 1 S.C.R. 494.

[51] Hogg, *op. cit.*, pp. 309-10.

[52] *Reference re The Agricultural Products Marketing Act* (1978), 19 N.R. 361 (S.C.C.).

[53] T.B. Smith, "Chickens and Eggs: Marketing and Trade and Commerce Power," in *The Constitution and the Future of Canada*, pp. 135-60, Special Lectures of the Law Society of Upper Canada (Toronto: Richard De Boo).

[54] *Shannon* v. *Lower Mainland Dairy Products Board*, [1938] A.C. 708 (P.C.); *Home Oil Distributors, Ltd.* v. *A.-G. B.C.*, [1940] S.C.R. 444; *Carnation Co. Ltd.* v. *Quebec Agricultural Marketing Board*, [1968] S.C.R. 238.

[55] Hogg, *op. cit.*, p. 310.

[56] R.S.C. 1970, c. T-10.

[57] *MacDonald* v. *Vapour Canada Limited* (1976), 66 D.L.R. (3d) 1 (S.C.C.).

[58] The term "extraprovincial" will be used to refer both to interprovincial works and undertakings connecting two or more provinces and those extending beyond the limits of one province but not into another.

[59] *C.P.R.* v. *A.-G. B.C.* (*Empress Hotel Case*), [1950] A.C. 122 (P.C.).

[60] (1905), 36 S.C.R. 596.

[61] *Ottawa Valley Power Co.* v. *A.-G. Ont.*, [1936] 4 D.L.R. 594 (Ont. C.A.) per Fisher, J.A., at p. 623; and *The B.C. Power Case*, *op. cit.*

[62] C.M. McNairn, "Transportation, Communication and the Constitution," *Canadian Bar Review* 47 (1969), p. 360.

[63] *Campbell-Bennett Ltd.* v. *Comstock Midwestern Ltd. et al.*, [1954] S.C.R. 207. The matter has probably now been settled by *Fulton* v. *Energy Resources Conservation Board, op. cit.*, which held that an electrical distribution system could be within section 92(10)(*a*).

[64] *B.C. Electric Railway Co. Ltd. and C.P.R.* v. *C.N.R., North Fraser Harbour Commissioners and Prov. of B.C.*, [1932] S.C.R. 161; *S.M.T. (Eastern) Ltd.* v. *Ruch*, [1940] 1 D.L.R. 190 (N.B.S. Ct. Ch.); *A.-G. Ont.* v. *Winner*, [1954] A.C. 541 (P.C.); *Kootenay and Elk Ry.* v. *C.P.R.*, [1974] S.C.R. 955; *cf. Luscar Collieries Ltd.* v. *McDonald*, [1927] A.C. 925 (P.C.), p. 932.

[65] *Re Westspur Pipe Line Co. Gathering System* (1957), 76 C.R.T.C. 158; and see Crommelin, *op. cit.*, pp. 111-12 and McNairn, *op. cit.*, pp. 373 and on. See also the *Fulton Case, op. cit.*

[66] McNairn, *op. cit.*, p. 361.

[67] *Ibid.*

[68] *Bonanza Creek Gold Mining Co.* v. *The King*, [1916] 1 A.C. 566 (P.C.).

[69] *Ibid.*

[70] Compare the discussion in Hogg, *op. cit.*, pp. 348-51, rejecting the functional limitation, with the opinions referred to in McNairn, *op. cit.*, accepting it. The Supreme Court of Canada in the *Fulton Case, op. cit.* since this was written appears to have settled this issue in support of my submission.

[71] McNairn, *op. cit.*, p. 359 and statutes therein cited where the federal Parliament has set out declarations concerning "undertakings."

[72] [1905] A.C. 52.

[73] Bora Laskin and A.S. Abel, eds., *Canadian Constitutional Law*, rev. 4th edition (Toronto: Carswell, 1975), p. 480.

[74] *In the Matter of the Incorporation of Companies in Canada* (1913), 48 S.C.R. 331, per Duff, J.; *Luscar Collieries* v. *McDonald*, [1925] S.C.R. 460, per Mignault, J.; *Reg.* v. *Thumlert* (1960), 20 D.L.R. (2d) 335 (Alta. C.A.).

[75] LaForest and Associates, *op. cit.*, pp. 13-14.

[76] See note 64.

[77] Thompson and Eddy, *op. cit.*, p. 74.

[78] *Canadian Industrial Gas & Oil Ltd.* v. *Govt. of Sask. et al. and A.-G. Can. et al.* (1977), 18 N.R. 107 (S.C.C.).

[79] *Ibid.*, p. 129.

[80] [1957] S.C.R. 198.

[81] *Ibid.*, pp. 204 and 205.

[82] *The Brooks-Bidlake Case, op. cit.*

[83] *Smylie* v. *The Queen, op. cit.*

[84] Crommelin, *op. cit.*, p. 123.

[85] *Carnation Company Ltd.* v. *Quebec Agricultural Marketing Board*, [1968] S.C.R. 238.

[86] Crommelin, *op. cit.*, p. 127.

[87] *Ibid.*, p. 121.

[88] [1967] S.C.R. 792.

[89] C. Martin, "Newfoundland's Case on Offshore Minerals: A Brief Outline," *Ottawa Law Review* 7 (1975): 34-61; see also I.J. Head, "The Canadian Offshore Minerals Reference," *University of Toronto Law Journal* 18 (1968): 131-57.

[90] See *Toronto and Niagara Power Co.* v. *Toronto*, [1912] A.C. 834 and Stats. Can. 1902, c. 104, s. 2.

[91] *Jorgensen* v. *A.-G. Can.*, [1971] S.C.R. 725.

[92] R.S.C. 1970, c. A-19, s. 17.

[93] *Pronto Uranium Mines Ltd.* v. *Ontario Labour Relations Board*, [1956] O.R. 862; *Denison Mines Ltd.* v. *A.-G. Can.*, [1973] 1 O.R. 797.

[94] *The Jorgensen Case, op. cit.*

[95] *Reference re Waters and Water Powers*, [1929] S.C.R. 200.

[96] *Interprovincial Co-operatives Ltd. et al.* v. *The Queen*, [1976] 1 S.R.C. 477.

[97] G.V. LaForest, "Interprovincial Rivers," *Canadian Bar Review* 50 (1972): 39 and references therein.

[98] *Ibid.*, p. 43.

[99] Lord Cairns, in *Swindon Waterworks Co.* v. *Wilts and Berks Navigation Co.* (1875), L.R. 7 H.L. 697, at p. 704. See also *Re Burnham* (1894), 22 O.A.R. 40 and Laskin, *Resources for Tomorrow, op. cit.*, p. 213.

[100] Zimmerman, "Inter-Provincial Water Use in Canada: Suggestions and Comparisons," in *Constitutional Aspects* of *Water Management, Vol. 2*, edited by D. Gibson (Winnipeg: University of Manitoba, Agassiz Center for Water Studies, 1969).

[101] This matter really is beyond the scope of the present paper and has not been looked into in detail.

[102] *Hydro Que. et al* v. *A.-G. Nfld. et al*, Aug. 5, 1977, Montreal Sup. Ct. (unreported), since writing reversed by the Quebec Court of Appeal (21 February 1980) and leave to appeal granted by the Supreme Court of Canada (2 June 1980).

[103] *Royal Bank of Canada* v. *The King*, [1913] A.C. 283 (P.C.); and *Interprovincial Cooperatives, op. cit.*

[104] *Nova Scotia Board of Censors* v. *McNeil* (1975), 55 D.L.R. (3d) 632 (S.C.C.).

[105] *Reg.* v. *Ontario Labour Relations Board: Ex p. Dunn* (1963), 39 D.L.R. (2d) 346 (Ont. High Ct.); *Retail, Wholesale and Department Store Union* v. *Reitmer Truck Lines Ltd.* (1966), 57 W.W.R. 104 (B.C.S.C.); *Reg.* v. *Manitoba Labour Board, Ex Parte Invictus* (1968), 65 D.L.R. (2d) 517 (Man. Q.B.).

[106] *Reference re Anti-Inflation Act* (1976), 68 D.L.R. (3d) 452 (S.C.C.).

[107] H. Marx, "The Energy Crisis and the Emergency Power in Canada," *Dalhousie Law Journal* 2 (1975): 446-54.

[108] See McNairn, *op. cit.*, pp. 373-93; and *Re Westspur Pipe Line Co. Gathering System* (1957), 76 C.R.T.C. 158, pp. 178-79. See also the *Fulton Case, op. cit.*

[109] See note 59.

[110] See note 47.

[111] See McNairn, *op. cit.*, p. 377 and references therein.

[112] See note 105.

[113] *Regina* v. *Ontario Labour Relations Board ex p. Dunn, op. cit.*

[114] *Reference re Validity of the Industrial Relations and Disputes Investigation Act*, [1955] S.C.R. 529.

[115] *Montreal* v. *Montreal Street Ry.*, [1912] A.C. 333 (P.C.).

[116] *Luscar Collieries, op. cit.*, p. 932. No reference was made in this decision to the previous *Montreal* case.

[117] *B.C. Electric Ry. Case, op. cit.*

[118] J.B. Ballem, "Constitutional Validity of Provincial Oil and Gas Legislation," *Canadian Bar Review* 41 (1963), p. 224.

[119] *The Queen* v. *Board of Transport Commissioners* (1967), 65 D.L.R. (2d) 425 (S.C.C.). No reference was made here to the *Montreal* case or the *B.C. Electric Ry. Co.* case with *Luscar Collieries* being relied upon.

[120] *Re Tank Truck Transport Ltd.* (1961), 25 D.L.R. (2d) 161; *aff'd* [1963] 1 O.R. 272; see also *R.* v. *Cooksville Magistrate's Court, Ex Parte Liquid Cargo* (1965), 46 D.L.R. (2d) 700 (Ont. High Ct.).

[121] *Reg.* v. *Manitoba Labour Board, Ex Parte Invictus*, *op. cit.*

[122] *Agence Maritime Inc.* v. *Conseil Canadien des Relations Ouvrières*, [1969] S.C.R. 851.

[123] See note 47.

[124] LaForest and Associates, *op. cit.*, p. 51 and references therein.

[125] *Ibid.*, pp. 53-54.

[126] *A.-G. Ont.* v. *Winner*, [1954] A.C. 541 (P.C.).

[127] *Ibid.*, p. 581.

[128] See the comment in J.D. Whyte and W.R. Lederman, *Canadian Constitutional Law* (Toronto: Butterworths, 1977), pp. 11-18, where the authors point out that such criticism may be unjustified in light of the administrative inconvenience for the operator of having divided jurisdiction.

[129] *In re Capital Cities Communications Inc.* (1977), 18 N.R. 181 (S.C.C.).

[130] See note 65.

[131] *Ottawa Valley Power Case*, *op. cit.*, p. 623.

[132] *Ibid.*

[133] *Campbell-Bennett Ltd.* v. *Comstock Midwestern Ltd. et al.*, [1954] S.C.R. 207.

[134] At the present time the export of power is regulated federally but not transfers between provinces. (See the *National Energy Board Act*, R.S.C. 1970, c. N-6, ss. 40-43.) The *Fulton Case*, *op. cit.* is relevant here.

[135] See the discussion of the *CIGOL Case*, notes 78 to 85.

[136] See notes 38 to 40.

[137] *The Ontario Egg Reference*, *op. cit.*

[138] See notes 121-122.

[139] LaForest, *op. cit.*, pp. xii-xiii.

[140] See note 126.

[141] For a discussion of issues arising under a scheme whereby authority is delegated from both levels of government to a single body, see Rowland J. Harrison, ''The Offshore Mineral Resources Agreement in the Maritime Provinces,'' *Dalhousie Law Journal* 4 (1978), p. 265.

[142] See section 56 of the proposed Constitution Act, 1981, attached as Schedule B to the consolidation of the proposed resolution and amendments as approved by the Special Joint Committee on the Constitution.

215-79

Chapter Thirteen

The Distribution of Resource Rents: For Whom the Firm Tolls

by
Leonard Waverman*

7230
3230
7210 Canada

INTRODUCTION

To discuss the distribution of resource rents in Canada requires two major ingredients—some concept of what constitutes "rent," as well as criteria for evaluating alternative distribution schemes. Examining the stated positions of the various actors in the drama indicates wide diversity of opinion on both issues.

Estimates of the size of the pie vary widely. On one side are industry spokesmen who argue that little if any rents are available, once one considers the need to develop expensive reserves for the future. On the other side are spokesmen such as Eric Kierans who argue that most of the return to the firms in the resource sector is rent, unnaturally appropriated from the people of Canada (Kierans, 1973). Opinion on the correct distribution of rents varies all the way from Alberta to Ottawa, some arguing that all rents accrue to residents of resource-rich provinces, others arguing that these rents belong to all Canadians.

Opinion is naturally going to range widely on these issues. It is impossible to think that all Canadians have the same opportunity costs or identical concepts of the socially correct rate of resource development and extraction. Some would prefer to see fast extraction, combined with high export and income levels—Canadians who own shares in resource-based firms or employees of these firms. Other Canadians might prefer to see a slower extraction rate with minimal export levels. The amount and flow of rents depend on the extraction profile. Even were all Canadians to agree on extraction rates, the actual measurement of rent is, as we will see, difficult, and designing fiscal instruments to efficiently and equitably collect rent, impossible. Nor is it obvious which Canadians should share in these rents.[1] Should all Canadians no matter where they reside receive equal portions of these rents? Should all Canadians no matter their income level receive a

255

proportionate share? Intermingled in this whole discussion is the role of the foreign investor and the share of rents accruable to him.

The scheme of this paper is as follows. First, the concept of rent is defined, both in terms of the short run (one to three years, the period where new additions to supply cannot be made) and the long run (five to ten years, a period long enough to allow supply responses to changes). Second, the question of who "ought" to receive the rents is discussed. Included is a brief survey of the constitutional basis for rent distribution in Canada. There is, however, little guide from economic theory to attempt to answer the question of who "ought" to receive rents. Then, alternative fiscal instruments used by governments are examined and evaluated. The evaluation relies on two criteria: the first is the economic efficiency of the instrument; the second is its ability to distinguish rent from the opportunity cost of a factor of production. In the final section, some thoughts, conclusions, and a brief summary are presented. Included is an estimate of the revenue from the oil and gas industry taxed by governments or directed towards consumers via the price ceiling on oil and natural gas.

THE CONCEPT OF ECONOMIC RENT

In the nineteenth century, David Ricardo examined the issue of the returns (rent) to factors of production (land, labour, capital) that follow *from* market-determined prices (rather than factor costs that *determine* market prices). The usual interpretation of his argument is as follows. Consider two vineyards from a total of n vineyards in a competitive wine industry. The price of wine is determined by the equilibrium market demand and supply. In Figure 1a, industry quantity Q is produced at price p. The supply schedule for wine is upward sloping because grape fields differ in their productive capacity. Assume that all vineyards are exactly one acre in size, but they differ continuously in the yield of grapes per acre. For the two vineyards shown in Figures 1b and 1c respectively, vineyard 1 has the lowest average and marginal costs (average costs including a normal return on capital) and vineyard n, the highest costs of all producing fields. Field n is the "marginal" field and earns no rent. Field 1, however, earns "rent" of area $ABCD. Let us assume, for simplicity, that the alternative use for field 1, were it not used to grow grapes, would be to lie fallow. Therefore, the land-owner of field 1 receives no return if the land is pulled out of wine production and a return of $ABCD when the field is used to produce wine grapes (*rent* is $ABCD). The government, however, could tax area $ABCD and leave the production of wine from field 1 unchanged, since rent is price determined, not price determining.

This form of analysis, plus additional hypotheses that the supply of land was fixed in total, meant that increased demand for agricultural products would generate increased land rents. Therefore, *rents* were proposed as major

Figure 1
THE RICARDIAN VINEYARD CASE

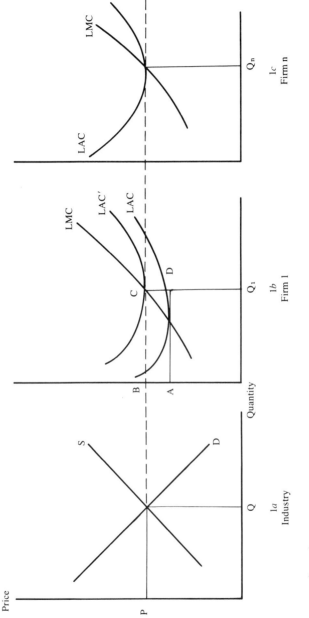

S: supply
D: demand
LAC: long-run average costs (including a normal return to capital)
LMC: long-run marginal costs
LAC': long-run average costs after sale of vineyard (rents BADC become capitalized)
Firm 1: intramarginal vineyard (close to market)
Firm 2: marginal vineyard (far from market)
P: market price

sources of revenue. In the report of the Royal Commission on the Transfer of the Natural Resources of Manitoba in 1929, it was stated that ''at Confederation it was decided to make the natural resources the cornerstone of provincial finance.''

Were the world as simple as the Ricardian rent model. Reality perturbs theory so that the concept and measurement of rent are not as transparent as in the above example. Assume that the factors of production (including land) have differing alternative uses and that land and resource deposits are traded freely in asset markets. Moreover, the supply of factors (including land) is not fixed and the production process for some factors is highly uncertain.

The effect of differing alternative uses yields a variable opportunity cost for individual units of some factor. Variable supply makes the specific return to the factor consist of cost and rent elements. Uncertainty of supply means that the cost of capital to the sector contains an unknown risk premium. How then in the *real* world is rent to be distinguished from factor costs and risk premiums?[2]

The concept of rent is far easier to formulate in the vineyard case than for natural resource reserves. Vineyards have, for all intents and purposes, an infinite life, while natural resources are depletable. As a result, the simple static market equilibrium shown in Figure 1 cannot hold for natural resources. Oil producers, for example, must not only contemplate supply and demand conditions today, but also supply and demand conditions in the future. For any single oil producer, holding back one unit of oil today would allow him to sell one more unit tomorrow. If the producer feels that oil prices will rise by more than the rate of interest, it is in his interest to withhold production today and sell tomorrow. Conversely, if oil prices are expected to rise by less than the rate of interest, producers will expand production today at the expense of tomorrow's output.[3] The theory of depletable resources therefore suggests that, in equilibrium, the net price (or market price minus extraction costs) will increase at the rate of interest. Part of the price then for any depletable resource is its scarcity value or user cost. For purposes of rent collection, this scarcity value should be acknowledged.

What then is *rent* in natural resource production? We must distinguish between two cases—quasi-rents (short run) and true rents (long run). Once an oil field or ore mine, for example, is fully developed and capital, in place, most costs are fixed. Since costs of shutting down production are high, the return required to keep production flowing is that the well-owner receive his variable costs. If the well-owner shuts down, he incurs the carrying charges on the fixed capital plus the costs of shutting down. If he maintains production, he incurs the carrying charges on the fixed capital alone.[4] Thus, *all* revenue above variable cost could be taxed away without affecting current production (the taxing of quasi-rents), but no new mine or oil field would be developed, since, in this example, taxes take a large share of the long-run opportunity costs—not just long-run rent.

When we are discussing rent extraction in the natural resource sector, we are implicitly discussing the supply curve (assuming that there is no monopoly) of the industry, the response of production and reserve additions, in both the short and long runs. The short-run, capital-sunk case can be considered as anything up to three or four years. The long run, accounting for potential changes in the rate of exploration, is likely over five years. Only in the case where all economic deposits are fully developed will fiscal policy be able to extract quasi-rents without affecting the flow of present production, the discovery rate of new reserves, and the asset value of all deposits.

As the earlier vineyard example showed, extracting true long-run rents will never affect supply or price. Available fiscal instruments are not, however, flexible enough to tax rent without also taxing returns to factors. As a result, actual tax policy will generate shifts in supply and price, worsening allocative efficiency (creating dead-weight losses). In analysing fiscal policy, rent collection instruments, and the distribution of resource rents, we must therefore examine the following:
1. Whether the rents are quasi or real long term
2. The cost of capital to the resource industry, including risk return
3. The effects of the fiscal or tax instruments on both the rate of present production and the rate of future exploration and development.
4. The dead-weight losses of the fiscal instruments in a static context and for the dynamic development of reserve discovery over time.

WHO OUGHT TO RECEIVE RESOURCE RENT?

There are many possible recipients of resource rents—the firm, the firm's factors of production, that is, labour or capital, the land-owner, or the state. Let us assume that firms purchase all their factors of production except land or mineral rights from competitive factor markets. Industry spokesmen argue that to raise sufficient capital, investors must receive most of the rent. There are three arguments presented. The first argument is that the industry itself must generate the necessary capital from internal sources. The market price of any resource (assuming competitive markets) reflects the costs of exploration, development, and production at the margin. For example, marginal oil fields today are more expensive to develop. The industry then suggests that the differential return between intramarginal (the oil fields found in the 1960s), and marginal fields (say, the Yukon or offshore Labrador) must be retained for investment in new production. The fallacy in this argument is the assumption that returns *above* the cost of capital to existing producers are necessary to finance new production. However, existing producers can borrow external funds from the capital market at the cost of capital and develop these same reserves. The corporate capture of all rent for internal funding of resource projects is unnecessary from a public viewpoint.

A second argument suggests that the resource industry is so risky that the rents are required to compensate investors for the extra uncertainty. The issue of the risk and uncertainty in natural resource production is still unsettled. Many writers, notably academics, argue that the risk of discovery for any single, exploratory drilling effort is high. However, for a large integrated company able to diversify its risks over many exploratory efforts, uncertainty is no greater than in many other industries.[5] When federal and provincial taxes began to rise in early 1974, so that the return on new investments in the resource sector seemed to fall below the level obtainable elsewhere, drilling rigs moved out of Canada. The resource sector does seem to require some risk premium; however, the exact amount is likely to be small. A third and related argument made by the industry is that the rents are not only required, but also were expected. While the petroleum sector, for example, has earned approximately a 10 per cent rate of return on its invested capital in the 1946−1973 period,[6] industry spokesmen suggest that this rate of return was far lower than required. It was the anticipation of rents that generated the investment in the industry even though the rates of return were so low. That the Canadian oil industry could have foreseen in 1956, for example, the OPEC price rise following the Arab-Israeli war of 1973 seems unfathomable. In fact, with some reduction in federal and provincial taxation, drilling rates have increased to new highs in this country.

Who ''should'' receive these rents? The essential point in discussing who should benefit is that economics has little to say. Since true resource rents are price determined, not price determining, *economic efficiency is not influenced by the distribution of the rent*.[7]

Imagine a world without taxation. Potential mineral developers lease likely mineral-bearing land from landholders. In a competitive world with perfect foresight, land-owners could appropriate the entire rent, since competing well drillers would bid amounts up to the amount of rent, which is known. In a similar competitive world with imperfect foresight, land-owners would not appropriate all rents if all would-be developers were risk averse, since the exact amount of rent is *a priori* unknown. ''Lucky'' developers would earn positive economic profits, which would become capitalized in the value of their firms. Note that we cannot say who ''ought'' to receive these rents. Since all factors are earning their marginal products (the condition of efficiency in factor markets), there is no economic justification for enacting legislation to force developers instead of land-owners to receive the rents.

What, if in this world, the state is the land-owner? To what extent do rents belong to all citizens? This is, of course, the argument made by Alberta, that the resources are gifts from nature to the people of the province and, as a result, Albertans, but not Canadians in general, should share in this providence of Mother Nature. Let us examine the constitutional basis for this claim, but the reader must remember that there is no conclusive efficiency argument that states that ''rent'' (properly measured) ought to belong to the land-owner, the state, or the firm.

The Legal Basis of Resource Ownership in Canada

In examining the legal basis for resource ownership in Canada, three separate jurisdictions must be examined—federal lands; the provinces of Quebec, Ontario, Nova Scotia, New Brunswick, Prince Edward Island, British Columbia, and Newfoundland; and the provinces of Manitoba, Saskatchewan, and Alberta.

In feudal England, the power of the Lord and King rested on his ability to raise money through ownership and taxation of land. Therefore, '' . . . in Canada, the sovereign owned all ungranted lands and (subject to certain qualifications in Quebec) had prerogative rights and privileges similar to those in England, and owned any revenues derived therefrom'' (La Forest, 1969, p. 11).

By Confederation, the four original founding provinces of Canada (Ontario, Quebec, Nova Scotia, New Brunswick) had been passed the entire control, management, disposition, and proceeds of revenues from Crown lands. '' . . . At the time of the Union, the entire control, management and disposition of the Crown lands . . . were confided to the executive administration of the province, provincial government . . . so that the Crown lands . . . to all intents and purposes [were] the public property of the respective provinces in which they were situated. . . . ''[8] The dominion government received at Confederation only the public works and property related to the jurisdiction of the government itself.

Prince Edward Island, British Columbia, and Newfoundland, upon entering Confederation, received the same rights in the ownership of natural resources as the four founding provinces. Manitoba, Saskatchewan, and Alberta, which were formed out of Rupert's Land, and the Northwest Territories did not receive these full prerogative rights until 1930. In the *British North America Act* of 1930, the federal government transferred to the three Prairie provinces all rights to Crown assets transferred to the other provinces by section 109 of the *British North America Act* of 1867.[9] Included expressly in the provincial control over Crown lands is the right to royalties from these lands. In the Northwest Territories and the Yukon, the federal government has express control over land, minerals, and royalties.

In terms of the distribution of resource rents, it is clear that the provinces (and the federal government in the Yukon and the Northwest Territories) can exact royalties to earn revenue for their respective purses.[10] La Forest has defined a royalty as ''the dues based on production payable for the privilege of working a mine . . . those casual revenues of the Crown derived from the royal prerogative . . . '' (La Forest 1969, p. 79).

The *B.N.A. Act* of 1867 gives the federal government exclusive rights to levy indirect taxes, while both the provincial and federal governments can levy direct taxes. While the distinction between direct and indirect taxes is unclear, indirect taxes are paid by the consumer (rather than the producer) and *determine* the price. Since *rent* is by definition price *determined*, taxation of rent *cannot* determine the price. As we have seen, however, in reality, rent is difficult to measure. As a result, attempts to tax rent may determine, in part, the price.[11]

The overlapping of tax jurisdictions is clearly evident in the progressive changes in both provincial and federal tax laws following the rapid escalation of world commodity prices after 1973. When several provinces enacted new royalty legislation aimed at capturing a significant portion of commodity price increases, the federal authorities quickly made these royalty payments non-deductible for federal income tax purposes (details below in the section on "Income Tax Policy"). The federal argument revolved around two issues. The first issue was the equitable division of resource rents among Canadians, not just those who happened to be resident in any province that had the providence of resource wealth. The second issue was the federal concern that much of the royalty increase was a disguised income tax.

Equity

The arguments on whether an *equitable* division of resource rents requires division among *all* Canadians or among just those resident in resource-rich provinces centre around two points. First, the resource-rich provinces argue that royalties were to be the corner-stone of provincial finance (as enunciated as we saw in Manitoba). These natural resources are a shrinking asset of the province, and the province must therefore maximize the present value of the depletable reserves. The federal counter-argument is that much of the development in the resource sectors, and a good deal of the provincial income from these sectors, has come about because of federal subsidization. Federal tax provisions (depletion allowance, current expensing, etc.) allowed resource-rich provinces to expand their discovered resource base and income simultaneously at the expense of federal tax collections. One can therefore argue that resource-rich provinces have gained at the expense of resource-poor provinces. The second point is whether equity considerations require resource deposits to be, in effect, equalized among all Canadians.

I feel there is good merit to the federal case on both equity and tax grounds. The resource sector in each province does owe much of its present size and profitability to federal tax advantages. While the province is the land-owner and does have the constitutional rights to set royalties, some recognition of federal assistance is warranted. To allow each province to levy royalties or direct taxes that restrict federal taxation potential could quickly diminish all federal tax income. The federal government has to establish the obligation of all provinces in Canada to pay their share of federal tax revenue no matter the special nature of provincial income. A more important concern rests on the grounds of distributional equity. Why should those who happen to be resident of some arbitrary jurisdiction be the sole beneficiaries of some fortuitous event, not related to their efforts. Let us assume that a technological development is made that makes the tides off the Bay of Fundy capable of producing all electricity for Canadians, at half the present price. Should all rents accrue to Nova Scotians? In my opinion, rents should not solely accrue, in this example, to the "tidal-rich" provinces.

Who Does Share in Resource Rents?

It is impossible, in fact, to divide the issue of who "ought" to share from who does share. Table 1 indicates the distribution of *oil industry*

Table 1
PROVINCIAL OIL REVENUES—1947 – 1978
(thousands of dollars)

	ALBERTA			TOTAL (Alta., Sask., B.C., Man.)		
	Land Sales	Rentals	Royalties	Land Sales	Rentals	Royalties
1978	609,884	66,972	2,717,651	809,964	89,531	2,991,854
1977	579,743	61,744	2,133,661	715,540	81,210	2,385,576
1976	160,154	55,653	1,762,710	212,486	73,260	1,992,600
1975	105,991	100,654	1,352,768	118,740	116,464	1,597,068
1974	58,252	76,287	964,083	85,176	92,229	1,171,643
1973	43,586	80,563	392,636	66,537	93,019	437,057
1972	25,727	75,195	196,284	50,914	87,969	226,972
1971	23,922	68,844	167,721	48,614	82,112	197,428
1970	25,688	56,111	136,417	45,858	69,214	164,439
1969	102,180	61,224	113,219	128,690	76,075	140,974
1968	93,391	55,235	101,321	113,672	70,114	127,802
1967	87,721	53,507	92,714	111,440	76,926	118,399
1966	99,120	53,220	75,266	122,156	67,024	99,728
1965	119,662	57,408	67,971	149,227	69,002	87,994
1964	84,820	42,771	60,854	106,328	54,186	80,886
1963	46,647	37,792	55,139	59,085	48,042	72,466
1962	33,154	38,476	47,990	48,283	42,157	61,521
1961	44,631	30,861	35,431	53,685	41,580	45,318
1947 – 60	564,382	263,868	245,887	634,633	340,988	278,161
TOTAL	2,908,655	1,336,385	10,719,723	3,681,028	1,671,102	12,277,886

Source: Oilweek, Feb. 12, 1979

revenues over the 1947–1978 period. Provincial coffers have received $12.3 billion or some 10 per cent of total oil industry revenue in this thirty-one-year period. Until 1970, the federal government took a far lower percentage of petroleum and coal products', and mineral fuels' profits through the corporate profit tax than of the manufacturing sector (Table 2). Note that the

Table 2
CORPORATE INCOME TAXES AND ADJUSTED* BOOK PROFIT,
1970–1976
(millions of dollars)

	Mineral Fuels	Petroleum and Coal Products	All Manufacturing	Total Non-Financial
Adjusted book profit before taxes (PROFIT)				
Corporate income taxes (TAX)				
Corporate tax/book profit (TAX/PROFIT)				
1970				
PROFIT	213.2	250.5	2306.2	5449.3
TAX	33.2	72.1	1182.5	2384.3
TAX/PROFIT	.156	.288	.513	.438
1971				
PROFIT	273.0	460.0	2841.0	6061.1
TAX	37.9	112.0	1352.6	2594.9
TAX/PROFIT	.138	.243	.478	.428
1972				
PROFIT	297.4	481.9	3669.6	7344.1
TAX	61.1	103.4	1500.2	2996.6
TAX/PROFIT	.205	.215	.425	.408
1973				
PROFIT	581.6	885.1	5909.7	12428.1
TAX	122.1	195.0	1927.8	3929.9
TAX/PROFIT	.210	.220	.326	.316
1974				
PROFIT	991.8	1447.5	8050.4	16601.6
TAX	318.8	378.8	2613.6	5589.7
TAX/PROFIT	.317	.331	.325	.337
1975				
PROFIT	1638.2	1345.9	6945.5	15496.5
TAX	638.2	502.1	2482.2	5803.9
TAX/PROFIT	.390	.373	.357	.375
1976				
PROFIT	665.9	811.8	4953.7	10406.2
TAX	201.9	243.9	1843.2	3879.9
TAX/PROFIT	.303	.300	.372	.373

* Book profit before taxes less net capital gains, non-taxable dividends received, and prior years losses applied.
Source: Statistics Canada, *Corporation Taxation Statistics, 1969–70*, Cat. No. 61-208 (Ottawa: Minister of Industry, Trade and Commerce).

calculations show only the direct government share. Not included are taxes on the increased incomes of shareholders in resource-based industries, taxation of income induced by resource availability, and so forth.

It is impossible to say whether the total governmental share is ''equitable'' or not, since no one has been able to estimate the actual rents in the resource sector. In the next section, the various instruments used to collect rents are discussed and evaluated. These instruments, in general, cannot distinguish rent from opportunity cost and, as a result, their use entails some resource misallocation.

FISCAL INSTRUMENTS

Royalty

The royalty is the instrument common to all provincial resource-rent collection. It has two serious defects, however, as a fiscal instrument. First, either in a per unit or *ad valorem* form, it is a quantity or total revenue tax that cannot distinguish between resources flowing from high rent-earning reserves and resources flowing from marginal reserves. All producers pay the same per unit royalty no matter the ''rents'' that their mineral rights earn. Second, the royalty becomes a cost to the producer and, as a result, will lead to inefficiencies in production. In Figure 2*a*, the unit costs of all ore mined is c. Both asset and flow markets are in equilibrium when the net price (price minus cost) grows at the rate of interest. In Figure 2*b*, the resources are depleted at time t as the price rises to choke off demand. A royalty of t per unit of output adds to average costs and marginal costs and, as a result, increases initial price, reducing the quantity sold today and lengthening the period of extraction (dotted line in Figure 2*a*). When the resource is not of uniform quality, but where grades are intermingled at any one site, the per unit tax causes the producer to increase the cut-off grade. Because the per unit tax adds to the cost of all units, high and low grade, the producer will mine ore of a grade higher than he would mine were there no royalty. As a result, the total quantity of ore recovered would fall (Bucovetsky, 1976).

The virtues of the royalty lie in its administrative simplicity and its cheat-proof nature. Changes in royalties in the three western provinces in the mid 1970s have attempted to make the tax more sensitive to differential rent.[12] Albertan oil royalties, for example, which were at 16.67 per cent of well-head prices before 1973, are now on a three-tier basis. The general royalty rate was increased to 22.9 per cent in January 1974, and differential rates applied to the well-head price increases were agreed upon between the provincial and federal authorities. The ''average'' Alberta royalty is now in the region of 36 to 39 per cent.

There are several additional problems, however, with the differential royalty as used in Alberta. First, it still cannot distinguish between rent and

Figure 2
EQUILIBRIUM FOR A NATURAL RESOURCE OVER TIME

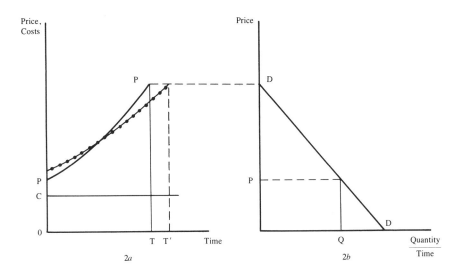

Note: OC is the constant costs of extraction. Given the market demand curve DD in 2*b*, the natural resource deposit is depleted at time t, price (PP) rising to choke off demand. The dotted line represents changes in the quantity and timing of production after the imposition of a royalty of t per unit of output.

the necessary return to the producer. In taxing some 50 per cent of the price increases, since 7 July 1975, the Albertan government is in fact arguing that rent is half of these price increases, which may or may not be true. Taxing "windfall gains" may in fact be just and equitable; taxing the opportunity cost of factors is not only inequitable but inefficient.

Besides the issue of how royalties "tax" existing production, a crucial issue is their impact on reserve development. The royalty truncates the supply function by affecting the decisions of producers to search for new reserves. When the price of oil is $12.75, not $6.50 per barrel, Canadian welfare is maximized when the marginal field is developed whose costs are $12.75 per barrel. But at the end of 1978, the price net of royalties to Albertan producers was roughly $6.00.[13] As a result, only fields whose expected costs were in the region of $6.00 would be developed, not fields whose expected costs were in the region of $12.75 per barrel. Society is best off when fields are developed that are marginal at $12.75, that is, yield no rent (see Figure 1). But with royalty schemes, marginal fields are discouraged, since all production, marginal or not, must pay the same royalty rate.[14]

The question is not whether there are adequate incentives after all the royalties and taxes are deducted from the price. The royalty is economically inefficient. Royalties do affect costs and output decisions and, therefore, there is a dead-weight loss in their use. We can attempt a "back-of-the-envelope" type evaluation of the dead-weight loss of Albertan oil royalties by examining simple supply and demand functions. We assume an arc elasticity of demand of −0.3 in the relevant range.[15] For simplicity, we assume that the royalty is per unit rather than *ad valorem* and that all oil produced in Canada is sold at $12.75 per barrel. Albertan producers received $8.00 per barrel before federal and provincial income taxes for 584 million barrels of oil sold in 1978.[16] The Albertan government received royalties of approximately $5.00 per barrel. If the market price were $8.00 rather than $12.75, consumers would purchase an additional 65 million barrels of oil.[17] Albertan producers would receive $522 million for this incremental production (at $8.00 per barrel). The dead-weight loss can be calculated as some $261 million or 10 per cent of the actual royalty revenue itself.[18]

In recognition of the likelihood of royalties truncating the supply function, several provinces have selected royalty schedules whose effective rates depend on the volume of funds plowed back into exploration and development. While the emphasis is on increasing reserves, the connection with rent collection is crude at best—perhaps the argument being that firms that earn no rent must be paying out all revenues to factors of production and therefore will have no funds to reinvest in exploration. Of course, firms with no rents earned could be borrowing for exploration and development.

The incentive for exploration and development varies by province. Tax changes in British Columbia in 1975, for example, led to producer returns on natural gas production varying from a loss in old gas, with no reinvestment of revenues, to a return of $0.42 (out of a well-head price of $0.90) on new gas where $0.20 was reinvested in British Columbia (Tuschak, 1975, p. 171). The Manitoba *Royalty and Tax Act*, 1 January 1975, has a number of features that are aimed at making the royalty a more efficient tax collection device. First under the act, the profit base for each company is determined as 18 per cent of the company's investment base (the company's assumed *normal* rate of return). Fifteen per cent of this normal rate of return is payable as tax, while any return over 18 per cent is taxable at 35 per cent. The investment base itself is calculated in real dollars and thus the inflationary aspect is removed. Tax incentives also exist to process the oil in Manitoba. The effectiveness of the legislation in capturing rent depends on whether 18 per cent before profit taxes is a rate of return sufficient to generate the required investment. Close examination of the act indicates that it likely will not be a sufficient rate of return. The reason for this pessimism is that the asset base on which the rate of return is calculated is defined as the cost of mining and service assets. Exploration funds are not, however, included. This is a legacy of the Kierans report written in February 1973 (Kierans, 1973). By ignoring exploration costs incurred in ''non-productive'' drilling, Kierans was able to state that a Manitoban government public mining corporation could earn a 35 per cent rate of return on its invested capital.

In November 1977, the Supreme Court of Canada ruled *ultra vires* the 1974 taxes on oil revenue enacted by the Province of Saskatchewan. These taxes, aimed at capturing 100 per cent of the differential between the Canadian price of oil and the basic Saskatchewan well-head price were held by the Supreme Court to be indirect taxes aimed at interprovincial trade.

It is evident from the above discussion that royalties are not the best means of capturing resource rents, except as they mimic progressive income taxes. In all cases, even where the royalty does correspond to a progressive income tax, the problem exists of actually measuring the rent for any one industry. Because *average* rates of return must be used for all firms, it appears that even progressive or differential royalty schedules have failed to distinguish rents and are capturing part of factor costs for some fields. In addition, the presence of these royalty schemes means that marginal fields go undiscovered.

Bonus Bidding

In addition to the royalty, provincial governments receive resource income from land sales and bonus payments. In theory, in a competitive resource market with perfect foresight, firms would be willing to bid in advance for mineral rights an amount equal to the present value of all rents (the bonus payment). The bonus payment could then, in theory, represent an alternative superior form of rent collection. In practice, however, its usefulness is questionable for three reasons.[19] First, with imperfect foresight and risk aversion, firms will underbid. Second, any monopoly elements in bidding will decrease the rent collected. Third, because the bonus payment is made ''up front,'' its effects on the firm's risk and total cost of capital are significant. In addition, small firms may be at a significant disadvantage in bonus bidding, since the ''up front'' payment may be difficult if not impossible to make. In short, the bonus bidding scheme places all risk on firms while yielding all potential rents to governments. But governments are better able to spread risk, especially compared to smaller, undiversified firms. Rent collection could therefore increase were the government willing to accept some of the industry's risks. Royalties are a form of risk taking by the government.

An alternative method of bonus bidding is to have firms bid, before exploration, the rate of royalties they would pay. While this form of *ex ante* bidding removes the problem of up front payments, numerous problems remain. Because of high uncertainty of actual finds, while the average bonus bid might be high, renegotiations, once production comes on stream, are all too likely. Take the case of the Mackenzie Delta gas fields. Prior to exploration, firms might have been willing to bid large royalties. Now that the structures appear fractured and of low productivity, high-bid royalties would ensure that production could *not* take place. On the other hand, a reservoir that turned out, *ex post*, to be at the high end of the distribution of possible size outcomes would have *ex ante* too low a bid and the government would be tempted to renegotiate (North Sea). Bonus bidding in either form—lump sums or the rate of royalties paid—is unlikely to prove to be an effective rent collection device. Governments, therefore, have been reluctant to concentrate on bonus bids as primary income-generating devices, especially for virgin territories.[20] Once a find is made, and surrounding lands take on new more definitive value, bonus bids have proved valuable.

Income Tax Policy

As Table 2 has indicated, the amount of revenue taken from the resource sector by government levels in the form of income taxes has been smaller than for all other industries in Canada, at least until 1974. Until the early 1970s, governments were convinced by industry arguments that specific factors (risk, etc.) required a higher ratio of after-tax to before-tax profits left with the industry.

Income tax policy of the federal government was not aimed at rent collection, but aimed at increasing the after-tax profitability of the resource sector.[21] Resource industries received two tax advantages unavailable to other industries. First, many items that were truly capital costs could be written off as current expenses. These included all exploration and development costs. Second, because the major asset of the mine was its depletable reserves, the resource owner could write off some depletion allowance against income. Because depletion is a difficult concept to measure, the firm was allowed to deduct a depletion allowance equal to some stated percentage of net income. Following the large price increases in the petroleum industry, the federal tax policy changed drastically. As of 23 June 1975, lease rentals and provincial royalties were made fully non-deductible against federal income tax payable. Exploration costs remained fully deductible against current income. Earned depletion amounted to the lesser of 25 per cent of production profits or $1 for every $3 of eligible drilling and exploration costs. As a partial compensation for the non-deductibility of royalties and lease rentals, the federal government enacted a special resource allowance equal to 25 per cent of production income.

In Tables 3 and 4 are reproduced some calculations by T.S. Tuschak, Senior Adviser, Department of Energy, Mines and Resources, Ottawa. As of 31 December 1973, the federal government received roughly 19 per cent of revenue produced by the petroleum industry. Changes in provincial royalty and tax schedules announced by the Alberta government in early 1974 reduced the federal share of oil industry revenue to 9 per cent. The federal budget of 18 November 1974 increased the federal share to 24 per cent. The 23 June 1975 federal budget reduced that share to 19 per cent. [*Editor's note:* The federal National Energy Program of 28 October 1980 would significantly increase the federal share of revenues from oil and natural gas production. This increase is largely at the expense of industry, as the provincial percentage share would decrease only marginally.]

Table 3
CASH FLOW GENERATION IN THE CANADIAN
PETROLEUM PRODUCING INDUSTRY, 1976−1979
(In constant 1975 dollars)
Cash flows under 23 June 1975 Budget rules[1]

Calendar Year	Gross Revenues Less Direct Operating Costs[2]	Deduct All Payments to Prov. Governments[3]	Deduct Federal Income Tax	Net Industry Cash Flow Available for Investment and Financial Costs[4]
	— $ Millions rounded to nearest $25 million —			
1976	7,425	3,200	1,400	2,825
	(100.0%)	(43.1%)	(18.8%)	(38.1%)
1977	8,625	3,750	1,700	3,175
	(100.0%)	(43.5%)	(19.7%)	(36.8%)
1978	9,150	3,975	1,825	3,350
	(100.0%)	(43.4%)	(19.9%)	(36.6%)
1979	9,400	4,075	1,875	3,450
	(100.0%)	(43.4%)	(19.9%)	(36.7%)
Totals	34,600	15,000	6,800	12,800
	(100.0%)	(43.4%)	(19.6%)	(37.0%)

Notes:
[1] The resource allowance of 25% of gross revenues less operating costs less CCA is combined with a 36% federal tax rate (net of provincial abatement) and a weighted average 11.5% provincial tax rate applied to taxable income as computed for federal purposes (specifically, royalties are non-deductible for all provinces).
[2] Oil prices are assumed to rise over the decade in response to domestic and international economic conditions. Gas prices are assumed to rise towards Btu parity with oil in consuming markets.
[3] It is assumed that (consistent with recent provincial policy announcements and actions) the provinces will adjust provincial royalty levies (or prices paid) to reflect changes in producer net-backs resulting from real cost increases or increased taxation burdens (resulting from simultaneous price and royalty increases). Estimates of provincial rebates are included.
[4] Only Crown royalties are reflected in provincial payments. Freehold royalties between parties in the private sector are included in industry cash flows.
[5] Land costs are included, for presentational purposes, in provincial revenues. The amounts shown for exploration and development are therefore exclusive of such costs. Total Canada expenditure forecasts are reflected.
[6] The lower end of the expected expenditure range has been used to calculate federal tax revenues and industry cash flows. Increments above the base levels could provide a tax shelter and somewhat increased industry cash flows available to meet such expenditure requirements.
Source: T.S. Tuschak (1975), p. 167.

Table 4
RATES OF RETURN AND REVENUE SHARES IN THE
PETROLEUM PRODUCING INDUSTRY

Situation As Of	1975 Constant Dollar Rates of Return 1947−2010[1,2]	Shares of Net Operating Income, 1975−1979[1]			
		Industry	Federal Government	Provincial Governments	Total
December 31, 1973 (provincial royalties largely unchanged from historical levels)					
— $ Millions (rounded)	— —	23,115	7,215	9,920	40,250
— %	13.7−14.3	57.4	17.9	24.7	100.0
April 30, 1974 (provincial royalties increased in response to higher prices anticipated)					
— $ Millions	— —	16,850	3,660	19,740	40,250
— %	12.2−13.0	41.9	9.1	49.0	100.0
May 6, 1974 Federal Budget					
— $ Millions	— —	8,280	9,680	22,290[3]	40,250
— %	9.7−11.1	20.6	24.0	55.4	100.0
November 18, 1974 Federal Budget					
— $ Millions	— —	10,480	7,695	22,075[3]	40,250
— %	10.5−11.6	26.0	19.1	54.9	100.0
June 23, 1975 Budget (incorporating all provincial royalty and tax rebate responses to federal actions of 1974)					
— $ Millions	— —	15,400	7,675	17,175	40,250
— %	12.1−13.0	38.3	19.1	42.7	100.0

Notes: [1] The range shown for rates of return reflects different treatment of frontier expenditures in calculating income taxes payable. Revenue distributions are those resulting from the use of frontier expenditures to shelter western Canadian earnings.

[2] The rate of return uses current dollars for 1947−1974. If a rate of return of 9−11% were used to gross up the past capital base (9−11% being consistent with risk adjusted rates of return available over that period) the calculated rate of return for the period 1975 to the exhaustion of the basin would be considerably higher than the rates shown here. For example, at 10% applied from 1947−1974 the rate of return for the 1975−2010 period is 19.6% (with no tax shelter from frontier expenditures reflected).

[3] Since the western provinces are agreeing provinces for corporate tax collection purposes, the federal budgets also redefined taxable income for provincial purposes.

Source: T.S. Tuschak (1975), p. 165.

In Table 4, the estimated cash flow through 1979 is given for the oil sector. The goal of the federal government was to maintain roughly the same share of revenue as before the vast price increases in petroleum products.

Income tax policy is not a completely effective means of rent collection. Part of the reason lies in the general inability of the corporate profit tax, as presently constituted, to distinguish between the required return to capital (including risk premium) and "pure" profits. In so far as the resource sector is concerned, were it taxed similarly to all other sectors, distortions similar to other sectors would appear, compounded by the unique problem of distortions in the profile of consumption and production over time. After-tax rates of return for corporate endeavours generally would lie below returns in the more lightly taxed non-corporate sector.[22] As a result, too few inputs would be devoted to corporate production in general and resource production in particular.

But the resource sector does receive special income tax advantages. These advantages, as others have pointed out, likely cause too great a rate of development (Hyndman and Bucovetsky, 1974).[23] The corporate tax advantages then also have social costs in terms of their impact on the exploitation of resources over time.

That the federal government was willing to accept a lower share of revenue from the petroleum and natural resource sectors than manufacturing industries, for example, reflected the basic belief that the provinces were the landholders and entitled to the major share of resource revenue. Note that after including royalties, the effective tax burden on the resource sector is now far higher than for all other industries in Canada. In combining a corporate profit tax with the relatively large royalty (albeit perhaps a two- or three-tier progressive royalty), fiscal instruments are predominately output based rather than profit based.

Distribution of Rents to Consumers

The objective of rent gathering by governments is to redistribute income from one group (shareholders) to "others." Normally, the "other" group is hard to determine, since rent revenue is thrown in with all other tax receipts. One could determine, in theory, the group of citizens who would have had to pay increased taxes (income or sales) were the government required to increase its tax receipts in lieu of rent collection. In fact, one could do this exercise for Alberta, where income taxes are lower than they would be were there no oil.

The Canadian federal government has used another instrument to redistribute income from shareholders to consumers. Beginning in late 1973, the price of oil sold in Canada has been kept below the comparative price of imported crude. In 1978, this subsidy amounted to $2.95 billion—$1.03 billion of direct subsidy to consumers who purchase imported crude and an implicit subsidy of $1.923 billion to consumers of domestic crude.[24]

East of the Ottawa valley, because of the National Oil Policy, refiners used only imported crude until early in 1974; imported oil was less costly than domestic crude (Debanné, 1974). With the rapid escalation in world oil prices, refiners east of the Ottawa valley received direct subsidies since late 1973. In late 1978, these subsidies reduced the $17.90 landed price of imports to the $13.65 constrained domestic price ($12.75 Edmonton well-head price plus transport costs). Because consumers of domestic crude receive oil at $13.65, worth $17.90 at its opportunity cost, an indirect subsidy is paid.

A number of authors have criticized the domestic price ceiling (Waverman, 1975; Grubel and Sydney Smith, 1975). Their arguments are that the effect of the price ceiling is to create shortfalls—the increment to supply is truncated while conservation of demand is inhibited. The domestic price ceiling is, therefore, inefficient.

Besides the issues of the size of the subsidy and its efficiency, the concept of equity is important. It has been shown (Waverman, 1975) that, with two minor adjustments, direct energy expenditures as a percentage of family income fall as income increases. While this pattern was not monotonic and varied considerably from region to region, a subsidy in the form of federal control of energy prices was equitable.

The price of natural gas is also constrained below its opportunity cost—long-run Btu parity with oil; natural gas city gate prices are currently set at 85 per cent of the corresponding oil price. But with oil being sold below its opportunity cost, the world price, Canadian domestic natural gas consumers receive, as it were, a double subsidy, amounting to $900 million in 1978.[25]

The Canadian government continually announced that the prices of both these energy supplies would reach world price levels in the late 1970s. This did not materialize, however. One cannot neglect the important effects that the domestic price ceiling (along with the export tax described below) have on producer incentives. The domestic price ceiling offers producers a price well below the opportunity cost of the oil—leading to underinvestment in energy production.[26]

Export Taxes

With the price of Canadian crude maintained below world oil price levels, foreign oil consumers of Canadian oil would earn windfall rents unless Canadian producers were allowed to charge market prices for their exports. Rather than allowing these producers to make up the difference between the domestic price of crude and foreign oil prices, the Canadian federal government levied a series of export taxes on oil (crude and products) shipments outside the country.

The export tax, by driving a wedge between the market price and the price to producers, must affect long-run supply and hence have a dead-weight loss. Grubel and Sydney Smith (1975) estimated the dead-weight loss of the two-price system in oil products to be approximately $250 million, assuming unitary elasticity of demand and supply. Using alternative measures of elasticity (-0.3 for demand, -0.7 for supply) yields an estimate of $65 million for the dead-weight loss in 1978 (oil exports having been reduced since 1975), roughly 10 per cent of the dead-weight loss of the Albertan royalty system.

State Participation

If the rents in resource extraction are largely unknown and conventional rent collection instruments inefficient, it has been argued that direct government participation can be an effective rent-collection device. But the inherent problems are still present unless 100 per cent participation is planned.

Participation can take two forms—a share of physical output or a share of equity in the firm. Many countries have opted for the former rather than a share in profits, since profits can be disguised. Relatively little cheating is possible when the government takes some percentage of actual output. The Canadian government participates in 50 per cent of the profits of Syncrude.

With both forms of participation, however, what percentage take is correct? This calculation, of course, requires that we know the rents earned! But participation has been suggested as an effective policy precisely because rents are difficult to measure. Taking a percentage of profits is just another form of income tax. A percentage of physical output is, again, not a function of the rents inherent in production. Therefore, participation does not overcome any of the difficulties in effectively calculating rent and taxing it accordingly.

SUMMARY

If Henry George's world did exist—assets in fixed supply with all the competitive assumptions fulfilled—rent collection would be easy. We could then argue "simply" about who should receive rents and utilize the most efficient instrument.

In fact, the world of Henry George is absent. Natural resources are not in fixed supply. Their production involves risk and investment over time. Also, their depletable nature makes optimal resource extraction a concern for public policy. Over and above these issues lies the constitutional bases for rent collection, with the provinces having the power to tax natural resources but limited to direct taxes.

Public policy must then be concerned with the appropriate amount of rent collection and the effects that tax instruments have on producer and consumer incentives.

Before the rapid price increases in natural resources beginning in 1973, the resource sector was lightly taxed. These tax advantages were unwise. Besides allowing too large a share of revenue to accrue to shareholders,[27] the tax advantages likely promoted too rapid an expansion of the sector. The sector now finds itself in the position, especially in the petroleum industry, where a large percentage of the field price is taken by provincial governments in the form of a royalty.

Provincial governments have the constitutional authority to set royalties on resource production. The federal government has the authority to set direct and indirect taxes. I feel there is sound justification for a significant federal share of resource rents. Part of the sector's well-being is due to past federal subsidies. Moreover, in a federation concerned with equity of income distribution, it is unacceptable that resource-rich areas attempt to tax their way out of Confederation.

The rapid rise in taxes imposed on the resource sector has generated a number of problems. First, the disorderly transition period did not inspire investor confidence. Second, the taxation schemes used have significant inefficiencies—most drive a wedge between the market price and the net back to the producer, truncating supply and imposing dead-weight losses on society.

Five major forms of rent collection are presently used in Canada— income taxation, royalties, bonus bidding, export taxes, and domestic price ceilings. A sixth form—state participation—has begun. Table 5 lists the instruments used to collect rent from the Canadian oil sector in 1978, the amounts collected and the dead-weight losses entailed, excluding the rent generated by state participation.[28] In 1978, total oil and gas industry tax revenues taken by governments or directed towards consumers was *$9.516 billion*. Total industry revenue *net* of these explicit and implicit taxes was $5.1 billion ($10 billion total revenue (*Oil Week*, 12 February 1979), minus royalties, income taxes, bonuses, land sales, and rental payments). Without belabouring the point, there is no evidence to judge whether $9,516 billion represents the bulk or over 100 per cent of long-run rents. The major single item in rent collection was the domestic price ceiling ($3.85 billion). Federal revenue through both income and export taxes represented 38 per cent of total government take from the petroleum sector. The inefficiencies generated by the direct taxes (royalties, export tax) led to a dead-weight loss, at a minimum, of $365 million.

Given the constitutional underpinnings of resource ownership and taxation, dead-weight losses will continue to be imposed. The amount of rent generated is very company specific. Unless vast quantities of firm data are gathered, taxation based on revenues generated will be an ineffective means of capturing actual rents.

Table 5
RENT COLLECTION IN PETROLEUM AND NATURAL GAS—1978
(millions of current dollars)

	Amount Collected	Dead-weight Loss
Income Taxes[a]	1,000	?
Royalties[b]	2,991	299[f]
Bonuses, Land Sales Rentals[c]	900	?
Export Taxes[d]	775⎱	65[g]
Domestic Price Ceiling[e]	3,850⎰	
State Participation	?	?

Calculated as 10 per cent of royalty income.
a. *OilWeek*, Feb. 12, 1979, p. 58, estimate
b. *OilWeek*, Feb. 12, 1979, p. 56
c. *OilWeek*, Feb, 12, 1979, p. 56
d. Author's estimate
e. See section of this paper entitled "Distribution of Rents to Consumers"
f. See section of this paper entitled "Royalty"; calculated as 10% of royalties
g. See section of this paper entitled "Export Taxes"

NOTES

* The original version of this paper was commissioned by the Ontario Economic Council in 1976. I am grateful to Peter Nemetz for many helpful suggestions. All errors are my responsibility.

[1] A complete discussion of this issue appears in Scott (1976).

[2] An assumption of perfect asset markets suggests that any purchaser of field 1 in the above example would pay the present value of the future time stream of rents ABCD in the purchase price. Therefore, the purchaser's average cost curve would not be LAC as in Figure 1b but LAC', the same as field n. Capitalization of differing rents, therefore, makes explicit taxation of rents difficult.

[3] For a very lucid description of the operations of resource markets, see Solow (1974).

[4] Note that the well owner would have to be certain that the price of oil would not rise by more than the interest rate, for then it would pay him to shut down.

[5] See Hyndman and Bucovetsky (1974), Millsaps *et al.* (1974).

[6] This return is book profits to book investment on an after-tax basis, deducting the depletion allowance. Since the depletion allowance is largely judgemental, true economic profits (ignoring much of the depletion allowance) could be much higher.

[7] Scott (1976) argues that two economic efficiency arguments (resource revenues are benefit taxes and should accrue to the government, which provide the benefits; resources are depletable and resource taxes should be used to pay off the burdens arising from depletion) provide limited justification for rent distribution.

[8] *Mercer* v. *Attorney General of Ontario* (1881) 5 S.C.R. 538, Act 633.

[9] The interest in mineral rights depends on the land titles acts in each Prairie province and the degree to which rights to land include subsurface rights to minerals. This is of particular importance in the Prairie provinces where subsurface rights were *not* generally given in land grants to homesteaders before or after 1930. "Because settlements of the Prairie provinces proceeded from east to west, homesteaders acquired patents and grants without reservations of mines and minerals to the Crown in greater numbers in the east than in the west, so that the percentage of privately-owned or "freehold" mineral rights increases from west to east. In Manitoba, approximately 75% of the mineral rights in the southwestern part of the Province where conditions are suited to oil and gas production is privately owned." D.E. Lewis and A.R. Thomson, *Canadian Oil and Gas*, Vol. 1, 4.29 (Toronto: Butterworths). In Alberta, mineral rights to 80 per cent of the land are owned by the Province of Alberta.

[10] The issue of rent collection in offshore waters is currently still being disputed.

[11] Paus-Jenssen (1979) argues, incorrectly, I feel, that the most onerous of the new provincial oil revenue taxation schemes (Saskatchewan, 1974) could not be price determining.

[12] There have been few major changes in federal or provincial taxation of oil and gas revenues since 1975 (except for Saskatchewan; see note 11 above and the discussion in the section of this paper entitled "Royalty"). [*Editor's note:* The most recent change has been the federal National Energy Program of October 1980, which followed the writing of this paper.]

[13] This is a *very rough* calculation based on a large diversified firm reinvesting $3.80 per barrel of current oil production.

[14] Recent developments in Alberta suggest that the royalty scheme, normally applicable, can be adjusted for special circumstances.

[15] While point elasticities are likely in the -0.6 to -0.2 range, the elasticity for a large price change is likely at the low end; hence the choice of -0.3 in the example.

[16] This ignores Albertan royalty policies differentiating old and new oil and exaggerates the royalties received. The estimate of Alberta production is taken from *Oilweek*; the producers' net take used assumes maximum reinvestment of resource revenue. Federal income taxes would be roughly $3.00 and provincial income taxes $0.70.

[17] This includes increased export demand.

[18] The dead-weight loss is calculated as one-half the gain to consumers from lowering the price. Were the calculations done only on the basis of domestic Canadian consumption of Albertan oil, the dead-weight loss would be $140 million or 10 per cent of actual royalties on domestically consumed Albertan oil.

[19] Bucovetsky (1976) has many arguments, some of which are summarized in the text below.

[20] With the exception of Alaska, which generated large bids in its sale of oil and gas rights.

[21] A related issue not discussed in this paper is the impact of provincial resource revenue on the relative level of equalization payments received by resource-poor as compared to resource-rich provinces.

[22] This is a general argument concerning the allocation of all society's inputs between corporations and other businesses (unincorporated, partnerships), and so on. Were one sector (e.g., corporate) taxed more highly than the non-corporate sector, competition for inputs, based on after-tax returns (which must be equal, compensating for risk, etc.) would lead to "too many" inputs in the more lightly taxed non-corporate sector. Equalization of taxation would see migration of factors to the corporate sector generally. If natural resource firms are generally corporations, differential income taxation would ensure "too few" inputs devoted to natural resource production, all other things equal. But as the following paragraphs in the text and note 23 show, the income of natural resources corporations was until lately taxed far more lightly than other corporations.

[23] This reduction in corporate profit taxation likely reduced the effective profit tax rate below that of the unincorporated sector.

[24] The calculation of this subsidy is as follows. Consumption of domestically produced oil in 1978 was 427.38 billion barrels, imports were 228.64 billion barrels, and exports were 156.48 billion barrels. It is difficult to calculate the "average" differential between the Canadian price and the landed import price because of differences in the types of crude and transport costs. One method is to examine the export tax, set by federal officials to be equal to the difference between a barrel at the Canadian-controlled well-head price delivered in Chicago (now St. Paul), as compared to U.S. "new" oil or Saudi Arabian oil. The average export tax in 1978 was $4.95. A second method is to compare prices for a certain type of Canadian crude with the price for an equivalent imported crude, both delivered to the same point. In December 1978, light imported crude (comparable to Edmonton crude used as the base for well-head price controls) was $17.90 (Canadian) in Montreal. Edmonton crude at that date delivered to Montreal was $13.65—the difference in price was $4.25. The same calculation for 1 January 1978 yields a differential of $5.20. I have used $4.50 as the differential in all calculations.

[25] Domestic natural gas consumption was 1475 bcf in 1978. The price differential between the controlled price and a "free-market" price was calculated to be $0.61/mcf—85 per cent of the $4.50 price differential for oil assuming an mcf of gas has 1 million Btu's and a barrel of crude, 6.3 million Btu's.

[26] The *net* effect of all government policies, taxes, subsidies, and so forth, on the rate of exploitation of Canadian energy resources is then unknown. Some policies such as the income tax concessions prior to 1974 likely led to too rapid a rate of extraction. The effect of the domestic price ceiling, tending to reduce the speed of extraction, cannot be viewed as a simple counterweighing influence to corporate tax concessions. First, there is no reason why the net effect of these two policies is the optimal rate of extraction. Second, each policy should be correct in its own right; new policies should not be designed to offset old policies. Instead, the old policies should be improved.

[27] This "too large" a share would consist of retained earnings and elements of cash flow, such as depletion, not paid out to factors of production.

[28] The recent expansion of the publicly owned Petro-Canada into many facets of the industry *may* generate rent if the return on the ventures is greater than the cost of capital.

BIBLIOGRAPHY

Bucovetsky, M. (1976) "The Design of Mineral Tax Policy." Toronto: University of Toronto.

Debanné, J.G. (1974) "Oil and Canadian Policy." In *The Energy Question: An International Failure of Policy*, Vol. 2, *North America*, edited by E.W. Erickson and L. Waverman. Toronto: University of Toronto Press.

Grubel, H.G. and Sydney Smith, S. (1975) "The Taxation of Windfall Gains on Stocks of Natural Resources." *Canadian Public Policy* 1 (Winter): 13-29.

Hyndman, R.M. and Bucovetsky, M. (1974) "Rents, Renters and Royalties: Government Revenue from Canadian Oil and Gas." In *The Energy Question: An International Failure of Policy*, Vol. 2, *North America*, edited by E.W. Erickson and L. Waverman. Toronto: University of Toronto Press.

Kierans, E. (1973) "Report on Natural Resources Policy in Manitoba," prepared for the secretariat for the Planning and Priorities Committee of Cabinet, Ottawa.

La Forest, G.V. (1969) *Natural Resources and Public Property Under the Canadian Constitution*. Toronto: University of Toronto Press.

Millsaps, S.W.; Spann, R.H.; and Erickson, E.W. (1974) "Tax Incentives in the U.S. Petroleum Industry." In *The Energy Question: An International Failure of Policy*, Vol. 2, *North America*. Toronto: University of Toronto Press.

Paus-Jenssen, A. (1979) "Resource Taxation and the Supreme Court of Canada: The Cigol Case." *Canadian Public Policy* 5 (Winter): 45-58.

Scott, A. (1976) "Who Should Get Natural Resource Revenues." In *Natural Resource Revenues: A Test of Federalism*, edited by A. Scott. Vancouver: B.C.: Institute for Economic Policy Analysis.

Solow, R.M. (1974) "The Economics of Resources or the Resources of Economics." *American Economic Review* 64 (May): 1-14.

Tuschak, T.S. (1975) "A Federal Perspective on the Tax Treatment of the Petroleum Industry." In *1975 Conference Report*, pp. 157-73. Toronto: Canadian Tax Foundation.

Waverman, L. (1975) "The Two Price System in Energy: Subsidies Forgotten." *Canadian Public Policy* 1 (Winter): 76-88.

Chapter Fourteen

Environmental Protection in the United Kingdom

7220
UK

by
David Williams

The protection of the environment is especially important in a small and heavily industrialized country such as the United Kingdom. England alone, in an area of 50,000 square miles, has a population of about 46 million people, and it is "more densely populated than any other country in Europe with the exception of the Netherlands."[1] Close proximity to the mainland of Europe and the "generally more equable climate and terrain" have contributed to the concentration of population in England; and "the south-east of England in particular which possesses in London the political, administrative and commercial centre of the United Kingdom, has been so naturally attractive to population and industry as to require the adoption of special measures to order and restrain their growth."[2] Inevitably, there have been serious environmental problems. Throughout the United Kingdom, however, there are special problems and demands in planning and the control of pollution, often exacerbated by the legacies of the industrial revolution in the eighteenth and nineteenth centuries. The Aberfan disaster of 1966 was a vivid and tragic reminder of how the ravages of the past can take their toll.[3]

In the post-war years, the complicated system of town and country planning has provided a focus for environmental controls. The principal statute is now the *Town and Country Planning Act 1971*, under which the planning authorities at the local level have considerable responsibilities in the preparation of overall schemes and in considering individual applications for development. Central direction is exercised through the Department of the Environment, which is also the ministry with the most extensive powers in relation to pollution generally.[4] The cost and complexity of the planning system have given rise to much disquiet over the past decade, and there have been several reappraisals of its effectiveness.[5] Changes in the system are not infrequent, always subject to the aim of securing "a broadly acceptable balance between adequate examination of the often complex issues involved, the need to take account of local opinion, and the need to produce reasonably speedy decisions.''[6] Planning administration is costly, and heavy expenses are incurred also by applicants for planning permission.

One of the underlying purposes behind the law and practice on planning is that of involving the public. Strenuous, though not always successful, efforts have been made to ensure public participation at the early stages of town and country planning: these are, in part, a response to ''a growing demand by many groups for more opportunity to contribute and for more say in the working out of policies which affect people not merely at election time, but continuously as proposals are being hammered out and, certainly, as they are being implemented.''[7] Local planning authorities are under a duty to inform and consult the public during the preparation of what are termed ''structure plans'' and ''local plans'' for their respective areas,[8] though at least one leading authority has eloquently set out his reservations concerning ''all the hubbub and brouhaha'' of ''citizen participation in the sophisticated world of town planning control.''[9]

Public participation is most visible at the stage of an appeal against refusal of planning permission, in those cases where an appellant has asked for a public local inquiry.[10] A public local inquiry, which is a device used in numerous statutory contexts, is a stage in the decision-making process of a government department. In the context of planning, such an inquiry is held unless it is agreed to pursue an appeal to the Secretary of State for the Environment by written representations alone.[11] A further feature of planning inquiries is that the inspector, who conducts the proceedings and would otherwise make a recommendation to the minister, is nowadays entrusted on a transferred basis with the responsibility of actually deciding the appeal in the majority of cases.[12] This would seem to suggest that wider considerations of national policy are less obtrusive in such cases; and amid the many factors that enter into the planning process as a whole, problems of pollution as such are perhaps often given a low priority.[13]

Nevertheless, the initiation of a major industrial development necessarily brings problems of pollution before the planning authority; and in such circumstances, the Secretary of State for the Environment may give directions (under section 35 of the *Town and Country Planning Act*) requiring the application for development to be referred to him for the initial determination. Before making a decision, he has to allow a public local inquiry, if so requested either by the applicant or the local planning authority.[14] Two recent inquiries, where the decision had been referred to the Secretary of State, demonstrate that a public local inquiry held in such circumstances is likely to be different in nature from the bulk of inquiries that arise at the appellate stage in the planning process. These were the Windscale Inquiry of 1977 and the Belvoir Inquiry of 1979−1980.

The Windscale Inquiry arose from an application by British Nuclear Fuels Ltd. for outline planning permission for ''a plant for reprocessing irradiated oxide nuclear fuels and support site services'' at their Windscale and Calder Works in Cumbria. The environmental implications were obvious, especially as the application was lodged less than a year after the publication of the sixth report of the Royal Commission on Environmental Pollution that was concerned with nuclear power;[15] and it was a reflection of the importance attached to the application that after it was called in under section 35, a High Court judge (the Honourable Mr. Justice Parker) was appointed to sit as the planning inspector along with two distinguished assessors.[16] A preliminary meeting was held in the Civic Hall in Whitehaven on 17 May 1977 to settle procedural matters, and the inquiry itself then extended over 100 days from 14 June to 4 November 1977. Evidence was taken from 146 witnesses and, in addition, some 1,500 documents (including many books) were submitted; and during the course of the inquiry, certain tests and research work were undertaken at the inspector's request.[17] The inspector's report was presented to the Secretary of State on 26 January 1978.[18] It ranged over several wide issues, including the basic question as to whether oxide fuel from United Kingdom reactors should be reprocessed in this country at all, and several ''conventional'' planning issues such as the effect on the amenities of the area. In the words of the then Secretary of State, the Windscale Inquiry ''showed that a planning inquiry could range over a very wide field, so that it could take in major national and international issues, as well as questions of need and environmental concern.''[19] Such a proceeding, in his view, is helpful in providing information and reassurance to the public. But in order to ensure that the matter should finally be debated fully in Parliament, he took the unusual step of formally rejecting the recommendation (which favoured the application by British Nuclear Fuels) and then allowing permission by means of a special development order under section 24 of the *Town and Country Planning Act*.[20]

The Belvoir Inquiry arose after the Secretary of State ''called in'' three applications by the National Coal Board to mine coal and undertake related surface development in the Vale of Belvoir area of Leicestershire, Lincolnshire, and Nottinghamshire. A leading Queen's Counsel, Mr. Michael Mann, was appointed as inspector and he, like Mr. Justice Parker, sat with two assessors.[21] The formal statement made by the Secretary of State, in which he gave his reasons for calling in the applications, indicated that he would, in reaching his decision (on the basis of the inspector's recommendation), ''take into account the environmental factors which would be involved, the extent to which coal will otherwise be economically available and all other planning aspects of the proposed development'';[22] and particularly with reference to the effects on the local and regional community, he expressly identified such matters as employment opportunities, the effect on agriculture, and the suitability of the area for tourism and recreation. After a preliminary meeting to consider procedural steps, the

Belvoir Inquiry finally began on 30 October 1979,[23] and it lasted well into 1980.[24] There was strong opposition to the proposed development, not least when the tenth Duke of Rutland appeared to denounce any further destruction of the English countryside: as owner of Belvoir castle, he warned that if the landscape were ruined through the proposed development, then a new ghost of Belvoir would forever haunt the philistines of the National Coal Board.[25]

These new types of inquiry are entrusted with the task of inquiring into issues of national as well as regional and local policy; and the political, economic, social, legal and environmental considerations are raised and discussed in a forum that many might see as merely a variant of a court of law. Because of the publicity of the proceedings, some matters, of course, cannot be raised adequately: an outstanding example in the Windscale Inquiry was the issue of terrorism and civil liberties "upon which the evidence which could be tendered . . . was very limited."[26] There are inevitable difficulties involved wherever it is sought to adjudicate on any wide issues of national policy under a procedure that "is in general of an adversary nature,"[27] bearing in mind that there is a "disparity between the resources available to the applicants and to the objectors."[28] Inquiries such as those at Windscale and Belvoir have stimulated interest in new methods of exploring some of the broader issues of policy in advance of the public adversary proceedings.[29] Particular attention has been directed to the feasibility of an environmental impact analysis (EIA),[30] for which the British substitute is currently the lengthy public inquiry: Mr. Justice Parker, for instance, said that he was "satisfied that all matters which might or would have been included in an EIA were properly investigated at the Inquiry," though he added that it is "possible but by no means certain that, had there been such an analysis, it would have saved time at the Inquiry."[31] A relatively detailed proposal for a "Project Inquiry" was made in a report published in 1979. Aiming at a form of home-made EIA, the Project Inquiry would be reserved for development proposals with substantial national and international implications, provided that the proposals are so complex and controversial as to demand a thorough, impartial, and public investigation.[32] Against the background of such an investigation, and the public debate it will have generated, the report envisaged a much shorter, site-specific planning inquiry as a prelude to a final decision by the Secretary of State.[33]

The inherent difficulty of attempts to separate the broad and specific issues relating to a proposed development is that of timing and motivation. Many people will not be alerted at all until the "executioner's block" is in view; in other words, until their own locality and amenities are directly affected by the proposed siting of a new development. Motorway inquiries in the past few years have brought out some of the intractable problems of separation of issues. A national policy may be laid down for a network of motorways, but the implications of that policy are rarely appreciated by individuals until construction work comes close to their homes. After a series of contentious public local inquiries into particular stages of motorway development,[34] the government issued a *White Paper on Transport Policy* in 1977[35] and a *White Paper on Roads* in 1978,[36] in part to ensure that highway inquiries could "now take place against the background of the statement of national roads policy, in the context of transport policy as a whole."[37] A further government publication of 1978 concentrated on the procedures at a highway inquiry, which, as was admitted, "undoubtedly forms the focus of interest not only for those likely to be directly affected by the particular proposal under consideration but also for those who are totally opposed to any major new road construction."[38] In order to confine the inquiry in future to local rather than national issues, the government promised earlier and more detailed information as part of a deliberate effort to increase openness in government,[39] and also suggested a number of procedural improvements designed to produce "less of a court-room atmosphere," thereby diminishing "the adversarial flavour" at an inquiry itself.[40]

Whatever the good intentions or logical persuasiveness of the government's arguments, however, it remains difficult and controversial to identify what may be raised and to what extent it should be raised at an inquiry. This is evident from two recent decisions in the English courts. The first, *Bushell* v. *Secretary of State for the Environment*,[41] concerned a local inquiry (which lasted one hundred days) into schemes for two, fifteen-mile stretches of motorway through rural areas to the south and southeast of Birmingham. Although the inquiry was completed in January 1974, the inspector did not report until June 1975 (his report consisted of 450 pages), and the Secretary of State's decision came in August 1976. Legal proceedings then ensued, in which the decision was challenged partly on the ground that objectors had been denied an opportunity of cross-examining officials on the methodology of traffic forecasting. The Secretary of State was upheld at first instance[42] and then reversed in a Court of Appeal presided over by Lord Denning MR[43] before finally being upheld in a majority decision in the House of Lords in 1980,[44] leading one of the majority to state that "the history of this lengthy and expensive litigation shows in my opinion the desirability of ministers having power, for the exercise of which they would be responsible to Parliament, to limit the matters which may be discussed at a local inquiry. If the need for a particular motorway can be discussed at every inquiry held in consequence of objections to a scheme to construct a part of it, the time it takes to deal with these matters is bound to be extended."[45] In the second case, *Lovelock* v. *Minister of Transport*,[46] the Court of Appeal rejected an attempt to quash or suspend compulsory purchase orders made for a stage of the M25 motorway in Essex. The appellate hearing had been expedited because the M25 had been under discussion for ten years and was, in Lord Denning's words, "deserving of the highest priority."[47] The "very determined lady'' who had brought the action had been involved in many public inquiries and in previous litigation, all related to the M25[48]: now, declared Lord Denning, it was "match point" and her case failed. Such a case serves to illustrate the persistence, even against all the odds, of those who are fundamentally opposed to particular developments on environmental grounds; and the judgment of Lord Diplock in the *Bushell* case shows how difficult it is to mould and ventilate the complex issues either in the law courts or earlier at the public local inquiry. The Council on Tribunals, a body responsible to the Lord Chancellor and charged with exercising a measure of oversight in respect of administrative tribunals and inquiries, has itself been closely involved in the special problems of highway inquiries as a result of complaints submitted to it; and the possibility of further procedural changes remains constantly under discussion.[49]

The Council on Tribunals is principally concerned with procedural fairness in tribunals and inquiries under its jurisdiction, and it is only in respect of procedure that it thus touches upon environmental controls. In the aftermath of the Windscale Inquiry, for instance, it considered some of the matters raised by Mr. Justice Parker about the arrangements for large-scale planning inquiries,[50] and it subsequently expressed "particular interest" in the Belvoir Inquiry.[51] More generally, as its annual reports indicate, it looks at procedural issues affecting inquiries in a wide range of statutory contexts; and on occasions it has issued special reports.[52] Environmental problems often emerge, but it would be misleading to assume that environmental protection can be secured exclusively through planning law and its attendant procedures. Public local inquiries serve to publicize, even to dramatize, environmental problems. Outside planning law altogether, however, there is a bewildering collection of executive or advisory bodies responsible for a continuous examination of means of preventing pollution and safeguarding the environment; and it is to these we shall now turn.

POLLUTION CONTROL

There are numerous statutes on general or particular aspects of pollution, and many of these are watched over in their operation by executive or advisory bodies at the fringe of government. Such bodies are themselves constantly under review, as becomes clear from the *Report on Non-Departmental Public Bodies*, which was issued in 1980.[53] One well-known advisory body, the Clean Air Council, has recently been abolished.[54] It had originally been set up by the *Clean Air Act 1956*, Section 23, for the purpose of "keeping under view the progress made . . . in abating the pollution of the air in England and Wales" and of "obtaining the advice of persons having special knowledge, experience or responsibility in regard to prevention of pollution of the air."[55] The stimulus for new legislation on clean air had come from the urgent investigations undertaken after the London smog of

1952[56] and general concern about domestic and industrial air pollution;[57] but legislation on clean air dates from the nineteenth century, and even today one has to take into account a number of different statutes and their enforcement through both central and local authorities.[58] Relevant statutes include the *Alkali etc. Works Regulation Act 1906*, the *Health and Safety at Work etc. Act 1974*, the *Clean Air Acts of 1956* and *1968*, the *Control of Pollution Act 1974*, and various *Public Health Acts*.

Given the complexity of the law—which could at least be eased if there were some effort at consolidation—it seems a pity to have abolished the one advisory body charged with overall responsibility for considering air pollution. The Noise Advisory Council, by contrast, which was set up in 1970 to advise the Secretary of State for the Environment, remained in existence, and it continued to produce reports on problems associated with noise. A report issued in 1971[59] formed the basis for extended provisions on noise in the *Control of Pollution Act 1974*,[60] and other reports have dealt with traffic noise,[61] noise in public places[62] and, on several occasions, aircraft noise: a report of 1977 considered problems associated with the Concorde,[63] and another of 1980 had returned to the topic of a third London airport.[64]

The year 1970, in which the Noise Advisory Council was set up, witnessed a significant new departure in governmental concern with pollution and the environment. The government set up a standing Royal Commission[65] ''to advise on matters, both national and international, concerning the pollution of the environment; on the adequacy of research in this field; and the future possibilities of danger to the environment.'' This represented a shift in emphasis from the fragmented approach of the past. It would now be possible to find out on a broader basis ''what is happening to our physical environment and to inform the British public about trends in pollution and needs for research and development.''[66] Many standing advisory bodies would remain in existence for relatively restricted purposes, but this would not prevent the Royal Commission from undertaking studies in the areas for which they were responsible.

Under its first chairman, Lord Ashby, the Royal Commission on Environmental Pollution proceeded to identify priorities in its work, looking broadly at the state of the natural environment and making a number of preliminary recommendations.[67] A relatively brief second report[68] looked into three selected issues in industrial pollution (including the disposal of toxic wastes on land), and a fuller third report[69] was devoted to pollution in some British estuaries and coastal waters. A number of the recommendations in the third report concerned administrative action, legislative changes, international implications, and economic factors, all of which demonstrated that the Royal Commission's remit was more than merely scientific. Although the three chairmen of the Commission to date—Lord Ashby, Lord Flowers, and Sir Hans Kornberg—have all been scientists, the members have included economists, trade unionists, land-owners, planners, politicians, and others experienced in public affairs, as well as scientists in a variety of disciplines. Individual appointments have been for a period of three years, subject to extension, and this has allowed for flexibility in composition at various stages in the Royal Commission's work.

The fourth, fifth, and sixth reports were prepared under the chairmanship of Lord Flowers. The fourth report[70] was in the nature of a reassessment of the Commission's role, taking into account recent changes in the environment, the structure of pollution control in the United Kingdom, and the training and manpower requirements for pollution control staff. The report appeared at a time, December 1974, when the United Kingdom needed to brace itself to face a "major reappraisal of available sources of energy and of energy use";[71] and the Royal Commission also took note of the increasing international discussion of its work, reflected both in membership of the European Economic Community (from January 1973) and in progress made with the United Nations Environment Programme. Moreover, two major statutes had been enacted in 1974: the *Control of Pollution Act* and the *Health and Safety at Work etc. Act*. The former, to which the Commission drew particular attention, represented "the evolutionary development of previous controls over a wide field, covering the disposal of wastes on land, pollution of water and air, and noise (though not noise from air or road traffic)."[72] The latter, which has important environmental aspects, derived from the recommendations of the *Report of the Committee on Safety and Health at Work*;[73] it provides for safety at places of work, administered by the Health and Safety Commission and its operating agency, the Health and Safety Executive.[74]

Some of the jurisdictional and administrative difficulties of pollution control were vividly brought out in the fifth report of the Royal Commission on the subject of air pollution.[75] One of its principal recommendations related to the provision in the *Health and Safety at Work etc. Act* transferring the Alkali Inspectorate, which has been prominent in controlling air pollution since the nineteenth century, from the Department of the Environment to the Health and Safety Executive: it was strongly urged, with a view to seeking a unified approach to pollution problems, that the inspectorate should return to the department.[76] Also as part of the move towards unification, the Royal Commission urged that there should be consolidation of the confusing statutory provisions on air pollution, stating that the legislative differences "appear to us to be inimical to the flexibility in control arrangements that we should like to see. The system has developed piecemeal as the need for action to deal with different aspects of air pollution was appreciated."[77] A further area of examination was the link between planning and pollution; and several recommendations were designed to achieve a better and more consistent recognition by planners of issues of pollution at all stages of the planning process.

The sixth report,[78] on the subject of nuclear energy, has achieved much publicity and was used extensively at the Windscale Inquiry. It was concerned with radioactivity and radiobiology, nuclear power, international and national control arrangements, reactor safety and siting, security and the safeguarding of plutonium, radioactive waste management, and energy strategy and the environment. The report contained a considerable amount of technical material, providing a base for numerous recommendations of a technical and wider nature. One of its recommendations, which has been implemented, was for the setting up of a body that is called the Radioactive Waste Management Advisory Committee (under the Department of the Environment); and it also proposed the setting up of ''a high level independent body to provide advice on energy strategy, taking account of economic, social, technical and environmental considerations.''[79] The government subsequently set up the Energy Commission in 1977 to advise on energy policy issues, under the aegis of the Department of Energy and, in 1978, an allied standing body known as the Commission on Energy and the Environment acting for both relevant departments ''to advise on the interaction between energy policy and the environment.'' Of these, the Energy Commission has failed to survive the ''quango-culling'' operations of 1979,[80] but the Commission on Energy and the Environment has been engaged since 1978 in a comprehensive study of ''the longer term environmental implications of future coal production, supply and use in the United Kingdom looking to the period around and beyond the end of this century, including likely new technologies and conversion to other fuels and raw materials.''[81] The members of the Commission are or have been involved with several other environmental bodies including the Royal Commission itself, the Clean Air Council, the Energy Commission, the Nature Conservancy Council,[82] the Natural Environment Research Council,[83] the Countryside Commission,[84] and the Advisory Council on Energy Conservation.[85] This list is perhaps a sufficient indication of the reliance in the United Kingdom on advisory bodies, many with overlapping jurisdictions and some with overlapping membership.[86] Indeed, the first chairman of the Commission on Energy and the Environment was Lord Flowers, who had chaired the Royal Commission from 1973 to 1976.

The seventh report of the Royal Commission,[87] now under the chairmanship of Sir Hans Kornberg, appeared in 1979 and was devoted to the subject of agriculture. This looked into problems and dangers of pollution associated with pesticides, nitrogen fertilizers, and farm wastes, and it also covered the effects of pollution on agriculture and aspects of planning. Specific topics included aerial spraying, pesticides control, eutrophication, slurry banks, sheep dipping, fish farms, and intensive livestock units. The unexpected implications of concern for the environment are shown in the treatment of a legal case, *Smedleys Ltd.* v. *Breed*,[88] in which the House of

Lords considered a prosecution under the *Food and Drugs Act 1955* arising from a caterpillar found in a tin of peas: the Commission recommended a widening of the statutory defences available in order to ''reduce somewhat the pressures for the unnecessary use of pesticides.''[89]

Work on the eighth report of the Royal Commission—on the subject of oil pollution at sea—is well under way, touching upon international as well as national problems. The extensive coastline of the United Kingdom is exposed to oil spillages from tankers and other ships using busy routes to and from Europe; and few people in the country will be unaware of at least some of a series of events stretching from the *Torrey Canyon* sinking in 1967[90] to spillages associated with such tankers as the *Amoco Cadiz*,[91] the *Betelgeuse*,[92] the *Christos Bitas*,[93] the *Eleni V*,[94] the *Esso Bernicia*,[95] the *Litiopa*,[96] and the *Tarpenbek*,[97] to say nothing of blow-outs such as that at the *Ixtoc 1* exploration well in the Gulf of Campeche.[98] The problems of oil pollution have been a stimulus to inquiry by international bodies, parliamentary bodies, government departments, independent advisory committees, and a variety of pressure groups; and there is already a considerable literature in the field.[99]

The growing literature on all aspects of the environment and pollution comes from a surprising variety of sources. At government level there are, for instance, regular reports issued by such agencies as the Central Unit on Environmental Pollution in the Department of the Environment: the Pollution Papers have ranged over general matters on pollution control[100] and controversial specific topics such as lead and the environment,[101] and chlorofluorocarbons and their effect on stratospheric ozone.[102] Some of the problems are transient, others are persistent. Sometimes discussion arises from a publicized event, such as the grounding of an oil tanker, or from a build-up of concern over some real or imagined nuisance or danger,[103] or when Parliament undertakes some new or consolidating legislation with environmental implications. At other times, the discussion arises from long-term scientific research or from protracted international negotiations. The effect of discussion—whether it stimulates governmental action—depends on a number of variables, including financial restraints and competing social and economic demands. Pressures for environmental safeguards are so many and so different in nature and seriousness that there are bound to be problems of timing, emphasis, and priority.

The approach to environmental control can be based on subject-matter, taking into account—as the chapter headings in Pollution Paper No. 9[104] suggest—air pollution, fresh water pollution, marine pollution, waste management, radioactivity, pesticides, and noise. But under each heading the issues are scientific and technical, political and administrative, social and economic, and legal both at the national and international levels. The purpose of this article has been to illustrate the complexity of the issues and to give some indication of how, over the past decade in particular, there has been a growing public awareness of the importance of environmental controls. It has to be recognized, at the same time, that terms such as ''environment'' and ''pollution'' can be interpreted loosely and that it would be wrong to assess possibilities of further controls except against a political and economic background. In its fourth report, the Royal Commission on Environmental Pollution had this to say:

> The last few years have seen mounting public concern and debate about global problems arising from the scale of man's activity. Industrial and population growth, and the concomitant problems of pollution, food production, consumption of non-renewable materials and energy supplies, are seen by many as leading inexorably to crisis for mankind. Some believe that this can be averted only by fundamental changes in the structures and values of industrial society, by the abandonment of expectations of continuing economic growth, and even by reversion to some form of pre-industrial community. Others see no limit to man's resourcefulness, expressed through continuing technological development and through his capacity to adapt to changing circumstances, to solve the problems he faces within the existing social and economic framework.[105]

These are wide issues, but as summarized in the fourth report, they remind us of the contexts in which pollution problems have to be considered. Pollution, in the words of the Royal Commission, ''is only one component of the inter-related problems facing mankind,'' and it cannot be studied in isolation.

NOTES

[1] Report of the Royal Commission on the Constitution, Cmnd. 5460 of 1973, para. 181. At the 1971 census, the population was 45,870,062. By contrast, Scotland had 5,227,706 people in an area of 30,000 square miles (*ibid.*, para. 65); Wales had 2,723,596 people in an area of just over 8,000 square miles (*ibid.*, para. 108); and Northern Ireland had a population of 1,536,065 in an area of about 5,000 square miles (*ibid.*, para. 144).

[2] *Ibid.*, para. 183. An isolated example of an effort to deter the drift to the southeast is seen in the *London Government Act 1963*, s. 73(2), which provides—in the context of giving publicity to the amenities and advantages of Greater London—that nothing in the section ''shall authorize the Greater London Council to give publicity in the United Kingdom, whether by advertising or otherwise, to the commercial and industrial advantages of Greater London.''

[3] The disaster was the slide of a coal tip on 21 October 1966: 144 people, mainly young children, died. See Lord Robens, *Ten Years Stint* (London: Cassell, 1972), chap. 12.

[4] On departmental responsibilities for pollution, see *Pollution Control in Great Britain: How It Works* (Pollution Paper No. 9, Department of the Environment (DOE), Central Unit on Environmental Pollution), 2d ed. (HMSO; 1978), chap. 1 and Annex A.

[5] See *Review of the Development Control System* by George Dobry Q.C.; Interim Report 1974 (HMSO, 1974) and Final Report 1975 (HMSO, 1975); *Planning Procedures* (8th Report from the Expenditure Committee of the House of Commons, Session 1976−77, HC 395-I, 19 May 1977); *Planning Procedures* (White Paper, Cmnd. 7056 of January 1978).

[6] Cmnd. 7056, para. 6.

[7] *People and Planning* (Report of the Committee on Public Participation in Planning, chaired by A.M. Skeffington M.P.) (HMSO, 1969), para. 7.

[8] Structure plans and local plans may be equated with strategic and tactical plans. See generally, Neal Alison Roberts, *The Reform of Planning Law* (London: Macmillan, 1976).

[9] Sir Desmond Heap, *The Land and the Development (or The Turmoil and the Torment)*, Hamlyn Lectures for 1975 (London: Stevens), pp. 34-35.

[10] See, on public local inquiries, R.E. Wraith and G.B. Lamb, *Public Inquiries as an Instrument of Government* (London: George Allen and Unwin, 1971) and W.A. Robson, "Public Inquiries as an Instrument of Government," in *Politics and Government at Home and Abroad* (London: George Allen and Unwin, 1967), chap. 7. In *Bushell* v. *Secretary of State for the Environment* [1980] 2 All ER 608, 612, Lord Diplock said that "local inquiry" was "an expression which when appearing in a statute has by now acquired a special meaning as a term of legal art."

[11] See Annual Report for 1964 of the Council on Tribunals, para. 76.

[12] See HC 395-I (note 5 above), para. 20.

[13] See, for example, Fifth Report of the Royal Commission on Environmental Pollution, Cmnd. 6371 of 1976; *Air Pollution Control: An Integrated Approach*, chap. 11, especially paras. 334-36.

[14] Alternatively, the Secretary of State could set up a planning inquiry commission under s. 47 of the act. This method of proceeding, which was allowed for in the wake of the non-statutory Roskill inquiry into a Third London Airport, 1968−70 (on which see *The Big Public Inquiry* (Council for Science and Society, Justice, the Outer Circle Policy Unit, 1979, chap. 3)), has never in fact been used. See also Olive Cook, *The Stansted Affair* (London: Pan Special, 1967).

[15] *Nuclear Power and the Environment*, Cmnd. 6618 of 1976. It was noted in the report that there are "few subjects in the field of environmental pollution to which people react so emotionally as they do to radioactivity" (para. 5).

[16] The assessors were Sir Edward Pochin and Professor Sir Frederick Warner.

[17] The legal profession was prominent at the inquiry, and several Queen's Counsel appeared for various parties.

[18] *The Windscale Inquiry*. Report by the Honourable Mr Justice Parker (HMSO, 1978).

[19] From a speech at Manchester on 13 September 1978, as set out in DOE Press Notice (No. 488) of 13 September 1978.

[20] For an account of the Windscale events and particularly the parliamentary proceedings after publication of the report, see Jeremy Bugler, "Windscale: A Case Study in Public Scrutiny," *New Society* (27 July 1978), pp. 183-86.

[21] See DOE Press Notice (No. 26) of 26 January 1979. The National Coal Board had announced in mid-1976 that a major coalfield had been discovered in the Vale.

[22] DOE Press Notice (No. 26) of 26 January 1979, addendum. Such a statement has to be made under Rule 6(1) of the Town and Country Planning (Inquiries Procedure) Rules 1974 (S.I. 1974, No. 419).

[23] See *Daily Telegraph* (29 October 1979), p. 10, and (31 October 1979), p. 10.

[24] No decision has yet been reported. The inquiry ended on its eighty-third day: see *The Times* (23 April 1980), p. 2.

[25] *The Times* (28 March 1980), p. 5.

[26] See the Windscale report, para. 7.1, and generally chap. 7. On the general issue of terrorism and civil liberties (including problems of security clearance), see the Sixth Report of the Royal Commission on Environmental Pollution (note 15 above), chap. 7; the White Paper in response to the Sixth Report, Cmnd. 6820; *Plutonium and Liberty* (some possible consequences of nuclear reprocessing for an open society), Justice Report 1978 (based on evidence presented to the Windscale Inquiry).

[27] Windscale report, para. 15.12.

[28] *Ibid.*, para. 15.8.

[29] See, especially, David Pearce, Lynne Edwards, and Geoff Beuret, *Decision Making for Energy Futures* (A Case Study of the Windscale Inquiry) (London: Macmillan, 1979).

[30] See *Environmental Impact Analysis* (a study prepared for the Secretaries of State for the Environment, Scotland and Wales), by J. Catlow and C.G. Thirlwall (DOE Research Report 11, 1976). See also J.F. Garner, *Environmental Impact Statements in the United States and in Britain* (the Denman Lecture, 1979, Department of Land Economy at Cambridge University).

[31] Windscale report, para. 14.9(i).

[32] *The Big Public Inquiry* (note 14 above).

[33] The device of a Special Development Order, so as to permit parliamentary debate, is envisaged as the means by which planning permission would finally be granted.

[34] For an account by one of the leading opponents of motorway developments, see John Tyme, *Motorways versus Democracy* (London: Macmillan, 1978), which is sub-titled ''Public inquiries into road proposals and their political significance.''

[35] Cmnd. 6386 of 1977.

[36] *Policy for Roads: England 1978*, Cmnd. 7132 of 1978. See also the Report of the Advisory Committee on Trunk Road Assessment (HMSO, 1978).

[37] Cmnd. 7132, para. 54. The paragraph went on: ''Potential objectors will be able to prepare for an inquiry with fuller information than they have had in the past; the lack of this has been used as a ground for contesting the holding of inquiries or to protract their course.''

[38] Report on the Review of Highway Inquiry Procedures, Cmnd. 7133 of 1978, para. 12.

³⁹ *Ibid.*, para. 31. On open government, see White Paper on Reform of Section 2 of the Official Secrets Act 1911, Cmnd. 7285 of 1978, referring (at para. 42) expressly to greater openness at highway inquiries. See, generally, a Green Paper (consultation paper) on *Open Government*, Cmnd. 7520 of 1979.

⁴⁰ Cmnd. 7133, para. 42.

⁴¹ [1980] 2 All ER 608, HL.

⁴² (1978) 76 Local Government Reports 460, Sir Douglas Frank Q.C.

⁴³ (1980) 78 Local Government Reports 10. This was a 2-1 decision of the Court of Appeal: see *Daily Telegraph* (31 July 1979), p. 3 ("Denning stops M-way after 100-day inquiry"); *The Guardian* (31 July 1979), p. 1 ("Denning halts motorway projects"); and Peter Levin, "Public Inquiries: The Need for Natural Justice," *New Society* (15 November 1979), pp. 371-72. Lord Denning commented at one point that there "has been a deplorable loss of confidence in public inquiries."

⁴⁴ *The Times* (8 February 1980), p. 3 ("Law Lords overturn ruling for reopening of motorways inquiry").

⁴⁵ [1980] 2 All ER at 623-24, *per* Viscount Dilhorne.

⁴⁶ Times Law Report for 11 June 1980 in *The Times* (12 June 1980), p. 9, CA. The first instance decision (Willis J.) was reported in the news columns of *The Times* (12 March 1980), p. 5.

⁴⁷ See also *Policy for Roads: England 1978*, (note 37 above), para. 123.

⁴⁸ See *Sunday Times* (2 March 1980), p. 6 ("Shy secretary stops a motorway"). An earlier case, in which she failed before the Court of Appeal, was "*Lovelock* v. *Secretary of State for Transport*," *Journal of Planning Law* (1979): 456.

⁴⁹ See *The Functions of the Council on Tribunals*, Special Report by the Council, Cmnd. 7805 of 1980, para. 3.11 and Appendix 3(b) setting out its work on highway inquiries between 1974 and 1980.

⁵⁰ See Annual Report of the Council on Tribunals for 1977−78, HC 74 of Session 1978−79, para. 7.24 to 7.28.

⁵¹ See Annual Report for 1978−79, HC 359 of Session 1979−80, para. 6.13 and 6.14.

⁵² See, for example, Special Report on Stansted Airport, Cmnd. 3559 of March 1968, which preceded the Roskill inquiry (note 14 above).

⁵³ Cmnd. 7797 of January 1980. The inquiry that led to this report resulted from recent controversy over the proliferation of so-called "quangos" (i.e., quasi-autonomous non-governmental organizations): see para. 17 of the report.

⁵⁴ *Ibid.*, page 71.

⁵⁵ Its creation followed on a recommendation of the Committee on Air Pollution, which reported in 1954 (Cmd. 9322 of 1954); see para. 119.

[56] See the Fifth Report of the Royal Commission on Environmental Pollution (note 13 above), para. 44.

[57] See Cmd. 9322 (note 55 above).

[58] See *Pollution Control in Great Britain: How it Works*, (Pollution Paper No. 9, DOE, Central Unit on Environmental Pollution, 1976), chap. 2 ("Air Pollution").

[59] *Neighbourhood Noise* (Report by the Working Party on the Noise Abatement Act) (HMSO, 1971).

[60] See generally on noise legislation, Pollution Paper No. 9 (note 58 above).

[61] *Traffic Noise: The Vehicle Regulations and their Enforcement* (HMSO, 1972).

[62] *Noise in Public Places* (HMSO, 1974). The chapter headings in this report include planning controls, licensing laws, amplified sound, motor sports, miscellaneous nuisances (acoustic bird scarers, chain saws, model aircraft, clay pigeon shooting, burglar alarms, and refuse collection vehicles), and noisy behaviour (including the slamming of car doors and the revving of engines late at night). An interesting appendix (Appendix A) sets out a list of 'model byelaws'—that is, byelaws drafted by the Home Office to be adopted if local authorities so choose—dealing with noise nuisance ranging from music in the vicinity of hospitals to noisy animals.

[63] *Concorde Noise Levels* (HMSO, 1977).

[64] *The Third London Airport* (HMSO, 1980). Since this chapter was written, the government has announced its intention to disband the Noise Advisory Council. See HC, Vol. 998, cc. 423-4, 13 February 1981.

[65] Almost all Royal Commissions are set up on an *ad hoc* basis for a particular inquiry.

[66] First Report of the Royal Commission on Environmental Pollution, Cmnd. 4585 of 1971, para. 2.

[67] First Report, Cmnd. 4585 of 1971.

[68] Cmnd. 4894 of 1972.

[69] Cmnd. 5054 of 1972.

[70] *Pollution Control: Progress and Problems*, Cmnd. 5780 of 1974.

[71] *Ibid.*, para. 17. See also *Energy and the Environment* (Report of a Working Part set up jointly by the Committee for Environmental Conservation, the Royal Society of Arts, and the Institute of Fuel, 1974).

[72] Cmnd. 5780, para. 19. The act, which is still not fully in force, adopted some previous recommendations of the Royal Commission. For an indication of the implementation of the act as at 1 July 1978, see *Pollution Control in Great Britain: How It Works* (Pollution Paper No. 9, DOE Central Unit on Environmental Pollution), 2d ed. (HMSO, 1978), Appendix B.

[73] Cmnd. 5034 of 1972. The terms of reference of this committee specifically excluded consideration of questions of general environmental pollution.

[74] See the Commission's report on *The Hazards of Conventional Sources of Energy* (HMSO, 1978).

[75] *Air Pollution Control: An Integrated Approach*, Cmnd. 6371 of 1976.

[76] The proposal was immediately countered by the Health and Safety Commission: see *Health and Safety. Industrial Air Pollution 1976* (HMSO, 1978), especially para. 12. The Royal Commission's recommendation has not been adopted as yet.

[77] Cmnd. 6371, para. 199.

[78] *Nuclear Power and the Environment*, Cmnd. 6618 of 1976.

[79] *Ibid.*, para. 515. The committee was appointed in 1978. The government's response to the Sixth Report appeared as a White Paper, Cmnd. 6820.

[80] See Cmnd. 7797, p. 68 (para. 3). When it was set up in 1977, the membership of the Energy Commission covered workers and management in the energy industries and conservation, environment, and consumer interests. See *Energy Policy* (A Consultative Document), Cmnd. 7101 of 1978 (foreword by Secretary of State for Energy); reference is made in Annex 7 to a number of listed Energy Commission papers.

[81] Commission on Energy and the Environment: Review of Activities 1978−79 (HMSO, 1979), p. 7.

[82] This body was established by statute in 1973, and is responsible for the conservation of flora, fauna, and geological and physiographical features throughout Great Britain.

[83] This body was established by Royal Charter under the *Science and Technology Act 1965* to cover research in physical and biological sciences relating to man's natural environment.

[84] This is an independent statutory body responsible for the promotion of and for advising on the conservation of the natural beauty and amenity of the countryside and on improving facilities for enjoying the countryside.

[85] This body was set up to advise the Secretary of State for Energy in carrying out his duty of promoting economy and efficiency in the use of energy.

[86] See *Decision Making for Energy Futures* (note 29 above), Appendix 5 setting out the membership of the Energy Commission, the Commission on Energy and the Environment, and the Royal Commission on Environmental Pollution.

[87] *Agriculture and Pollution*, Cmnd. 7644 of 1979.

[88] [1974] AC 839, HL. See also *"Greater Manchester Council v. Lockwood Foods Ltd,"* *Criminal Law Review* (1979): 593, DC, the case of a black beetle in a can of strawberries, on which a news report appeared in the *Daily Telegraph* (10 August 1979), p. 13.

[89] Cmnd. 7644, para. 3.86.

[90] The tanker stranded on 18 March 1967 near the Scilly Isles. See report on *Coastal Pollution* from the Select Committee on Science and Technology, HC 421-I of Session 1967−68 and the

government's reply to the Select Committee at Cmnd. 3880 of 1969. See also White Paper on *The Torrey Canyon*, Cmnd. 3246 of 1967.

[91] See, generally, Second Report from the Expenditure Committee, HC 105-I of Session 1978–79, on "Measures to prevent collisions and strandings of noxious cargo carriers in waters around the United Kingdom." See also the government's response to the Expenditure Committee's Report at Cmnd. 7525 of 1979.

[92] See the recent report in the Republic of Ireland on the circumstances in which the French tanker *Betelgeuse* exploded on 8 January 1979 alongside the Gulf offshore terminal in Bantry Bay: see *Sunday Times* (27 July 1980), p. 4.

[93] See *Christos Bitas—The Fight at Sea Against Pollution* (Department of Trade, HMSO, 1978), an account of the measures taken to combat oil pollution after the tanker was damaged off the coast of South Wales on 12 October 1978.

[94] See Fourth Report from the Select Committee on Science and Technology, *Eleni V*, HC 684 of Session 1977–78, on the events after the Greek-registered tanker was in collision off the coast of Norfolk on 6 May 1978. See the government's reply to the Select Committee at Cmnd. 7429 of 1978.

[95] The tanker collided with part of the jetty at the Sullum Voe terminal in the Shetland Islands on 30 December 1978.

[96] This involved a leak occurring in the discharge of oil from the *Litiopa* off the north coast of Anglesey on 10 October 1978.

[97] See *The Tarpenbek Incident* (Department of Trade, HMSO, 1979), on the events after the product tanker *Tarpenbek* was in collision off Selsey Bill on 21 June 1979.

[98] The well blew out and caught fire on 3 June 1979, and the spillage was widely reported in Britain. A less serious blow-out occurred at *Ekofisk Bravo* in the North Sea in 1977.

[99] See, for example, *Accidental Oil Pollution of the Sea* (Pollution Paper No. 8, DOE Central Unit on Environmental Pollution, 1976), which is an official report on oil spills and clean-up measures; *Improved Arrangements to Combat Pollution at Sea* (Department of Trade, HMSO, 1979); *Liability and Compensation for Marine Oil Pollution Damage* (Department of Trade, HMSO, 1979); and frequent reports from such private bodies as the Advisory Committee on Oil Pollution at Sea (ACOPS). A recent parliamentary debate was in the House of Lords, on oil spills in the North Sea, at HL, Vol. 410, cc. 1681-1713, 25 June 1980.

[100] See *The Monitoring of the Environment in the United Kingdom* (Pollution Paper No. 1, 1974); *Controlling Pollution* (Pollution Paper No. 4, 1975); and *Pollution Control in Great Britain: How It Works* (Pollution Paper No. 9, 1978, 2d ed.).

[101] See *Lead in the Environment and the Significance to Man* (Pollution Paper No. 2, 1974); *Lead in Drinking Water* (Pollution Paper No. 12, 1977); and *Lead Pollution in Birmingham* (Pollution Paper No. 14, 1978). The recent concern about lead pollution centres particularly on the so-called "Spaghetti Junction," a motorway interchange in Birmingham. A Working Party of the Department of Health and Social Security reported on *Lead and Health* (HMSO, 1980), which the Secretary of State for Social Services describes (in the foreword) as "a valuable and timely addition to international work on environmental exposure to lead."

[102] See *Chlorofluorocarbons and Their Effect on Stratospheric Ozone* (Pollution Paper No. 5, 1976), and Second Report (Pollution Paper No. 15, 1979). The United States has been especially anxious to secure international co-operation in relation to chlorofluorocarbons—see *The Times* (19 May 1980), p. 2.

[103] See, for instance, the renewed controversy about the herbicide 2, 4, 5-T (*The Times* (12 June 1980)). Other problems have included genetic engineering (see report of the Working Party on the Experimental Manipulation of the Genetic Composition of Micro-organisms, Cmnd. 5880 of 1975) or odours (see report of the Working Party on the Suppression of Odours from Offensive and Selected Other Trades, Warren Spring Laboratory (HMSO, 1974 and 1975)—in separate parts).

[104] HMSO, 1978, 2d ed.

[105] Cmnd. 5780, para. 13.

The Institute for Research on Public Policy
PUBLICATIONS AVAILABLE*
June 1981

BOOKS

Leroy O. Stone &
Claude Marceau

Canadian Population Trends and Public Policy Through the 1980s. 1977 $4.00

Raymond Breton

The Canadian Condition: A Guide to Research in Public Policy. 1977 $2.95

Raymond Breton

Une orientation de la recherche politique dans le contexte canadien. 1978 $2.95

J.W. Rowley &
W.T. Stanbury, eds.

Competition Policy in Canada: Stage II, Bill C-13. 1978 $12.95

C.F. Smart &
W.T. Stanbury, eds.

Studies on Crisis Management. 1978 $9.95

W.T. Stanbury, ed.

Studies on Regulation in Canada. 1978 $9.95

Michael Hudson

Canada in the New Monetary Order: Borrow? Devalue? Restructure! 1978 $6.95

W.A.W. Neilson &
J.C. MacPherson, eds.

The Legislative Process in Canada: The Need for Reform. 1978 $12.95

David K. Foot, ed.

Public Employment and Compensation in Canada: Myths and Realities. 1978 $10.95

W.E. Cundiff &
Mado Reid, eds.

Issues in Canada/U.S. Transborder Computer Data Flows. 1979 $6.50

G.B. Reschenthaler &
B. Roberts, eds.

Perspectives on Canadian Airline Regulation. 1979 $13.50

P.K. Gorecki &
W.T. Stanbury, eds.

Perspectives on the Royal Commission on Corporate Concentration. 1979 $15.95

David K. Foot

Public Employment in Canada: Statistical Series. 1979 $15.00

---

* Order Address: The Institute for Research on Public Policy
P.O. Box 9300, Station "A"
TORONTO, Ontario
M5W 2C7

Meyer W. Bucovetsky, ed.	*Studies on Public Employment and Compensation in Canada.* 1979 $14.95
Richard French & André Béliveau	*The RCMP and the Management of National Security.* 1979 $6.95
Richard French & André Béliveau	*La GRC et la gestion de la sécurité nationale.* 1979 $6.95
Leroy O. Stone & Michael J. MacLean	*Future Income Prospects for Canada's Senior Citizens.* 1979 $7.95
Douglas G. Hartle	*Public Policy Decision Making and Regulation.* 1979 $12.95
Richard Bird (in collaboration with Bucovetsky & Foot)	*The Growth of Public Employment in Canada.* 1979 $12.95
G. Bruce Doern & Allan M. Maslove, eds.	*The Public Evaluation of Government Spending.* 1979 $10.95
Richard Price, ed.	*The Spirit of the Alberta Indian Treaties.* 1979 $8.95
Peter N. Nemetz, ed.	*Energy Policy: The Global Challenge.* 1979 $16.95
Richard J. Schultz	*Federalism and the Regulatory Process.* 1979 $1.50
Richard J. Schultz	*Le fédéralisme et le processus de réglementation.* 1979 $1.50
Lionel D. Feldman & Katherine A. Graham	*Bargaining for Cities. Municipalities and Intergovernmental Relations: An Assessment.* 1979 $10.95
Elliot J. Feldman & Neil Nevitte, eds.	*The Future of North America: Canada, the United States, and Quebec Nationalism.* 1979 $7.95
Maximo Halty-Carrere	*Technological Development Strategies for Developing Countries.* 1979 $12.95
G.B. Reschenthaler	*Occupational Health and Safety in Canada: The Economics and Three Case Studies.* 1979 $5.00
David R. Protheroe	*Imports and Politics: Trade Decision-Making in Canada, 1968–1979.* 1980 $8.95

G. Bruce Doern	*Government Intervention in the Canadian Nuclear Industry.* 1980 $8.95
G. Bruce Doern & R.W. Morrison, eds.	*Canadian Nuclear Policies.* 1980 $14.95
W.T. Stanbury, ed.	*Government Regulation: Scope, Growth, Process.* 1980 $10.95
Yoshi Tsurumi with Rebecca R. Tsurumi	*Sogoshosha: Engines of Export-Based Growth.* 1980 $8.95
Allan M. Maslove & Gene Swimmer	*Wage Controls in Canada, 1975 –78: A Study in Public Decision Making.* 1980 $11.95
T. Gregory Kane	*Consumers and the Regulators: Intervention in the Federal Regulatory Process.* 1980 $10.95
Albert Breton & Anthony Scott	*The Design of Federations.* 1980 $6.95
A.R. Bailey & D.G. Hull	*The Way Out: A More Revenue-Dependent Public Sector and How It Might Revitalize the Process of Governing.* 1980 $6.95
Réjean Lachapelle & Jacques Henripin	*La situation démolinguistique au Canada: évolution passée et prospective.* 1980 $24.95
Raymond Breton, Jeffrey G. Reitz & Victor F. Valentine	*Cultural Boundaries and the Cohesion of Canada.* 1980 $18.95
David R. Harvey	*Christmas Turkey or Prairie Vulture? An Economic Analysis of the Crow's Nest Pass Grain Rates.* 1980 $10.95
Stuart McFadyen, Colin Hoskins & David Gillen	*Canadian Broadcasting: Market Structure and Economic Performance.* 1980 $15.95
Richard M. Bird	*Taxing Corporations.* 1980 $6.95
Albert Breton & Raymond Breton	*Why Disunity? An Analysis of Linguistic and Regional Cleavages in Canada.* 1980 $6.95
Leroy O. Stone & Susan Fletcher	*A Profile of Canada's Older Population.* 1980 $7.95

Peter N. Nemetz, ed.	*Resource Policy: International Perspectives.* 1980 $18.95
Keith A.J. Hay, ed.	*Canadian Perspectives on Economic Relations with Japan.* 1980 $18.95
Raymond Breton and Gail Grant	*La langue de travail au Québec: synthèse de la recherche sur la rencontre de deux langues.* 1981 $10.95
Diane Vanasse	*L'évolution de la population scolaire du Québec.* 1981 $12.95
Raymond Breton, Jeffrey G. Reitz and Victor F. Valentine	*Les frontières culturelles et la cohésion du Canada.* 1981 $18.95
David M. Cameron, ed.	*Regionalism and Supranationalism: Challenges and Alternatives to the Nation-State in Canada and Europe.* 1981 $9.95
Peter Aucoin, ed.	*The Politics and Management of Restraint in Government.* 1981 $17.95
H.V. Kroeker, ed.	*Sovereign People or Sovereign Governments.* 1981 $12.95
Heather Menzies	*Women and the Chip.* 1981 $6.95
Nicole S. Morgan	*Nowhere to Go? Consequences of the Demographic Imbalance in Decision-Making Groups of the Federal Public Service.* 1981 $8.95
Nicole S. Morgan	*Où aller? Les conséquences prévisibles des déséquilibres démographiques chez les groupes de décision de la fonction publique fédérale.* 1981 $8.95
Peter N. Nemetz, ed.	*Energy Crisis: Policy Response.* 1981 $10.95

OCCASIONAL PAPERS

W.E. Cundiff (No. 1)	*Nodule Shock? Seabed Mining and the Future of the Canadian Nickel Industry.* 1978 $3.00
IRPP/Brookings (No. 2)	*Conference on Canadian-U.S. Economic Relations.* 1978 $3.00
Robert A. Russel (No. 3)	*The Electronic Briefcase: The Office of the Future.* 1978 $3.00

C.C. Gotlieb
(No. 4)

Computers in the Home: What They Can Do for Us—And to Us. 1978 $3.00

Raymond Breton &
Gail Grant Akian
(No. 5)

Urban Institutions and People of Indian Ancestry. 1978 $3.00

K.A. Hay
(No. 6)

Friends or Acquaintances? Canada as a Resource Supplier to the Japanese Economy. 1979 $3.00

T. Atkinson
(No. 7)

Trends in Life Satisfaction. 1979 $3.00

M. McLean
(No. 8)

The Impact of the Microelectronics Industry on the Structure of the Canadian Economy. 1979 $3.00

Fred Thompson &
W.T. Stanbury
(No. 9)

The Political Economy of Interest Groups in the Legislative Process in Canada. 1979 $3.00

Gordon B. Thompson
(No. 10)

Memo from Mercury: Information Technology **Is** *Different*. 1979 $3.00

Pierre Sormany
(No. 11)

Les micro-esclaves: vers une bio-industrie canadienne. 1979 $3.00

K. Hartley, P.N. Nemetz,
S. Schwartz, D. Uyeno,
I. Vertinsky & J. Young
(No. 12)

Energy R & D Decision Making for Canada. 1979 $3.00

David Hoffman &
Zavis P. Zeman, eds.
(No. 13)

The Dynamics of the Technological Leadership of the World. 1980 $3.00

Russell Wilkins
(No. 13*a*)

Health Status in Canada, 1926 – 1976. 1980 $3.00

Russell Wilkins
(No. 13*b*)

L'état de santé au Canada, 1926 – 1976. 1980 $3.00

P. Pergler
(No. 14)

The Automated Citizen: Social and Political Impact of Interactive Broadcasting. 1980 $4.95

Zavis P. Zeman
(No. 15)

Men with the Yen. 1980 $5.95

Donald G. Cartwright
(No. 16)

Official Language Populations in Canada: Patterns and Contacts. 1980 $4.95

REPORT

Dhiru Patel   *Dealing With Interracial Conflict: Policy*
                                 Alternatives. 1980 $5.95

WORKING PAPERS (No Charge)**

W.E. Cundiff   *Issues in Canada/U.S. Transborder Computer Data*
(No. 1)   *Flows.* 1978 (Out of print; in IRPP book of same
                                 title.)

John Cornwall   *Industrial Investment and Canadian Economic*
(No. 2)   *Growth: Some Scenarios for the Eighties.* 1978

Russell Wilkins   *L'espérance de vie par quartier à Montréal, 1976:*
(No. 3)   *un indicateur social pour la planification.* 1979

F.J. Fletcher &   *Canadian Attitude Trends, 1960−1978.* 1979
R.J. Drummond
(No. 4)

** Order Working Papers from
The Institute for Research on Public Policy
P.O. Box 3670
Halifax South
Halifax, Nova Scotia
B3J 3K6

2818